Reality Without Realism

Arkady Plotnitsky

Reality Without Realism

Matter, Thought, and Technology in Quantum Physics

Springer

Arkady Plotnitsky
Literature, Theory, and Culture Program, Philosophy
and Literature Program
College of Liberal Arts
Purdue University
West Lafayette, IN, USA

ISBN 978-3-030-84577-3 ISBN 978-3-030-84578-0 (eBook)
https://doi.org/10.1007/978-3-030-84578-0

This Springer imprint is published by the registered company Springer Nature Switzerland AG
The registered company address is: Gewerbestrasse 11, 6330 Cham, Switzerland

If there is such a thing as a sense of reality, and no one will doubt that it has its justification for existence—then there must also be something that we can call a sense of possibility.

—Robert Musil, The Man Without Qualities

Preface

Any observation of atomic phenomena will involve an interaction with the agency of observation not to be neglected.

—Niels Bohr, "The Quantum Postulate and the Recent Development of Atomic Theory" (1927) (Bohr 1987, v. 1, p. 54).

This book considers quantum theory as a theory based on the new relationships among matter, thought, and experimental technology, as against those previously found in physics, such as classical physics and relativity. There, as Niels Bohr argued from the outset of his work on the epistemology of quantum mechanics (QM), "our … description of physical phenomena [is] based of the idea that the phenomena concerned may be observed *without disturbing them appreciably*" (Bohr 1987, v. 1, p. 53; emphasis added). By contrast, "any observation of atomic [quantum] phenomena will involve an *interaction[of the object under investigation]with the agency of observation* not to be neglected" (Bohr 1987, v. 1, p. 54; emphasis added). My emphasis reflects a subtle nature of this contrast, due to the fact that the interaction between the object under investigation and the agency of observation *gives rise* to a quantum phenomenon rather than disturbs it. Bohr became weary of using the language of "disturbing of [quantum] phenomena by observation" (Bohr 1987, v. 2, p. 64). It is possible to say that this interaction "disturbs" the object with which the measuring instrument interacts, but even this claim requires qualifications, discussed below and in detail later in this study. In any event, speaking of "an interaction [of the object under investigation] with the agency of observation," which cannot be neglected, expresses more precisely the irreducible role of experimental technology in the constitution of quantum phenomena.

The argument of this book is based on its title concept, "reality without realism" (RWR), grounded in this role, and the view, the RWR view, defined by this concept, of quantum theory, specifically QM and quantum field theory (QFT) in their standard versions, referring by QFT, unless qualified, to high-energy, relativistic versions of it. (Alternative theories of quantum phenomena, such as Bohmian mechanics or spontaneous collapse theories, will only be mentioned in passing.) The RWR view places a stratum of physical reality thus designated, in this study, in particular, the one ultimately responsible for quantum phenomena, beyond representation or knowledge, or even conception, and thus defines the corresponding set of interpretations, RWR-type interpretations, of quantum phenomena and QM or QFT. Although quantum phenomena and QM, or QFT, are commonly interpreted jointly, as they will, mostly, be in this book, quantum phenomena can be interpreted separately because they are independent of a theory accounting for them. The concept of reality without realism presupposes general concepts of *reality* and existence, discussed in detail in Chap. 2. Briefly, by reality I understand that which is assumed to exist without making any claims concerning the *character* of this existence. Such claims define realism, which, in most understandings of the term, assumes the possibility of forming an (idealized) representation or at least a conception of the reality responsible for the phenomena considered. By contrast, the absence of such a claim allows one to place a given reality or part of it beyond representation or knowledge, or conception, in accord with the RWR view.

The concept of reality without realism and the RWR view were introduced by this author previously, in several works cited later in this study. They will, however, be given a more radical form in this study by virtue of two new assumptions. The first, defining what I shall call the universal RWR (U-RWR) view, as against the quantum RWR (Q-RWR) view, applicable strictly in quantum theory, is that the ultimate constitution of nature, as responsible for *all physical phenomena* (rather than only quantum phenomena), is a form of reality without realism, in the strong version of this concept, which places this constitution beyond conception, rather than only beyond representation or knowledge. The U-RWR view, however, allows for a realist treatment of some physical phenomena, such as those considered in classical physics or relativity, for all practical purposes. The second assumption is that the concept of quantum object in the Q-RWR view, or any physical object in the U-RWR view, is an idealization applicable only at the time of observation. Nature has no physical objects. They are concepts or idealizations created by us in our interaction with nature.

Speaking of "the *idea* that the phenomena concerned may be *observed without disturbing them* appreciably" in classical physics or relativity does not imply, and Bohr never claimed so, that these phenomena are *independent of our means of observing them*, but only that they may be treated as such for all practical purposes within the proper scopes of these theories. Hence, Bohr speaks of "the idea" that these phenomena may be so observed, rather than that such is in fact the case. The grounding philosophical position of this book, leading to the RWR view, first the Q-RWR view and then to the U-RWR view, is based on the *assumption* that ultimately the observation of *any physical phenomenon* involves an interaction between the world and our agencies of observation, beginning with that of our bodies and brains. (The latter, too, may be seen as technologies, created by nature, of observing, representing, and thinking about the world.) This interaction ultimately defines all physical phenomena, without allowing us to be certain that these phenomena or theories accounting for them represent, however ideally or approximately, nature as it exists independently. In some cases, however, such as that of classical physics or relativity, assuming that a theory does so is, in the U-RWR view, a practically justified idealization. QM or QFT expressly do not in RWR-type interpretations, defined by the Q-RWR view. With quantum theory, the book argues, the character of thought in theoretical physics was radically changed by the assumption of the irreducible role of experimental technology in the constitution of the physical phenomena considered by a given theory. These phenomena were, accordingly, no longer defined by nature alone, apart from human participation in it or by allowing one to neglect this participation, as physical phenomena were in classical physics or relativity. Or so it appeared. Quantum physics helped us to realize that experimental technology, beginning with that of our bodies, irreducibly shapes all physical phenomena, and made us rethink the relationships among matter, thought, and technology in all of physics and beyond, although this book will only address this rethinking in physics.

In the U-RWR view, no representation, however idealized or approximate, of the ultimate constitution of nature by a physical theory or even any conception of this constitution is possible, *at least as things stand now*. I qualify because the possibility that a physical theory or a set of physical theories will be able to do so in the future cannot be excluded, which would compel even those who previously held the U-RWR view to abandon it in favor of realism. The phrase "the ultimate constitution of nature" (or of physical reality) which occurs frequently in this study requires qualification as well. A theory that is, or is assumed to be, correct, as far as things stand now, does not necessarily need to account for this ultimate constitution. Thermodynamics is an example of such a theory, although it is not a fundamental theory, a theory commonly (including in this study) defined as being concerned with the ultimate constitution of nature. A nineteenth-century example of such a theory was the kinetic theory of molecular or atomic constitution of nature, with atoms and molecules assumed to behave individually in accordance with the laws of classical mechanics. In fact, no current theory, specifically general relativity or QFT (two theories now considered as workable fundamental theories), account for the ultimate constitution of nature. It is commonly assumed that general relativity and QFT account for those aspects of this constitution that they consider governed, respectively, by gravity and three other fundamental forces of nature (electromagnetism, the weak force, and the strong force). The fact, however, that these theories are not in accord with each other implies that they do not account for how nature ultimately works. There are theories that aim at providing, at least eventually, such an account, but they are hypothetical without much consensus concerning their relative promise.

Unlike, at least in principle, in classical physics and relativity, in quantum physics the experimental technology of measuring instruments capable of observing quantum phenomena *expressly physically interferes* with the stratum of nature that is ultimately responsible, by interacting with this technology, for quantum phenomena. This interference and, correlatively, the specific character of quantum phenomena, as manifested in paradigmatic quantum experiments, such as the double-slit experiment, led to the view of quantum phenomena according to which the ultimate constitution of reality responsible for them is placed beyond representation or even conception, thus making this reality a reality without realism. I shall speak of this concept or this view as "weak" if this reality is assumed to only be beyond representation or knowledge and as "strong" if it is assumed to be beyond conception, as it will be in this study. Importantly, however, especially in the context of QFT and the question of elementary particles, discussed in Chap. 8, there is, in the present view, no assumption of a uniform character of the ultimate, RWR-type reality considered in either QM or QFT, a uniformity only manifested differently in quantum experiments. This assumption is in fact in conflict with the strong RWR view, adopted by this study, which precludes any concept of this reality, including that of its unified nature. While each time unknowable or even unthinkable, this reality is assumed to be each time different, thus, automatically, making each quantum phenomenon, as an effect of this reality, individual and even unique, manifesting the individual and unique nature of this reality each time one encounters it. I shall refer to this assumption as the quantum individuality, QI, postulate, which is, as explained below, accompanied by the quantum discreteness, QD, postulate, according to which quantum phenomena are irreducibly discrete relative to each other. Both postulates could be, and have been, interpreted in realist terms, including by assuming a continuous and possibly uniform nature of the reality ultimately underlying them, which would imply that this individuality and discreteness arise in our interaction with this reality. This assumption is in principle precluded in RWR-type interpretations, especially of the strong type, such as the one adopted by this study. That does not mean that one cannot, in RWR-type interpretations, relate, or relate *to*, quantum phenomena otherwise, in particular, by predicting, by means of QM or QFT (or possibly other theory), the probabilities of future quantum phenomena on the basis of already established ones.

As discussed in the Introduction, the RWR view has a longer history, extending even from the pre-Socratics and, in modern times, from Immanuel Kant's philosophy, which is, arguably, the most recognizable modern precursor of the RWR view. This view as such, however, was initiated by Bohr's 1913 atomic theory, considered in detail in Chap. 3. The RWR view was only partially applicable there, in defining the discrete transitions, "quantum jumps," between the so-called stationary states of the electrons in atoms. Stationary states were represented classically as orbits (along which the electrons moved around the atomic nuclei) and hence in a realist way. The theory, accordingly, has been sometimes called "semi-classical," the term that came to be applied, more generally, along with "the old quantum theory," to quantum theory preceding QM. This partial application of the RWR view does not diminish the revolutionary nature of Bohr's concept. This was an audacious step both physically, because it was incompatible with classical mechanics and electrodynamics alike, and philosophically, because it implied that it may not be possible to form a representation and even conception of how nature ultimately works.

It was expected at the time that Bohr's concept of quantum jumps, also the first instance of the QD postulate, was only a stopgap measure that would eventually be abandoned, and that some form of realism would be restored to all fundamental physics. It was, however, this concept that became central for Werner Heisenberg, who built on it by abandoning an orbital representation of stationary states as well. The RWR view, at the time in its weak form (precluding a representation of quantum objects and behavior) grounded, fully, Heisenberg's epistemology of QM in his discovery of the theory in 1925. Heisenberg also introduced a new form of mathematical formalism, based on a mathematics never previously used in physics, *in effect* that of Hilbert spaces over the field of complex numbers, $\mathbb{C}$. Heisenberg did not speak in these terms, introduced later by John von Neumann, and standard ever since. The main credit for bringing this type of formalism into physics, however, belongs to Heisenberg.

The interpretation of quantum phenomena and QM, or QFT, adopted by this book builds on these and related developments, and contains several new features. In particular, as noted, this interpretation redefines the concept of quantum object as an idealization applicable only at the time of observation or measurement, rather than as existing independently, in contrast to the ultimate constitution of the reality responsible for quantum phenomena, along with, in this interpretation, quantum objects. The concept amplifies the significance of experimental technology in defining the concepts of quantum theory, beginning with the concept of "quantum measurement" as understood in this study, following Bohr, eventually leading him to his concept of "phenomenon," as applied in quantum physics.

This concept makes the terms "observation" and "measurement," as conventionally understood, inapplicable in the case of quantum phenomena. These terms are remnants of classical physics or still earlier history, from which classical physics inherited them, beginning with the rise of geo-*metry* in ancient Greece. As understood here, a quantum measurement *does not measure* any property of the ultimate constitution of the reality responsible for quantum phenomena, which this constitution would be assumed to possess before, or even in, the act of observation. Hence, the concept of observation requires a different understanding as well. An act of observation in quantum physics *establishes* quantum phenomena by an *interaction* between the instrument and the quantum object, or in the present view, the ultimate constitution of the reality responsible for quantum phenomena and, at the time of measurement, quantum objects, as applicable idealizations (which are RWR-type entities as well). Then what is so observed, as data or information, can be measured classically, just as one measures what is observed in classical physics, where, however, what is so observed and measured could be associated with the object considered. As far as the emergence of the observed data or information is concerned, a quantum measurement is not a measurement of anything but a number or bit generator, something akin to a quantum computer created by our interaction with nature. In the case of quantum phenomena, there is thus, technically, a difference between an observation, which construct a phenomenon, and a measurement, which classically measures one or another physical property of this phenomenon. While keeping this qualification in mind, I shall, for the sake of economy, speak, in referring the whole process just described, of quantum measurement, unless qualifications are necessary.

The overall philosophical position of this book, stated above, extends the Q-RWR view to the view, the U-RWR view, of *all physical phenomena*, at least as things stand now, assuming that the concept of observational technology extends to our bodies and brain. Although the last assumption is applicable to phenomena beyond physics, indeed to all possible phenomena, this book is only concerned with the U-RWR view, or any form of the RWR view, such as the Q-RWR view, in physics rather than with possible applications of this view elsewhere. The extension of the Q-RWR view to the U-RWR view is automatic if one assumes that the ultimate constitution of nature is the same as that of the stratum of this constitution responsible for quantum phenomena. This extension need not, however, depend on this assumption because this constitution, while still inconceivable, may be different or contain strata other than those responsible for quantum phenomena. Given that most of my use and discussion of the RWR view will be those of the Q-RWR view, the RWR view will hereafter refer to the Q-RWR view, unless the context requires a specification of it as the Q- or the U-RWR view.

At the same time (in the RWR view, by definition), it is not experimental technology alone but nature and our interactions with it by means of this technology that are responsible for quantum phenomena and ultimately all physical phenomena, technology, like all technology, helped by our thought, a product of our brains and thus our bodies. No physical phenomena, any more than any other phenomena, are possible without thought. Nor is experimental technology possible without thought, which is necessary to create and use it. Our thought itself may be seen as, or, given that it is more than technology alone, as providing, an essential form of technology, however helped by other technologies, from the origins of humanity on. How we use these other technologies, from the oldest known tools on, still depends on thought, or society and culture, which have capacities for using technology for both enhancing and diminishing the creative power of thought. We need thought when we use our bodies or something already existing in nature, such as the Sun or clusters of galaxies for gravitational lensing (observing which, however, requires telescopes and other devices), as experimental technology. They are still made experimental technologies by our thought. It is possible, even in performing experiments in physics, to use our bodies or even tools apart from thinking, at least conscious thinking. Only thinking, however, makes possible any directed or effective use and, especially, any creation of technology. Galileo Galilei and Sir Isaac Newton, arguably, along with René Descartes, the greatest founding figures of modern theoretical physics were also great experimentalists and creators of new experimental devices, most, famously, new types of telescopes. While that of Galileo was a refinement of its Dutch prototype, Newton, in 1668, designed and built the first practical reflecting telescope, still known as the Newtonian reflector, part of the history of technological thought extending to the Hubble telescope and Large Hadron Collider (LHC). Reciprocally, however, we also think *with* technology, especially in physics, but not only there.

Experimental technology is a broader concept than that of measuring instruments. It would, for example, involve devices that make it possible to use measuring instruments, like the 26.7 km circular tunnel of the LHC under the Jura Mountains. It is difficult to give a strict definition of technology. There is no consensus concerning such a definition even when it is used in the more conventional sense of referring to tools and devices. Arguably, most generally, technology is a means, technique (the term originating in the ancient Greek word *technē*) of doing something, getting "from here to there," as it were, and helping us to do it more successfully than previously. Or, as Brian Arthur put it, a bit more formally, technology is *a means to fulfill a human purpose* (Arthur 2011, p. 20). He adds two other basic meanings, "an assemblage of practices and components" and, most broadly "the entire *collection of devices and engineering practices available to a culture*" (Arthur 2011, p. 20). Defining it as a technique of doing something serves best my purposes in this book, including in considering thinking, which, again, may be seen as technology or enabling technologies, for example, conceptual or mathematical thinking in theoretical physics, or conceptual thinking in mathematics itself. (Arthur only deals with technology in its conventional sense.) Bohr, in referring to the new mathematics of Heisenberg's matrix mechanics in 1925, spoke of "the mathematical *instruments* created by the higher algebra," in other words, the mathematical technology (Bohr 1987, v. 1, p. 51; emphasis added). The ancient Greek *technē* also referred to art, in the sense of making, creating an artwork. This meaning (and ancient Greek thinking concerning *technē*) served Martin Heidegger's critique of modern technology, also in its use in science, a critique that, I would argue, is only partially applicable to either technology or science, especially quantum physics, which, as I shall suggest, may be seen as in accord with ancient Greek thought (Heidegger 2004). This meaning is retained in the concept of liberal arts (*artes liberalis*), a term found in Cicero, and probably used before him. It is true that not all thought is technology, but much of it is.

The experimental technology of quantum physics, helped by our thought, enables us to understand how nature works, in the RWR view, strictly as manifested in effects observed in measuring devices, without allowing us to know or even to conceive of the ultimate constitution of the reality responsible for these effects as quantum phenomena. This reality or any physical reality is not technology; it is something technology helps us to discover, understand, and work with. However, it can become part of technology, beginning with the quantum parts of measuring instruments through which they interact with this reality, or as part of devices we use elsewhere, such as lasers, electronic equipment, or MRI machines. Most recent discoveries in fundamental physics, especially particle physics, such as that of the Higgs boson, were the result of the joint workings of three technologies:

(1) The experimental technology of particle accelerators, such as the Large Hadron Collider (LHC), only part of which is the technology of observation and measurement.
(2) The technology of physical and sometimes philosophical concepts used in physics, classical, relativistic, and quantum.
(3) The mathematical technology of quantum theory, in this case, specifically QFT.

Any quantum event, I argue, is made possible by the first technology and could only be treated as quantum by the joint workings of the second and third technologies. A more recent addition is digital computer technology, which is becoming increasingly more prominent and even indispensable in physics. There is yet another technology involved, that of science as a social and cultural project. Its role is, however, a separate subject, which will only be mentioned in passing in this book. None of these technologies, however, can exist apart from thought and the technology of thought itself, reciprocally helped by other technologies.

Only nature itself, as matter, may be assumed to exist independently of thought, which assumption implies that it had existed when we did not exist and will continue to exist when we will no longer exist. Generally, "matter" is a narrower concept than "nature," although when the ultimate *material* constitution of nature is considered both concepts merge. There are some exceptions to this view of the independent or even any other existence of matter, including views, such as, famously, that of Plato or his pre-Socratic predecessor Parmenides, or in modern times that of Bishop Berkeley, that deny the existence of matter or nature, or anything apart from thought, altogether. Such views are useful in suggesting that any conception of how anything exists, or even that it exists, including when assumed to be independent of human thought (as Plato's ideal reality was as well), belongs to thought. It need not follow, however, that something which such concepts represent, or to which they relate otherwise than by representing it, possibly placing it beyond representation or even conception, does not exist.

In the modern (materialist) view, which emerged roughly in the mid-eighteenth century and was developed in the nineteenth century, thought is as an activity of the brain and thus of our bodies, and as such is a product of our biological evolution and, thus, ultimately of nature. On the other hand, as Kant, an eighteen-century philosopher as well, realized the ultimate constitution of nature or matter, considered independently, *may be* beyond the capacity of thought to represent or know it. Going beyond Kant, who does not appear to have considered this more radical limit, this book argues that this constitution may even be beyond our capacity to conceive it, in accordance with the U-RWR view, at least in modern physics, to considering which this book is limited. It is possible that an RWR-type view could apply elsewhere in science, for example, by placing beyond representation or thought the ultimate workings of living matter in biology or neuroscience, or the workings of thought in psychology, but this is a separate subject. It could also apply, as Kant anticipated as well (even if, as explained below, not reaching the RWR view, even the weak RWR view), to mental reality, for example, in mathematics. Anything that we can conceive of, however, or the assumption that something is unconceivable, still belongs to thought.

Of course, one could (as against, the RWR view) assume, as many physicists and philosophers have done, that phenomenal entities considered in physics could ground representations, ideal or approximate, of nature by means of physical theories, and thus allow one to form viable concepts of nature, or some parts of it, considered as an independent physical reality. As, however, Kant, again, realized, it is difficult and perhaps impossible to guarantee that such mental entities, even the most basic ones, such as space, time, and existence, or our theories based on them correspond to how nature ultimately is, as, in Kant's terms, a thing-in-itself (or noumena) vs. any phenomenon, defined as an entity formed by our thought. A phenomenon may be assumed to represent or reflect the corresponding thing-in-itself, but is always a construction that may not correspond to this thing-in-itself, which could, according to Kant, also be a mental rather than only material entity. This constructive dimension of our thought, even in the case of our immediate perceptions or what so appears (since it is no longer assumed to be immediate but mediated by conscious or unconscious thought) distinguished Kant's philosophy from that of his predecessors, such as and in particular John Locke. This view makes Kant's philosophy an important precursor of the RWR view, in this case, the U-RWR view. The RWR view (either the Q- or the U-RWR view) is, however, more radical than that of Kant. While Kant defined objects or noumena, as things-in-themselves, as not knowable (vs. phenomena that are representable or knowable), he still allowed that they may be conceivable, even though he also appears to have assumed that there is no guarantee that such a conception is ultimately correct, as opposed to being workable in practice (Kant 1997, p. 115). By contrast, the *strong* RWR view places the ultimate character of the reality considered beyond conception, although, as explained below, this placement is only claimed here to be workable in practice as well. Kant's view is closer to, and may appear to be the same as, the *weak* RWR view. For the reasons explained in Chap. 2, however, Kant's view did not reach the weak RWR view either and remained a form of realism, moreover, classically causal in character. (I shall define the concept of classical causality, expressly precluded by the RWR view, below.)

The U-RWR view does, however, allow one to *treat*, for all practical purposes, certain observable physical phenomena as representing objects in nature and their behavior as independent of our means of observing them. Such objects and behavior, for example, paradigmatically, planets, and their motion around the Sun, are represented by physical theories, such as classical physics and relativity, in a suitably idealized and mathematized way, as, in the case of the solar system, by Newton's classical theory of gravity or by Albert Einstein's general relativity. This scheme works, within the proper scope of these theories, *at least up to a point*, for all practical purposes, in particular, for making (ideally) exact predictions of the outcome of experiments, observed as phenomena, which may be treated as, for all practical purposes, representing the corresponding objects and their behavior. I qualify for the following reasons. First, Newton's theory, while an excellent approximation, is incorrect even within its proper scope, for example, in the case of the aberrant behavior of Mercury, which requires general relativity to be accurately predicted. Secondly, general relativity, too, has its complexities, for example, those related to singularities, such as black holes. Of course, one still needed to create the mathematics enabling these representations and predictions, which was a major achievement of classical physics and relativity alike. This is why the proper names of Newton and Einstein accompany these theories, more so than Newton's or Einstein's realist assumptions, although these assumptions sometimes carry their signatures as well.

If such a representation of objects in nature on the basis of the observed phenomena is assumed to be possible, even in an idealized and approximate way, it becomes a form of realism. Realism has other forms, and this form of it has different versions and additional features, considered later in this study. In the U-RWR view, which applies to the ultimate constitution of nature, any such representation can only be understood as a provisional idealization. It is effective for practical purposes but is underlain by an RWR-type reality of the ultimate workings of nature that in principle precludes a representation or even conception. Throughout this study, by idealization, I refer to a workable conception of something or, as this is an idealization as well, a lack or impossibility of such a conception, as in the case in an RWR-type reality, in other words, something that may be different from what it idealizes rather than, as in some uses of this term, any form of approximation, defined by a proximity to what is idealized. In sum, if one gives the RWR view the U-RWR scope, the ultimate character of the reality considered in physics is always that of a reality without realism, but some phenomena allow the reality responsible for them to be treated in realist terms.

As discussed in Chap. 6, a parallel and in fact correlative situation is obtained in the case of causality, usually defined in terms of relationships between events and the laws of such relationships. The question is, however, what is the nature of these relationships and these laws. In the Q-RWR view, the essential and indeed the only relationships, when it is possible to ascertain them, between events, such as those of measurements, are probabilistic. Indeed, all events considered are those of observation or measurement, creating phenomena (some of which can be created by nature, but can only be events for us), underlain by the ultimate constitution of the reality responsible for them as a reality without realism. Accordingly, any form of causality that could be defined under these assumptions is probabilistic. I shall define such relationships in terms of "probabilistic causality." As discussed in Chap. 6, these relationships take a specific form, that of "quantum causality," in quantum theory, given the nature of quantum phenomena and the data observed in them, beginning with the fact that, even in the case of the simplest possible quantum systems, no predictions other than probabilistic are *in general* possible in quantum physics on experimental grounds, as things stand now, because identically prepared quantum experiments *in general* lead to different outcomes. (There are exceptional idealized cases, on which I comment below.) On the other hand, in either realist or (provisionally) the U-RWR view, the connections between certain events or phenomena allow the reality responsible for them to be treated in terms of the concept of "classical causality." This form of causality is not *fundamentally* probabilistic, although it may involve probability, as a practical expedient in the cases when we lack sufficient knowledge to access the architecture of the classical causal reality underlying the events considered. I have several reasons for adopting the term "classical causality," instead of just "causality," as is more common in designating this concept, including for founding figures considered here, beginning with Bohr, in most of his writings. (There is an important exception in one of his later works, in 1958, when Bohr adopts a different view of causality [Bohr 1987, v. 3, p. 4–5], which I shall discuss in Chap. 6, Sect. 6.6) In particular, first, although it has a much longer history, this concept has defined classical physics from its birth and was then extended to relativity, which introduced certain restrictions on it, discussed later. Secondly, there are alternative definitions of causality, including probabilistic in nature (which classical causality is not), discussed in Chap. 6 (Sect.6.6). The concept of classical causality, considered in detail in Chap. 2, is defined by the claim that the state, state X, of a physical system is determined, in accordance with a law, at all future moments of time once it is determined at a given moment of time, state A, and state A is determined in accordance with the same law by any of the system's previous states. It is a more complex issue, discussed in Chap. 2 as well, whether A could be seen as the cause of X. The fact that it is not necessarily the case and related considerations have compelled some, beginning with P. S. Laplace, to speak of determinism instead in referring to this concept. I shall use the term determinism differently, as an epistemological category referring to the possibility of predicting the outcomes of classically causal processes ideally exactly in accordance with laws that define them. In classical mechanics, when dealing with individual objects or small systems, both concepts coincide or rather (because they are different concepts) are correlatively applicable. On the other hand, classical statistical mechanics or chaos theory is classically causal but not deterministic in view of the complexity of the systems considered, which limit us to probabilistic or statistical predictions concerning their behavior.

In the U-RWR view, then, while certain phenomena may be treated as classically causal for all practical purposes within the scope of certain theories, the ultimate constitution of nature responsible for these phenomena may not be, because this constitution is assumed to be a form of reality without realism, which prevents classical causality. For one thing, this definition implies the concept of a law that governs the strictly determined (rather than probabilistic) relationships between events and thus at least a partial conception and even representation of the reality responsible for these events as defined by this law. Accordingly, in the U-RWR view, in doing any form of physics (which in this respect becomes similar to QM or QFT in the Q-RWR view), one *ultimately* deals only with measurements of and predictions concerning observed phenomena or events, rather than with representing the ultimate constitution of the reality responsible for them. Such a representation is, again, only allowed to be assumed for all practical purposes in considering certain types of phenomena. In the U-RWR view, the combined workings of all three—nature, thought, and experimental technology (including our bodies)—are responsible for all physical phenomena, and define our measurements and, through them, justify, when possible, our predictions concerning physical phenomena. This remains the case even when these predictions are predictions with probability one. Such predictions are sometimes possible in dealing with quantum phenomena, although, as will be seen, their nature acquires new complexities. The possibility of exact predictions does not mean that we have an access to the ultimate constitution of nature, and in the U-RWR view (or Q-RWR view) we do not. While real as part of our thought, predictions are never part of a physical reality. Only the outcomes of observation and measurements are, which still leaves the ultimate constitution of reality responsible for them beyond representation or conception. Nature does not assign probabilities and makes no measurements or predictions. Only we do.

These relationships between reality and probability (in the absence of classical causality) in a RWR view, either a Q- or U- one, are reflected in my epigraph to this book, from Robert Musil's great unfinished novel, *The Man Without Qualities*: "If there is such a thing as a sense of reality, and no one will doubt that it has its justification for existence—then there must also be something that we can call a sense of possibility. Whoever has it, it does not say, for instance: Here this or that has happened, will happen, must happen; but he invents: Here such and such might, could, or ought to happen" (Musil 1995, p. 11). That, Musil adds, does not mean that one "shrinks from reality, but sees it as a project, something yet to be invented" (Musil 1995, p. 11). In the present view, only part of reality is open to such a project or invention. Besides, there is "the touch of reality" that can surprise us (Musil 1995, p. 93), sometimes in response to our own touch, for example, as in our experiments in physics, especially in quantum physics. None of this diminishes the significance of either reality that which we invent and that, a reality without realism, which we cannot invent. The RWR-type reality that we cannot invent may help us to invent the reality or realities that we can invent, which is what happens in quantum theory in the RWR view. Importantly, Musil only speaks about "a sense," an assumption of reality, thus, consistently with the view adopted here. The RWR view (of either U or Q type) is, too, assumed by this study to be *an idealization*, a product of thought, which is only practically justified, given quantum phenomena and theories predicting them, such as QM or QFT. The existence of these phenomena is crucial for the RWR view, inferred from nature on the basis of them, as their interpretation, rather than merely invented or assumed on some external grounds.

Assuming that the RWR view is an idealization, ultimately only practically justified, also precludes one from definitively claiming that the ultimate constitution of the reality responsible for quantum phenomena or, in the U-RWR view, the ultimate constitution (of the reality) of nature is unrepresentable or inconceivable. It may be or at least may be assumed to be ultimately conceivable or even representable, in accord with one or another realist view. At least this type of assumption or idealization might suffice for all fundamental physics, even for those who had previously adopted the Q-RWR view or even the U-RWR view, more resilient as the latter may be. The reason for this greater resilience in the U-RWR view is due to an extended view of technology, which includes our bodies and brains. Our concepts and language, or our thinking, defined by our evolutionary emergence as human animals have developed in our interaction with the world of objects consisting of billions of billions of atoms, and thus, on a scale very different from those, literally unimaginably small, considered in quantum theory, or at the opposite end of the available physical scales, from those at the scale of the Universe. Accordingly, there is no special reason to assume that this thinking should be able to conceive of the workings of nature on these scales, or even to relate these working otherwise, for example, in terms of probabilistic predictions, as in QM or QFT. It is true that our thinking has done reasonably well thus far, in particular, our mathematical thinking and our thinking that created our experimental technology. But how far would our mathematics be able to go, given that, at least in the present view, it is still human, a product, refined as it may be, of human thought? In any event, it is possible that the U-RWR view will lose its raison d'être in physics. It may happen, for example, if there is no physical theory, like QM or QFT, that defies realist assumptions at least in some interpretations and thus suggests the possibility of the U-RWR view. The U-RWR view may continue to be held elsewhere or considered as a philosophical possibility, but physics will have no need for it. Assuming the Q-RWR view to be only an idealization also allows for the possibility that quantum phenomena and QM or QFT can be interpreted in realist terms, as they have been.

If, however, one adopts the U-RWR view as the primary idealization, underlying all other idealizations used by physical theories, any realist view is assumed to be provisional and ultimately inapplicable. The situation can be represented diagrammatically as follows:

> (U-RWR): the ultimate constitution of nature → physical objects (defined, as RWR-type concepts or idealizations, at the time observation) → observational instruments (some of which are observed as classical physical object) → phenomena (defined by what is observed in observational instruments).

In quantum physics, in the Q-RWR view, a parallel diagram applies, with the ultimate constitution of the reality considered only assumed to be responsible for quantum phenomena:

> (Q-RWR): the ultimate constitution of the reality responsible for quantum objects and quantum phenomena → quantum objects (defined as RWR-type concepts or idealizations at the time measurement) → measuring instruments (observed as classical physical objects) → quantum phenomena (defined by what is observed in measuring instruments).

That the concepts or idealizations of "physical objects" and "quantum objects" are only applicable in considering observations is the view specifically adopted by this study. It is not necessarily found in other versions of the RWR view, including that of Bohr, who assumed that quantum objects can be considered as idealizations applicable independently of observation. Either diagram would, then, need to be adjusted accordingly. Finally, in classical physics or relativity, because it is possible, for all practical purposes, to neglect both the interference of measuring instruments and the ultimate constitutions of the reality considered, the diagram converts into:

> (R): physical reality = objects = (or assumed to be represented by a theory on the basis of) phenomena.

(Technically, if one assumes the U-RWR view as underlying any realism, this diagram should be designated as R-RWR.)

As indicated above, the extension of the Q-RWR view to the U-RWR view need not assume that the ultimate constitution of nature is the same as that of the stratum of this constitution responsible for quantum phenomena. This constitution may be different and contain different inconceivable strata, responsible for different physical phenomena. This difference could, for example, account for the fact that general relativity cannot be brought in accord with QFT and the standard model, which may imply that a deeper theory or set of theories is necessary to do so.

RWR-type interpretations of quantum phenomena may be different within each set, weak or strong, as are, for example, Bohr's interpretation, in its ultimate version (developed by him in the late 1930s), and the one adopted in this study, although both are of the strong RWR type and the present interpretation follows that of Bohr in several key aspects. In particular, Bohr was, again, the first to ground his interpretation (in all of its versions) in the irreducible role of measurement technology in the constitution of quantum phenomena and, in its ultimate version, in the RWR view as applied to the ultimate constitution of the reality responsible for quantum phenomena. This constitution is commonly, including in Bohr's interpretation (in all of its versions), associated with quantum objects. The interpretation adopted in this study, however, takes a more stratified view, as embodied in the Q-RWR diagram:

(Q-RWR): the ultimate constitution of the reality responsible for quantum objects and quantum phenomena → quantum objects (defined as RWR-type entities at the time measurement) → measuring instruments (observed as classical physical objects) → quantum phenomena (defined by what is observed in measuring instruments).

In this interpretation, while what is considered as a quantum object in a given experiment is made possible by the ultimate constitution of the reality responsible for quantum phenomena in its interaction with measuring instruments, and while these two strata of reality are equally beyond conception, they are not the same forms of idealization. The ultimate constitution of the reality responsible for quantum phenomena is assumed to exist independently of our interactions with it. On the other hand, the reality idealized as quantum objects is, while still of the RWR type, assumed to exist only *at the time observation.* It follows that there is no assumption of quantum objects, such as electrons, photons, or quarks, existing independently in nature apart from our interaction with it by means of observational technology. Hence, one cannot speak of the behavior of quantum objects as independent of observation either. One can only consider this behavior as part of observations or measurement, again, understanding by measurement the construction of quantum phenomena by means of measuring instruments capable of interaction with the ultimate, RWR-type constitution of the reality responsible for these phenomena. (Technically, again, this construction is that of *observed* phenomena, while a measurement qua measurement is a classical measurement of one or another property of a phenomena.) This view, on this point following Bohr, makes it impossible to separate, extract, quantum objects from quantum phenomena observed in measuring instruments. Bohr spoke in this connection of the indivisibility or wholeness of quantum phenomena, if, again, without assuming the present concept of a quantum object and the stratified idealization of the reality responsible for quantum phenomena assumed here: the ultimate RWR-type constitution of the reality considered, idealized as existing independently of our interaction with it; quantum objects, idealized, as RWR-type entities, as existing only in measurement; and quantum phenomena, idealized as effects of measurements classically observed in measuring instruments.

On the other hand, Bohr appears to have assumed the U-RWR view by suggesting that no physical theory can represent how nature ultimately works even when, as in classical physics or relativity, the physical objects considered and their behavior may be treated as corresponding to the phenomena observed (e.g., Bohr 1987, v. 1, p. 18). This view also appears to be suggested by Bohr's reference to "the idea," in the sentence that I quoted at the outset: "our ... description of physical phenomena based of *the idea* that the phenomena concerned may be observed without disturbing them appreciably," (Bohr 1987, v. 1, pp. 53; emphasis added). In other words, the assumption, "the idea," that such is the case may be workable for all practical purposes, but it need not mean, and does not in the U-RWR view, or even in Kant's view that these phenomena correspond to how things (objects) actually are in nature.

Quantum phenomena, defined by effects, observed in measuring instruments, allow for a representational and thus realist treatment, as do, in the first place, measuring instruments themselves or, more accurately, their observable parts. Measuring instruments also have quantum strata through which they interact with quantum objects, as considered in Chap. 2. In Bohr's interpretation and, following it, the present interpretation, both the observable parts of measuring instruments and, hence, quantum phenomena are treated by means of classical physics. The *quantum* character of quantum phenomena is defined by the particular configurations of their classically observed features. Bohr, accordingly, spoke of quantum phenomena as "the idealization of observation" (Bohr 1987, v. 1, p. 55). Eventually, Bohr adopted the term "phenomenon," in considering quantum phenomena, as referring strictly to what is observed or, more precisely, what *has already been* observed, in specified setups, in measuring instruments, as effects of their interaction with quantum objects (Bohr 1987, v. 2, pp. 64, 71). That Bohr's concept refers strictly to what has already been observed, rather than only predicted, is an important aspect of his concept, the significance of which will be apparent throughout this study.

Each quantum phenomenon is always discrete in relation to any other quantum phenomena, the assumption defined here as the quantum discreteness, QD, postulate, accompanied by the quantum individuality, QI, postulate, which assumes each quantum phenomena to be strictly individual, unique. Bohr assumed both postulates beginning with the introduction of what he called the quantum postulate, a form of the QI postulate, from the outset of his interpretation of QM (Bohr 1987, v. 1, p. 53). This discreteness is, in the RWR view, beginning with that of Bohr, not the same as that of quantum objects, which, including elementary particles, are beyond representation or even conception, and hence cannot, in the RWR view, be assumed to be discrete, for example, corresponding to the idea of particles, or continuous, for example, corresponding the idea of waves. Initially, this discreteness, responsible for using the term "quantum," was understood as associated with quantum objects, first as quanta of energy and then as particle-like objects, as it has often been later as well. This shift of the concept of discreteness from quantum objects to quantum phenomena emerged in Bohr as correlative to his ultimate strong RWR-type interpretation. For, if it is no longer possible, as it is not in such interpretations, to know or even conceive of how each quantum phenomenon comes about, it follows that it is also impossible to assume them to be connected by means of a continuous and classically causal process of the type considered in classical physics or relativity. A different, realist interpretation of quantum phenomena and QM, or QFT, would be required for this assumption to be workable, and such interpretations have of course been offered.

The situation just outlined and the RWR view of it leads to what I define as "the quantum indefinitiveness postulate," one of the key postulate of this study, correlative to both the QI and QD postulates. The postulate *precludes making definitive statements of any kind, including mathematical ones, concerning the relationship between any two individual quantum phenomena or events, indeed to definitively ascertain the existence of any such relationship.* The postulate does allow making definitive statements concerning individual phenomena or events, statements related to measurements, which define such individual phenomena or events. It also allows statements concerning the relationships between multiple events, in this case statements statistical in nature. The postulate only concerns events that have already occurred, rather than possible future events, in which case one can make probabilistic or statistical statements concerning connections between events that have already occurred and possible future events.

While introduced by Bohr's 1913 atomic theory, as part of it, and extended to QM as a whole by Heisenberg in 1925, the RWR view was given its full-fledged conceptual form by Bohr's interpretation (in its ultimate form, by 1930s, of the strong RWR type) of quantum phenomena and QM in terms of complementarity, his most famous concept, considered in Chap. 6. Complementarity is defined by

(a) a mutual exclusivity of certain phenomena, entities, or conceptions, such as, and in particular, those of the position and momentum measurements, which can never be performed simultaneously in view of the uncertainty relations, and yet
(b) the possibility of considering each one of them separately at any given point and
(c) the necessity of considering all of them at different moments of time for a comprehensive account of the totality of phenomena that one must consider in quantum physics.

In Bohr's ultimate, strong RWR-type version of his interpretation, in place by the late 1930s, complementarity applies strictly to quantum phenomena observed in measuring instruments. Bohr introduced his interpretation in 1927 in the so-called Como lecture, where, as discussed later in this study, he retreated to a more realist view, which he, however, quickly abandoned in favor of the RWR view. He developed his interpretation, via several revisions, culminating in his ultimate version, over the course of the next decade, in part under the impact of his debate with Einstein on "epistemological problems in atomic physics," as Bohr characterized it in his 1949 account of this debate (Bohr 1987, v. 2, pp. 32–66). I shall discuss the Bohr–Einstein debate in the Introduction. Bohr's interpretation, in any of its versions, will be distinguished in this study from "the Copenhagen interpretation," because there is no single such interpretation, as even Bohr has changed his a few times. It is more suitable to speak, as Heisenberg did, of "the Copenhagen spirit of quantum theory" or, as a handier shorthand, "the spirit of Copenhagen," referring to certain common features of a group of interpretations, which may be different in their other features (Heisenberg 1930, p. iv). Some of these interpretations are not of the RWR type. The RWR view was given additional dimensions, discussed in Chap. 7, by Dirac's discovery of antimatter, a consequence of his equation for the relativistic electron in 1928. Quantum electrodynamics (QED), the first form of QFT, was introduced by Dirac in 1927. QFT in general, including a theory of nuclear forces, was developed in the 1930s.

Building on Bohr's 1913 atomic theory, Heisenberg's approach represented a radical change in the nature of theoretical physics as, in Galileo's title words, "a mathematical science of nature," from classical mechanics to relativity theory (Galileo 1991). With Galileo and Descartes, physics brought together experiment and mathematics. According to Heidegger, commenting on the rise of modern physics with

Descartes and Galileo, "modern science is experimental because of its mathematical project" (Heidegger 1967, p. 93). Experiment and mathematics not only relate to each other but also mutually define each other once one bases one's account of nature on measurable quantities. These relationships became further mathematically grounded, giving mathematics an even greater primacy in the project of modern physics, with the introduction of mathematical formalisms enabling one to represent and, as a result, predict the outcomes of experiments. All modern physics, again, only offers an idealized and, specifically, mathematized treatment of natural phenomena.

Throughout the history of modern physics prior to Bohr's 1913 atomic theory and then QM, physicists, including, at early stages of quantum theory, such figures as Max Planck and Einstein, had aimed at describing and developing (idealized) mathematical representations of the physical objects considered and their behavior. This imperative has never been relinquished and is still dominant in physics and philosophy alike. It was only let go, in part, by Bohr, leading him to his 1913 atomic theory, and, building on Bohr's theory, altogether by Heisenberg, leading him to his discovery of QM in 1925. At the same time, Heisenberg kept intact, on new grounds, the mutually defining reciprocity of the mathematical and experimental character of quantum physics. This reciprocity was possible by virtue of relating his formalism and the data considered in terms of probabilistic or statistical predictions, without representing the ultimate nature of reality responsible for these data, thus in accord with the (weak) RWR view. Indeed, according to Heisenberg, such a representation only did not appear to be possible at the time, which, he said, compelled him to adopt his approach (Heisenberg 1925, p. 265). The fact, however, that such a representation was not part of his theory was momentous.

It is true that QM was co-discovered, as wave mechanics, by Schrödinger on the basis of thinking defined by the realist imperative of finding a physical description and mathematical representation of quantum objects and processes, in this case in wave terms. As discussed in Chap. 5, however, Schrödinger could not sustain his agenda, especially as concerns the discreteness of observed quantum phenomena and the probabilistic nature of quantum predictions. Against his own grain, he made several mathematical moves that were in conflict with this agenda. It was this conflict that led to him to his, *correct*, theory, which Bohr called, "a gigantic advance over all previous forms of quantum mechanics" (reported in Heisenberg 1971, p. 76). Both formalisms were quickly proven to be mathematically equivalent.

The situation thus brought about and the new relationships among matter, thought, and experimental philosophy that it established has continued to shape quantum theory throughout its subsequent history. Far from being found satisfactory by many and even by most, this situation and these relationships have also defined the debate concerning quantum theory throughout its history, from the confrontation between Bohr and Einstein to our own time. In 1949, two decades after this confrontation commenced at the Solvay Conference in Brussels in 1927, Bohr spoke of "the dissent among the physicists themselves [as] the cause of skepticism about the necessity of going so far in renouncing customary demand as regards the explanation of natural phenomena," by this time as far as the strong RWR view of quantum phenomena (Bohr 1987, v. 2, p. 63). Einstein was foremost on Bohr's mind, especially given that this comment was made in Bohr's "Discussion with Einstein on Epistemological Problems in Atomic Physics," reflecting on their debates during these decades (Bohr 1987, v. 2, pp. 32–66). Einstein was the most prominent figure in and a symbol of this dissent. But it was hardly Einstein alone. Schrödinger, a co-creator of QM, quickly joined Einstein, followed by many, event most, physicists and philosophers.

While most of them, beginning with Einstein, recognized the extraordinary predictive capacity of QM and then QFT, their attitude toward this renunciation went beyond mere skepticism. It was shaped by the realist imperative of making theoretical physics a science aiming at a representation of physical objects and behavior considered by any physical theory, especially if fundamental. This type of project was pursued by Einstein, unsuccessfully (which fact did not diminish his faith in this imperative), in his attempt to unify gravity and electromagnetism for 40 years. Realism, in this case also known as "scientific realism," has sometimes assumed a more naïve form than the view held by Einstein. Einstein saw the practice of theoretical physics as that of the invention of new mathematized physical concepts, possibly quite far from or even incompatible with our phenomenal intuition. According to Einstein, one could only approach the ultimate architecture of physical reality by means of such concepts (Einstein 1949, p. 47). Schrödinger held a similar view, in part following Heinrich Hertz's and Ludwig Boltzmann's thinking concerning the nature of physical theories (Schrödinger 1935, pp. 152–153). Probability and statistics would only be allowed in considering systems whose mechanical complexity was beyond tracking their behavior, on the model of classical statistical physics, while the predictions concerning the behavior the ultimate individual constituents of such systems or of nature were required to be ideally exact. This demand posed major difficulties in quantum theory on experimental grounds, given that, as noted, identically prepared quantum experiments, no matter how simple the quantum systems considered, in general lead to different outcomes, which make our predictions concerning them probabilistic or statistical. There are special cases, such as those of the Einstein–Podolsky–Rosen (EPR)-type experiments, where these predictions can be ideally exact, but only special cases, while otherwise our predictions are not exact, again, even as concerns the simplest possible quantum systems. None of these factors deterred Einstein, who never abandoned his belief that a future theory of quantum phenomena (or more fundamental phenomena that underlay them) would return to realism.

Thus, while it was seen by Bohr and a few around him as the commencement of a new era in theoretical physics and the relationships between mathematics and physics (Bohr 1987, v. 1, p. 51), QM was not welcome news to most physicists and philosophers, any more than Bohr's 1913 theory was, notwithstanding the remarkable successes of both theories, especially QM. The realist imperative remains dominant in physics and philosophy alike, and the hope that the development of fundamental physics will eventually obey this imperative remains as alive as ever. A fulfillment of this hope would enable one to bring QM or QFT under the umbrella of this imperative or will replace them with realist alternatives. Some, again, argue for realist interpretations of both theories in their present form.

This book will address the reasons for thus dominance and will comment on approaches to QM and QFT based on this imperative. The book's argument, however, is in accord with Bohr's view that QM brought with it a new era of the relationships among matter, thought, and experimental technology in fundamental physics, and gives this argument a broader scope and new features. The RWR view is a product of this era. The book also argues that, far from limiting creativity in physics, as it might appear to some to do, the RWR view has helped this creatively. It has done so from Bohr's 1913 atomic theory throughout the history of quantum theory, and it might continue to do so in the future. It does, however, change the nature of thinking and knowledge, because that which is beyond thought, ultimately nature itself in its ultimate constitution, becomes an irreducible and shaping part of thought, and thus of the relationships among matter, thought, and technology. Before quantum physics, modern physics allowed one to avoid dealing with that which cannot in principle be thought. Quantum physics brought the unthinkable into our thinking in physics and perhaps made it unavoidable there, or at least created the possibility, even if not necessity, to interpret quantum phenomena and quantum theory in this way, fully in accord with the experimental data defining these phenomena and the conceptual and mathematical structure of quantum theory, QM and QFT. Beginning with Bohr's 1913 atomic theory, the idea, the *thought*, that nature or at least some part of nature may be beyond thought has created new possibilities for thought, and it might continue to do so. One cannot be certain that it will, especially given that the current state of fundamental physics is incomplete in its account of the ultimate constitution of nature. But it might. Nothing in the current state of fundamental physics appears to preclude it either.

West Lafayette, USA Arkady Plotnitsky

References

Arthur, B.: The Nature of Technology: What It Is and How It Evolves. Free Press, New York, NY, USA (2011)

Bohr, N.: The Philosophical Writings of Niels Bohr, Vols. 3. Ox Bow Press, Woodbridge, CT, USA (1987)

Einstein, A.: Autobiographical Notes (tr. Schilpp, P. A.). Open Court, La Salle, IL, USA (1949)

Galileo, G.: Dialogues Concerning Two New Sciences (tr. Crew, H., De Salvio, A.). Prometheus Books, Amherst, NY, USA (1991)

Heidegger, M.: What Is a Thing? (tr. Barton, WB, Jr., Deutsch, V). Gateway, South Bend, IN, USA (1967)

Heidegger, M.: The Question Concerning Technology, and Other Essay. Harper, New York (2004)

Heisenberg, W.: Quantum-theoretical re-interpretation of kinematical and mechanical relations. In: Van der Waerden, B.L. (ed.) Sources of Quantum Mechanics, 1968, pp. 261–277. Dover, New York, NY, USA, Reprint (1925)

Heisenberg, W.: The Physical Principles of the Quantum Theory (tr. Eckhart, K., Hoyt, F.C.). Dover, New York, NY, USA, Reprint 1949 (1930)

Heisenberg, W.: Physics and Beyond: Encounters and Conversations. G. Allen & Unwin, London, UK (1971)

Kant, I.: Critique of Pure Reason (tr. Guyer, P., Wood, A.W.). Cambridge University Press, Cambridge, UK (1997)
Musil, R.: A Man Without Qualities, vol. 1 (tr. Wilkins, S.). Vintage, New York, NY, USA (1996)
Schrödinger, E.: The present situation in quantum mechanics. In: Wheeler, J.A., Zurek, W.H. (eds.), Quantum Theory and Measurement, 1983, pp. 152–167. Princeton University Press, Princeton, NJ, USA (1935)

Acknowledgments

I would like to express my special gratitude to G. Mauro D'Ariano, Gregg Jaeger, Andrei Khrennikov, and Athanase Papadopoulos, for their knowledge and friendship, which were essential in my work on this book. I additionally thank G. Mauro D'Ariano, for many discussions that were invaluable in clarifying several key subjects discussed in the book. My additional thanks also to Andrei Khrennikov for inviting me to many conferences at the Linnaeus University, previously Växjö University, the longest running conference ever on quantum foundations, where many of the ideas of this book were initially presented. I am also grateful to many participants of these conferences for helping in my thinking. I would also like to thank Henry Folse, Christopher A. Fuchs, Lucien Hardy, Emmanuel Haven, Laurent Freidel, Jan-Åke Larsson, Theo Niewenhuizen, Masanao Ozawa, and Paolo Perinotti, for productive exchanges. Most of them were among the participants of Växjö conferences as well.

I am grateful to Purdue University for its support of my work through several research leaves and awards, most especially the great honor of Distinguished Professorship, which also funded part of my work on the book.

My personal thanks to Paula Geyh, Inge-Vera Lipsius, and Marsha Plotnitsky.

I am grateful to Springer Nature for publishing the book, and I would like to thank those with whom I worked there, especially my Editor Elena Griniary for so kindly shepherding this project to its publication, and Francesca Ferrari for her work on the final stages of preparing the manuscript. My thanks to Martin Whitehead for his help with editing the manuscript.

Earlier versions of portions of several chapters have been published previously as articles "Nature Has No Elementary Particles and Makes No Measurements or Predictions: Quantum Measurement and Quantum Theory, from Bohr to Bell and from Bell to Bohr," *Entropy* 2021, 23, 1197, https://doi.org/10.3390/e23091197; "'Something happened': on the real, the actual, and the virtual in elementary particle physics," *The European Physical Journal*, https://doi.org/10.1140/epjs/s11734-021-00075-3; "The Unavoidable Interaction Between the Object and the Measuring Instruments: Reality, Probability, and Nonlocality in Quantum Physics," *Foundations of Physics* (2020) 50:1824–1858. https://doi.org/10.1007/s10701-020-00353-5; "Reality, Indeterminacy, Probability, and Information in Quantum Theory," *Entropy* 2020, 22,

747; doi: 10.3390/e22070747; "Transitions without connections: quantum states, from Bohr and Heisenberg to quantum information theory," *The European Physical Journal,* https://doi.org/10.1140/epjst/e2018-800082-6; "The Heisenberg Method: Geometry Algebra, and Probability in Quantum Theory," *Entropy* **2018**, 20 (9) 656; https://doi.org/10.3390/e20090656; "On the Character of Quantum Law: Complementarity, Entanglement, and Information," *Foundations of Physics* 47 (2017) 47: 1115–1154. However, the arguments of these articles were significantly revised and most of them indeed changed, sometimes leading to different conclusions.

Contents

Chapter 1
Introduction

> *In quantum mechanics, we are not dealing with an arbitrary renunciation of a more detailed analysis of atomic [quantum] phenomena but with a recognition that such an analysis is in principle excluded.*
>
> —Niels Bohr, "Discussion with Einstein on Epistemological Problems in Atomic Physics," (1949) (Bohr 1987, v. 2, p. 62)

Abstract This book primarily belongs to the *philosophy* of quantum theory rather than the *history* of quantum theory, and the main reasons for considering the history of quantum theory in the book are physical and philosophical, putting history in the service of physics and philosophy. Conversely, however, neither physics nor philosophy has ever existed or could have been possible without history. Every theory, no matter how innovative, has a history and depends on it, a circumstance more often used than reflected on. By contrast, this book takes advantage of it, by considering the thinking of founding figures of quantum theory, most especially Bohr, Heisenberg, Schrödinger, and Dirac, as reflected in this, historically oriented, introduction. Section 1.1 offers a general perspective of the philosophy and history of quantum theory, and their relationships. Section 1.2 considers three junctures of the prehistory of the RWR view, defined by the ideas of Kant, Riemann, and Einstein. Section 1.3 sketches the emergence of the RWR view of quantum theory, inaugurated by Bohr's 1913 atomic theory. Finally, Sect. 1.4 places the argument of this book in the context of the Bohr–Einstein debate, which still shapes our own debate concerning quantum foundations and the nature of fundamental physics.

Keywords History · Philosophy · Philosophy of physics · Reality · Realism · Reality without realism · Thinking

1.1 From History to Physics and Philosophy

Historically, this book follows the development of the RWR view of quantum theory, from Bohr's 1913 atomic theory, which inaugurates it, to the introduction of QM, which gives it its full-fledged form in the corresponding RWR-type interpretations, such as and, in particular, that of Bohr to the emergence of QFT, which (again,

A. Plotnitsky, *Reality Without Realism*,
https://doi.org/10.1007/978-3-030-84578-0_1

in the corresponding interpretations) gives it new dimensions, and then to its role in present-day quantum theory and its interpretation. This history is outlined in Sect. 1.3 of this introduction. The present section offers a general perspective on the philosophy and history of quantum theory, and their relationships. Moving the book's historical inquiry about half a century back in time, Sect. 1.2 considers three junctures of the prehistory of the RWR view, defined by the ideas of Kant, Bernhard Riemann, and Einstein. Finally, Sect. 1.4 places the argument of this book in the context of the Bohr–Einstein debate on, in Bohr's title phrase of his account of this debate, "epistemological problems in atomic physics" and the viability of the RWR view in fundamental physics (Bohr 1987, v. 2, p. 34). This debate still shapes our own debate concerning quantum foundations and the nature of fundamental physics, which continues with an undiminished intensity and, it appears, no end in sight.

This book, thus, offers the reader a hefty dose of history, sketched in this, mainly historical, introduction, and discussed in later chapters. Nevertheless, as my more philosophical preface would suggest, the project of this book does not primarily belong to *the history of quantum theory*. It belongs to *the philosophy of quantum theory*, considered here as a manifestation of the new relationships among matter, thought, and experimental technology, grounding the RWR view of quantum theory, the Q-RWR view. This view is ultimately extended by the book to the U-RWR view, as the RWR view of all physical phenomena. The main reasons for considering the history of quantum theory in the book are physical and philosophical as well, putting history in the service of physics and philosophy. Conversely, however, neither physics nor philosophy, including the philosophy of physics, has ever existed apart from or could have been possible without history. Every theory, no matter how innovative, has a history and depends on it, a circumstance more often used, even if without realizing it, than reflected on, although not missed by Bohr or earlier Ernst Mach (e.g., Mach 1988, pp. xxii–xxiii). By contrast, I would like to take advantage of this circumstance by considering the thinking of some of the founding figures of quantum theory, most especially Bohr, Heisenberg, Schrödinger, and Dirac, and, as my limits permit, a few others, such as Einstein, Hermann Weyl, John Archibald Wheeler, and Julian Schwinger, with the aim of contributing to a deeper understanding of QM and QFT now and thus to our own thinking concerning these theories.

We arrive at new ideas and theories by engaging with this history, which helps us to understand earlier ideas and to create our own, especially when earlier ideas are those of the likes of Bohr, Heisenberg, Schrödinger, and Dirac, or Einstein, in whose shadow the thought of all these figures has developed. This shadow, now appearing as a spirit, now as a ghost, is enormous and has spread over the subsequent history of fundamental physics and continues to spread over it. All these figures arrived at their ideas by engaging with their history. Their engagement with the preceding history of quantum theory, sometimes in each other's works, as in that of Heisenberg with Bohr's thinking or that of Dirac with Heisenberg's thinking (both cases considered in this book), may be more apparent and more important. But their thinking was also shaped by a longer history of physics and philosophy, from their more immediate predecessors, such as James Clerk Maxwell or Ludwig Boltzmann, to as far back as Aristotle and Plato, or the pre-Socratics, such as the Pythagoreans, the inventors of the mathematical view of the cosmos, or Leucippus and Democritus, the inventors of the atomistic view of nature.

Physics is thinking about what is true or probable about nature or those aspects of nature that physics considers, and sometimes, especially in dealing with foundational questions, it is, more philosophically, also thinking about how to think about physics. The latter type of thinking is more common in the philosophy of physics but is not absent in physics either, especially at times of crisis, such that which precipitated the invention of QM. In modern physics, which emerged with Kepler, Descartes, and Galileo, what is true or probable is determined by means of mathematical structures or models by which physical theories relate to the experimental data considered and interpretations of these theories and physical phenomena containing these data. These models may do so either by using a mathematical representation of the processes assumed to be responsible for these phenomena or data and predicting them on the basis of this representation, as in classical physics or relativity, or by only using a mathematical formalism to predict these data in the absence of such a representation, as in QM or QFT, at least in RWR-type interpretations. How close we come to understanding nature, including in its ultimate constitution, depends on our interactions with nature by means of mathematical theories and experimental technologies. These interactions are ultimately part of nature, too, but a particular part of it, specific to us, to our thinking and the technologies that we use, beginning with that of our bodies and brains, responsible for our thinking.

The establishment of physics as a mathematical-experimental science of nature was a major change enacted by modern physics, vis-à-vis that of the preceding thinking about it, from Aristotle on. This is not to say that this change was entirely discontinuous. For example, while essentially different in its mathematical nature and physical laws (mathematically expressed), the concept of motion in classical physics is still Aristotelian as governed by both continuity and classical causality. Nevertheless, making physics a *mathematical*-experimental science, with mathematics, as stated in the Preface, defining this conjunction, was a momentous change. This change was given a new dimension in quantum theory by virtue of abandoning altogether, in the final break with Aristotle, the concept of motion at the level of the ultimate constitution of the physical reality considered. On the other hand, Heisenberg eventually saw this final break with Aristotle, as a return, in physics, especially via the concept of symmetry in QFT, to Plato, a more mathematically oriented philosopher. Heisenberg had long-standing connections to Plato's thinking, whose dialogue *Timaeus* he read, in particular, the part dealing with Plato's mathematical and thus mental, rather than physical (as in Democritus) conception of atoms, during his work leading to his invention of quantum mechanics (Heisenberg 1962, pp. 59–75, 1989, pp. 71–88; Plotnitsky 2011, pp. 471–472). And yet, at the same time, Heisenberg also used Aristotle's concept of "*potentia*" in his interpretation of QM in his later thinking, thus bringing Plato and Aristotle together and confirming the complexity of the interplay of continuities and breaks in this history of physics, from the pre-Socratics on (Heisenberg 1962, p. 44).

But then, most breaks with the immediate history are returns of one kind or another to an earlier history, which, however, need to merely be returns to the same (that of Heisenberg to Plato was not). They are returns to some previous configurations of thought transformed by new configurations of thought. The originality of thinking, such as that of Heisenberg, one of the most original thinkers in twentieth-century physics, is always defined by this combination of difference and sameness, break and continuity, and transformation and tradition. It is radically original and deeply traditional at the same time, admittedly a difficult and sometimes paradoxical combination, nevertheless managed by Heisenberg or Bohr, or other key figures considered in this study. One way to accomplish this is what Wheeler described as a "radical conservative" way of thinking, which he expressly associated with Bohr's thinking but which is applicable to other figures considered in this study, certainly Heisenberg. The phrase refers to thinking shaped by "adhering to well-established physical law (be conservative) but follow these laws to their most extreme domain (be radical), where unexpected insight into nature might be found. [Wheeler] attributed that philosophy to his own revered mentor, Niels Bohr" (Misner et al. 2009, p. 40). Radical conservative thinking, however, is thinking in the first place, a key way of *thinking* about physics for Bohr and Wheeler alike.

The primary focus on thinking in physics extends the approach adopted in this author's previous work (e.g., Plotnitsky 2016), in the present study with a greater role given to mathematical thinking. This focus makes this approach different from those grounded in, and primarily dealing with, the logical and propositional structures of quantum theory, either in their own right or in addressing ontological and epistemological questions. Approaches of the latter type are especially common in the analytic philosophy of physics. These questions are, of course, essential, and they will be addressed in this study, sometimes by engaging with literature in the analytic philosophy of physics, which made important contributions to these subjects, although primarily along realist lines. But they will be considered as part of our *thinking* concerning quantum physics, in particular, its concepts on which this study will primarily focus. This does not prevent one from pursuing these questions with the same *philosophical rigor* (I shall define this concept, as different from that of mathematical or scientific rigor, presently) as that of the analytic philosophy of physics.

It is true that this study relies less extensively on the technical aspects of the mathematics of quantum theory than most work in the latter field does, thus potentially making the book appeal to a broader audience, which might be interested in it because the philosophical situation defining quantum theory has important connections with and implications for other fields of inquiry. This lesser *reliance* will, hopefully, not disappoint some more mathematically inclined readers of the book, given its actual content. Its lesser reliance on the technical engagement with and expertise in mathematics does not mean that the book is not concerned with mathematics or that it does not contain it or even is about it, including some technical mathematics of QM or QFT. A major portion of the book (about 30% of its space) concerns both the workings of mathematics in QM and QFT, and the history and philosophy of mathematics, especially in considering the relationships between algebra and geometry. A section of the book is devoted to these relationships in twentieth-century mathematics itself. This section is part of Chap. 4, which offers a discussion of Heisenberg's work from the perspective of these relationships. Chapter 5, on Schrödinger, is largely mathematical in its content as well. There is also a discussion, admittedly, more conceptual than technical, of category theory in Chap. 9. In reality, this book engages with mathematics no less than most philosophical treatments of quantum physics, even if in part in a different way, in part, because sometimes it does so similarly as well. As I explained in the Preface, quantum theory emerges as a new *mathematical*-experimental science of nature, with mathematics defining this conjunction, as it does in most modern physics, from Descartes and Galileo on, because, to return to Heidegger's assessment, "modern science is experimental because of its mathematical project." This book is, accordingly, about mathematical thinking in physics and, via physics, in mathematics itself. In particular, it relates quantum theory to, and even sees it as form of, the Pythagorean thinking that emerged with the pre-Socratics and has accompanied mathematics throughout its history into our own time as, on the one hand, proto RWR thinking and, on the other, as thinking defined by the relationships between geometry and algebra (in the Pythagoreans, arithmetic). The book also offers a new perspective on mathematics' divorce from its connections to physics, which has defined the development of mathematics from roughly 1800, especially "modernist mathematics," from around 1900 on (a concept explained in Chap. 4). This perspective arises because, beginning with Heisenberg, quantum theory found a new way of using the abstract mathematics that emerged from this divorce, a way very different from that of classical physics or relativity, while at the same time reflecting the new relationships between algebra and geometry in modernist mathematics. This book is about the new type of the relationships between mathematics and physics that emerged with Heisenberg's discovery of QM, in which I comment in closing this Introduction.

The philosophical rigor, which I invoked above in connection with this book's project, is not the same as (technical) mathematical rigor in mathematics or physics. It is also worth keeping in mind that a rigorous use of mathematics in physics is not the same as mathematical rigor in mathematics itself, based on strict, or at least stricter, definitions, proofs, and so forth. The use of mathematics in physics has even been criticized by mathematicians and philosophers, although there are also defenders of this use even among mathematicians. This is, however, a separate subject. Most (it is difficult to think of exceptions) technically oriented books in the analytical philosophy of physics are not pursuing calculations which relate the mathematical formalism of physical theories to numbers observed in experiments or related to those observed in experiments. Without doing so, even if only potentially at some future states of a given theory, there is no mathematical rigor of physics.

This is, naturally, not a criticism of such projects in the analytic philosophy of physics. The philosophical rigor that I have in mind is something else. While it is, necessarily, the rigor of logic, it is, as understood by this study, most essentially the rigor of concepts, as discussed in detail in Chap. 2. I also argue, for the reasons explained in Chap. 2 as well, that the role of concepts and, thus, the rigor of concepts are equally essential to mathematics and physics, especially in their creative aspects, a role emphasized by Riemann and Einstein, both of whom the book follows on this point. It is also worth keeping in mind that, while requiring logical, conceptual, or other rigor, thinking in philosophy, mathematics, or physics is never reducible to rigor. This book will address this unavoidable supplement of rigor, to the degree possible, because this supplement does not readily, and sometimes not at all, lend itself to a (rigorous) analysis. One might, for example, surmise and plausibly discern some of the reasons for Heisenberg's invention, ultimately a guess, of his concept of matrix variables, one of the great mathematical inventions of quantum theory, with a mathematically rigorous connection to the experimental data he aimed to account for. It is, however, difficult to assume that one can ever really explain this guess, or the great guess of Schrödinger, his invention of the complex wave function and his time-dependent equation, discussed in Chap. 5, entitled "Schrödinger's Great Guess." The project of this study is shaped most essentially by the role of concepts, physical, mathematical, and philosophical (or concepts combining these types of concepts) and the rigor of concepts in quantum theory.

As Weyl said, in opening his classic book, *Space Time Matter*, a great example of an investigation focused on the role of concepts (a focus common to Weyl, and several of his main predecessors, in particular, Riemann, Hilbert, and Einstein), "Philosophy, mathematics, and physics have each a share in the problems presented [in the book]," problems defined by the rise of relativity theory as fundamentally different from classical physics (Weyl 1952, p. 2). This is also true, with equally or even more profound implications, for philosophy, mathematics, and physics, and their relationships in quantum theory, with concepts playing the main role in each of these fields of thought and their relationships.

1.2 Toward RWR Thinking, with Kant, Riemann, and Einstein

One can trace some among the ingredients of the idea of reality without realism to as early as pre-Socratic philosophy, in particular, the Pythagoreans and their discovery of incommensurable magnitudes (which we now understand in terms of irrational numbers), and in modern times, especially Kant's philosophy. The latter, as noted in the Preface, may be considered as the main modern precursor of the RWR view. There are also important connections between Kant's philosophy and the question of causality, central to the history of quantum theory. The question of causality and the role of Kant's philosophy in considering it will be discussed later. Before, however, I consider Kant, I would like to connect the concept of reality without realism to Riemann's view of physical reality (in this case that of space), which was both influenced by and departed from Kant. This view was offered in his famous Habilitation lecture, "The Hypotheses That Lie at the Foundations of Geometry" (Riemann 1854), given in 1854 but not published until 1868, 2 years after Riemann's death. As other works of Riemann, a uniquely important figure for the development of modern mathematics and physics, the lecture had a momentous impact on both fields. Einstein's general relativity is the most famous example of its significance in physics, but far from the only one.

My main reason for choosing Riemann's lecture as my starting point in this section is that Riemann's view of physical reality, in particular, that underlying what we experience as space, but by implication, in general, emerged from modern physics, rather than from philosophical considerations, although philosophy permeates Riemann's lecture and his related works. These works deal extensively with the relationships among physics, mathematics, and philosophy, and Riemann's lecture traces the genealogy of his ideas about geometry to thinking in all three fields. The lecture offered a radical reconceptualization of space and geometry. The concept of space considered by Riemann was physical, mathematically conceptualized in terms of his concept of manifold [*Mannigfaltigkeit*], which has played a major role in mathematics and physics ever since. Riemann spaces and geometry encompassed both Euclidean and non-Euclidean geometry (of both negative and positive curvatures). Euclidean space and geometry became merely special cases of these more general concepts, rather than the primary or even the only forms of them, as they had been before Riemann. Riemann's reconceptualization of geometry also placed non-Euclidean geometry (of negative curvature), discovered two decades earlier, in a properly comprehensive mathematical scheme. In addition, the lecture introduced spaces and geometries of positive curvatures, generalizing the concept of spherical geometry (which was only two dimensional) to three dimensions, and mathematically, to any number of dimensions.

The discovery of non-Euclidean geometry around 1830 already implied that physical space may not be Euclidean, which gave this discovery a revolutionary significance beyond mathematics. No measurements made at the time showed any deviation from Euclidean geometry in physical space in the immediate vicinity of the Earth or beyond. The possibility was momentous enough, however. The situation became even more dramatic with Riemann. His argument implied that the infinite number of possible types of manifolds and geometries could be associated with physical space, which also included space and geometries of variable curvatures, another major innovation of Riemann, central for general relativity. Any such association is a hypothesis, which was what "the hypotheses" of Riemann's title referred to. As such, it is subject to verification, qualification, refinement, and so forth. This process can rule out some among possible geometries as suitable for physics or require different geometries at different scales (Riemann 1854, p. 23). This is how the situation appears in modern physics and cosmology. Courtesy of Einstein's general relativity, we know well certain (more or less) local geometries, say, the one, curved, in the vicinity of the solar system or the earth, or more esoteric ones, for example, in the vicinity of black holes. It is, however, more difficult to be certain concerning the ultimate geometry of the Universe, although the current cosmological data suggests that it is, on the average, flat, as far as one can observe it.

In closing his lecture, Riemann used an astutely chosen phrase: "a reality underlying space" (Riemann 1854, p. 33). This phrase implies, on Kantian lines (although Riemann rejects Kant's view of the a priori nature of our intuition of space or time), that this reality may not be spatial in our phenomenal sense. It could, for example, have a dimension higher than three and may even (another remarkable insight of Riemann) be infinite-dimensional. The concepts of higher dimensional spaces, with different geometries, came into prominence in the wake of the publication of Riemann's lecture and have become ubiquitous in the twentieth- and twenty-first-century physics, more recently in string theory and other approaches to quantum gravity, or in mathematics itself. There, however, the term "space" refers to mathematical objects, such as Hilbert spaces used in QM and QFT. Riemann also stated that the reality underlying space could be discrete and thus have the topological dimension zero. This would imply a different physics, because it would not allow one to use differential calculus, which had grounded classical physics since Newton (Riemann 1854, p. 33). The ideas of discrete spaces or spaces of higher (if not infinite) dimensions, or the corresponding forms of reality underlying three-dimensional space, were entertained before Riemann, including, in the case of higher dimensions, by Kant in his early (1747) work, although Kant rejected the idea (Kant 2012; de Bianchi and Wells 2015). In part following Riemann's rethinking of the nature of geometry, the concept of discrete space was developed in mathematics and has been sporadically used in physics, including in quantum theory, for example, by Heisenberg in the 1930s in trying to deal with difficulties of QFT, and more recently in some attempts at developing quantum gravity. If the reality underlying space (or time), or what phenomenally appears as such, as a continuous manifold, is discrete, for example, at Planck's scale, making space an emerging property, this reality would imply a very different physics. In any event, the fundamental physics in discrete space (or time) will be essentially different, as Riemann surmised nearly two centuries ago. Riemann's comments on the nature of physics in the lecture have a great subtlety. He says, in particular:

> Still more complicated relations can occur if the line element cannot be represented, as was presupposed, as the square root of a differential expression of the second degree. Now it seems that the empirical notions on which the metrical determinations of space are based, the concept of a solid body and that of a ray of light, lose their validity in the infinitely small; it is therefore quite definitely conceivable that the metric relations of space in the infinitely small do not conform to the hypotheses of geometry; and in fact one ought to assume this as soon as it permits a simpler way of explaining phenomena. (Riemann 1854, p. 32)

Riemann initially invoked the immeasurably small [*Unmessbarkleine*], also in its direct sense of that which cannot be measured, the questions concerning which he considers important, as opposed to those concerning the immeasurably large [*Unmessbargrosse*], which "are idle questions for the explanation of nature [*die Naturerklärung*]" (Riemann 1854, p. 32). Riemann offers no justification for the last assessment, and I shall put it aside. These questions may not be as idle as Riemann thought given present-day cosmology, most of which, however, still deals with *measurably* large scales, extremely large as they may be. Riemann then shifts from the immeasurably small to "the infinitely small [*Unendlichkleine*]," the infinitesimally small. Riemann sees the mathematical representation of space, or time, or physical processes in space and time, offered by classical physics, as defined by the kinematical and dynamical principles established by the figures he mentions in the lecture—Archimedes, Galileo, and Newton (Riemann 1854, p. 32). Riemann also sees physics as based, mathematically, on the principles of differential calculus, which is an analysis of the infinitely small. This is not the same as the immeasurably small, but it provides the proper mathematical representation of the physical concepts just mentioned, which explains Riemann's shift from the immeasurably small to the infinitely small. Riemann, thus, envisions not only that space in the immeasurably small may not conform to the hypothesis of Euclidean geometry, but also that it may not conform to the axioms of geometry at all, for example, to those of Riemannian geometry, used by Einstein in general relativity. In the latter case, the concept of metric relations still applies, although these relations are, in general, non-Euclidean and allow for a variable curvature. Bringing together gravity and quantum theory may change this. Riemann suggests next that "the reality underlying space" may be "a discrete manifold":

> The question of the validity of the hypotheses of geometry in the infinitely small is bound up with the question of the basis for the metric relations of space. In connection with this question, which may indeed still be ranked as part of the study of space, the above remark is applicable, that in a discrete manifold the principle of metric relations is already contained in the concept of the manifold, but in a continuous one it must come from something else. Therefore, either the reality underlying space must form a discrete manifold, or the basis for the metric relations must be sought outside it, in binding forces that act upon it.
>
> An answer to these questions can be found only by starting from the conception of phenomena which has hitherto been approved by experience, and for which Newton laid the foundation, and gradually modifying it under the compulsion of facts that cannot be explained by it. Investigations like the one just made here, which begin from general concepts, can only serve to ensure that this work is not hindered by unduly restricted concepts and that progress in comprehending the connection of things is not obstructed by traditional prejudices. This leads us away into the domain of another science, the realm of physics, into which the nature of the present occasion does not allow us to enter. (Riemann 1854, p. 33)

If the reality underlying space is discrete, then the ground of the metric relation of the corresponding manifold is given by the *mathematical* concept of this manifold as such, while if it is continuous, this ground is given by the concept of phenomena justified by experience and, in physics, defined by the physical principles arising from the nature of these phenomena. Such principles may, however, need to be changed "under the compulsion of [new] facts." This argumentation explains and reveals deeper implications of Riemann's phrase "the reality underlying space" as referring to the difference between our phenomenal concept of space and the nature of physical reality underlying it. This difference, extrapolated to physical reality in general, suggests the possibility of the RWR view, especially in the immeasurably small, the physics of which was rethought by quantum theory, eventually giving rise to the RWR view and the corresponding interpretations of quantum phenomena and quantum theory.

I am not saying that Riemann ever adopted the RWR view and saw the reality underlying space as being beyond representation, discrete or continuous, flat or curved, three-dimensional or higher dimensional (all of which possibilities Riemann, remarkably, entertained in 1854), let alone beyond conception. While Riemann's view of physical reality, expressed in the phrase "the reality underlying space," may only be one step away from the RWR view, this step is not easy to make. It had, I argue in this study, not been made in physics before Bohr's 1913 atomic theory. If anybody, Kant might have been closer at least to the weak RWR view, still, as explained below, without quite reaching it. This proximity arises because Kant denied that space, or time, or (classical) causality or any phenomenal entities, conceptual or not (as space and time were assumed by him to be forms of our phenomenal intuition rather than concepts) represented the ultimate nature of reality, material, or mental, which he saw as a noumenal object, a thing-in-itself. Riemann does not appear to have held this type of view, and he rejected the a priori nature of our sense of spatiality or temporality. Kant's scheme, however, defined by distinguishing noumena or things-in-themselves, as *objects* (referring to the ultimate nature of reality that is beyond a representation or knowledge, if not a conception) and *phenomena* (which belong to human thought and can be known), might apply even if our intuition of space and time has emerged from experience. Kant was right to argue that our mind *actively* shapes or constructs, rather than merely perceives, the world we observe, even if it does not entirely define the world, because the world still affects this construction, both independently and via our bodies. This aspect of Kant's philosophy is strongly supported by modern neuroscience, which, on that score, follows Kant, as opposed to John Locke, who assumed that our knowledge of the world is the product of our accumulated experience, starting with the initial tabula rasa of thought. Hence, the technologies of our bodies and brains, which make possible our thought, create the world we experience rather than merely enabling us to perceive it. This does not mean that we cannot think about the reality underlying the world that we thus experience, or think about nature, especially in its ultimate constitution, as different from what we thus create, quite the contrary. The possibility that such may be the case was Riemann's or Kant's point. It is also the grounding assumption of the U-RWR view, which, in its strong form, places the ultimate constitution of nature beyond thought. This placement, however, is still a product of thought. Thought cannot escape itself even in thinking about that which is beyond thought. On the other hand, it is capable of thinking that something may be.

The strong RWR view is, then, manifestly more radical than that of Kant. While Kant places things-in-themselves beyond representation or knowledge, he allows that a conception of them could be formed and accepted as at least practically justified and even could in principle be true, although this truth cannot be guaranteed, a qualification that gave Kant's epistemology greater subtlety (Kant 1997, p. 115). Georg W. F. Hegel, in part building on Kant's view of the possibility of such a conception, and giving concepts the defining role in philosophy, appears to have allowed human thought a greater, even if never complete, power of forming a true conception of the ultimate nature of reality, through the formation of rigorous concepts (Hegel 2019, p. 48). Both Riemann and Einstein are closer to Hegel as concerns this possibility and in giving the primary role to the formation of concepts, in this case physical concepts, suitably mathematized, in this pursuit of a mathematically idealized representation of physical reality. For the moment, Kant's position (or that of Hegel's) is thus manifestly short of the strong RWR view, which places the ultimate character of the reality responsible for quantum phenomena beyond conception, although, as noted in the Preface, this view is assumed to be only practically justified as well.

While Kant's argumentation is closer to the weak RWR view, it does not amount to this view either. This is because, once such a conception is practically justified, a representation of things-in-themselves based on this conception becomes practically justified as well, even though it is impossible to speak of any knowledge of things-in-themselves. Accordingly, Kant's scheme may ultimately be seen as a form of realism, moreover, classically causal in character, which is excluded by the RWR view, because classical causality implies a conception, at least a partial one, of the ultimate character of the reality considered.

In the RWR view, as the Q-RWR view, what is practically justified is not any given representation or conception of the ultimate nature of reality responsible for quantum phenomena, but the *impossibility* of any such representation (in the weak RWR view) or conception (in the strong RWR view), either as things stand now, even if such a conception is in principle possible, or ever, in which case forming such a conception is precluded. The U-RWR view extends this situation to the ultimate constitution of nature and thus to the reality ultimately responsible for all physical phenomena, while allowing one to consider, with a practical justification, that some physical theories may be treated as representing the objects considered and their behavior. The absence of classical causality, again, follows automatically, while the probabilistic relationships between events are possible. If such a conception becomes possible at some point, at least as practically justified, then a representation based on this conception becomes also possible, again, at least as practically justified. This would bring the situation into accord with Kant's argumentation just described and thus realism. As indicated in the Preface, however, and discussed in detail in Chap. 2, the qualification "as things stand now" still applies, even if one assumes that no conception of the ultimate nature of reality is ever possible. It applies because even this assumption may become obsolete in physics, either as the Q-RWR view, applicable in quantum theory (because it may be replaced by a realist alternative in interpreting quantum theory in its present form or by an alternative realist theory of quantum phenomena), or as the U-RWR view, if there is no longer physics to justify this view. The impossibility of forming a conception of some stratum of reality will then only retain its value as a philosophical position, or will apply elsewhere. For example, the ultimate nature of the workings of our brain may prove to be beyond our capacity to conceive it.

Riemann and then Einstein went beyond merely assuming that a practical justification for a given theory is sufficient, although they would have admitted that sometimes it may be. In giving, as Hegel did, concepts the primary role in this process, they thought that a theory ideally approaching how nature ultimately works was, in principle, possible, by means of concepts. In physics, one only deals with a mathematically idealized representation of physical reality within the scope of a given theory, such as classical physics or relativity, which may disregard other aspects of nature, aspects that may not be open to such a representation. The question is, however, whether all physical reality could ever be made available to such a representation or whether the connections between different domains or scales so represented could or could not ultimately be established, an alternative sometimes known as that between "reductionism" and "emergence." Kant, again, would not deny the possibility of reaching, at least ideally or in part, the ultimate nature of reality by means of human thought either, although his overall scheme would make such a claim difficult for any given conception in any domain of investigation. Riemann, Einstein, and Kant, or Hegel, all, again, equally denied the empiricist view that our experience alone could be sufficient for understanding how nature works. For them, this understanding was only possible by means of concepts and thus required the invention of new concepts. Relating experimental evidence and theoretical thinking by means of concepts allows one to conclude that the geometry of space is not Euclidean and our physics, such as that of general relativity, is not Newtonian, as against Kant's view or at least his ultimate view.

Kant's argument (which is subtler than it might appear and especially than it was often portrayed) concerning the Euclidean and Newtonian nature of the physical world, extensively discussed in commentaries on Kant, is beyond my scope. I would like, however, to note that the possibility that the geometry of space is non-Euclidean does not disprove Kant's scheme, as it is sometimes contended. Were non-Euclidean geometry available to him, Kant might have been more open to this possibility than others, some *Kantians* (whose thinking should be distinguished from that of Kant) among them. Kant might have been incorrect in assuming that the ultimate nature of space is Euclidean or is governed by Newtonian physics, and both assumptions could be questioned, against Kant's own grain, even from within Kant's argumentation. He might, however, have been right insofar as our basic phenomenal spatial intuition is Euclidean (although this claim is under debate in present-day cognitive psychology) and that our basic phenomenal physical intuition is Newtonian. Or, in Bohr's and Heisenberg's more precise terms, Euclidean geometry and classical mechanics are mathematized refinements of our general phenomenal intuition of space, time, and motion.

As discussed in Chap. 2, however, one could only partially rely on this intuition already in special relativity, the first theory that showed the incapacity of our phenomenal intuition to grasp how nature works. One could no longer rely on this intuition at all (except as heuristic help) in developing the mathematical formalism of QM or QFT, or even part ("quantum jumps") of Bohr's 1913 theory. In the case of relativity, however, as a realist and classically causal (indeed deterministic) theory, the mathematical formalism still offered an idealized representation of the physical processes considered. In the case of QM, at least in RWR-type interpretations, one could only rely on the abstract mathematics of the formalism and its ability to relate to the data observed in quantum experiments in terms of probabilistic or statistical predictions. No other predictions concerning quantum phenomena are, again, possible.

1.3 The Rise of RWR Thinking

While one can trace the *prehistory* of the RWR view to earlier thinking, even from the pre-Socratics on, and in modern time, to Kant and post-Kantian thought, the *history* of this view as such begins with quantum physics, inaugurated by Planck's discovery of his black-body radiation law in 1900. It took, however, another decade before the difficulties and paradoxes of quantum theory led to the RWR view. It emerged as a result of three revolutionary discoveries—Bohr's discovery of his 1913 atomic theory, Heisenberg's discovery of QM, and Dirac's discovery of QED, especially his relativistic equation for the free electron in 1928, which led to the discovery of antimatter in 1931—and crystalized with Bohr's ultimate interpretation of quantum phenomena and QM by the late 1930s. I summarize each of these three discoveries.

In conflict with classical mechanics and classical electrodynamics alike (which could not, however, account for the manifested stability of atoms as electromagnetic systems, a fact that served as a major impetus for Bohr), Bohr's 1913 atomic theory introduced the concept of "quantum jumps," as discontinuous transitions between so-called stationary quantum states of electrons. In each quantum jump, discrete quanta of radiation, hv, was emitted (or absorbed) in accordance with Planck's light-quantum theory. Stationary states were represented by electrons orbiting nuclei, in accordance with the laws of classical mechanics, without radiating energy while in orbit, thus in contradiction to classical electrodynamics. By contrast, quantum jumps could not be so represented, and appeared to allow for no mechanical representation or even conception. Bohr's theory was a radical transformation of the nature of thinking in physics.

Building on Bohr's thinking, Heisenberg, in his 1925 approach to QM, abandoned an orbital representation of stationary states, replacing them with just energy levels of electrons, while retaining Bohr's concept of discontinuous transitions, quantum jumps, between them. Heisenberg's theory only predicted the probabilities of transitions between these energy levels, without providing any representation of electrons themselves and their behavior. These transitions, moreover, were only manifested as effects of the interactions between electrons and measuring instruments, reflecting the irreducible role of experimental technology in the constitution of quantum phenomena. Heisenberg's approach was grounded in the circumstance (still found in quantum physics) that no observation of quantum objects or their independent behavior is possible. Nobody has ever observed, at least thus far, an electron or photon as such, or any quantum object, no matter how large, and some, such as Bose–Einstein condensates, are large, although their quantum nature is defined by their microscopic constitution. It is only possible to observe traces, such as spots on photographic plates, left by their interactions with measuring instruments. The existence of quantum objects is inferred from these traces. Heisenberg also introduced a new form of mathematics, never previously used in physics, in effect (he did not use these terms), that of Hilbert spaces over $\mathbb{C}$ and noncommuting operators there. This mathematics enabled the probabilistic predictions of quantum events observed in measuring instruments, as effects of their interaction with quantum objects.

Dirac's introduction of QED, especially his discovery of his relativistic equation for the electron, in 1928, was the third key discovery in the history of the RWR view. Dirac's theory of the relativistic electron, by means of a new mathematical theory (while still using Hilbert spaces over $\mathbb{C}$ and noncommutative operators there), added a new dimension that of particle transformation and multiplicity to quantum physics. Unlike in low-energy quantum regimes handled by QM, or the whole preceding physics, it was no longer possible to maintain the identity of a quantum object, such as an electron, even in a single experiment. Dirac's equation was an equation for both the electron and the positron, although it took a few years before Dirac proposed the existence of the positron in 1931. In any high-energy experiments, dealing with quantum-electrodynamical processes, either an electron or a positron (or a photon or an electron–positron pair) could be observed, keeping in mind that, in the RWR view, this only means that the corresponding effect can be registered in measuring instruments. In high-energy quantum regimes and QFT, even in a single experiment, one can no longer deal with particles of the same type, but had to deal with particles of other types, transforming into each other. It is this circumstance that adds the irreducibly multiple to the irreducibly inconceivable found in RWR-type interpretations of low-energy quantum phenomena and QM, giving Dirac's equation and the discovery of antimatter their significance for the history of the RWR view.

I am not saying that other events did not play their roles in the history of the RWR view, including several preceding the discovery of QM, such as Louis de Broglie's 1923 theory of matter wave; Satyendra Nath Bose's discovery of the Bose–Einstein statistics in 1924; and the 1925 discovery of spin by Ralph Kronig, George Uhlenbeck, and Samuel Goudsmit. I shall discuss all three in Chap. 3. Then, of course, as arguably the most significant of these developments, there is Bohr's interpretation of quantum phenomena and QM, to which a large portion of this study is devoted. This interpretation was, however, not a discovery of a new theory, although it did introduce new physics, defined by the irreducible role of measuring instruments in quantum phenomena, a view anticipated by Heisenberg. Schrödinger's co-discovery of QM was important as well. His wave mechanics played little role in the history of the RWR view. As indicated in the Preface, however, in order to bring his equations (time independent and then time dependent) into accord with the experimental data, Schrödinger introduced certain mathematical features that were crucial to the mathematical formalism of QM. The importance of these features compels me to devote a separate chapter, Chap. 5, to his derivation, especially of his time-dependent equation. This equation had remarkable new features, in particular, a complex wave function. It also contained the first-order derivative in time, t, vs. the second-order derivatives in spatial coordinates (x, y, z), which required a complex wave function. Both features are essential to QM, including its probabilistic aspects. The first-order derivative in t was a defining feature of Dirac's equation, also important in establishing Schrödinger's time-dependent equation as the non-relativistic limit of Dirac's equation, a key aspect of Dirac's derivation of his equation.

Several subsequent developments were also significant in the history of the RWR view, including as concerns the idea, adopted here, that quantum objects are idealizations defined only at the time of measurement, rather than referring to something existing independently. Among them were Bohr and Léon Rosenfeld's 1933 analysis of measurement in QFT, updated in 1950 (Bohr and Rosenfeld 1933, 1950); the rise of the theory of nuclear forces in 1930 (with Heisenberg, again, playing a key role, although the main credit belongs to Hideki Yukawa, who was the first to consider the strong force as an independent force of nature); the development, specifically renormalization, of QED in the 1940s by Richard Feynman, Julian Schwinger, Shinichiro Tomonaga, and Freeman Dyson; the development of the standard model of elementary particle physics; and more recently the rise of quantum information theory, to which this study devotes a chapter as well (Chap. 9). In any event, as concerns the RWR view, none of the developments just listed was as important as the three discoveries described above.

In accordance with the history just outlined, this book has an overarching historical trajectory, proceeding from the prehistory of the RWR view, sketched in this Introduction, to Chap. 3, anchored in Bohr's 1913 atomic theory; then to Chap. 4, anchored in Heisenberg's discovery of QM and Chap. 5, in Schrödinger's discovery of his time-dependent equation; to Chap. 6, anchored in Bohr's interpretation of quantum phenomena and QM, and his concept of complementarity; to Chap. 7, anchored in the EPR experiment, and the question of quantum nonlocality; to Chap. 8, anchored in QFT; and finally, to Chap. 9, anchored in quantum information theory. (I shall explain my repeated use of "anchored" presently.)

As stated at the outset of this Introduction, however, this book belongs to *the physics and philosophyof quantum theory* rather than to *the historyof quantum theory*, and the main reasons for considering the history just outlined are, again, physical and philosophical. Each chapter offers a different angle, defined by one of the main discoveries in question and their implications (hence my use of "anchored" above), on the nature of quantum theory. Even as concern these perspectives themselves, it is, in each chapter, a matter of focus on one of them in relation to, rather than as separate from, the others. Chapter 2 is outside this historical sequence: it offers a general discussion of the RWR view and the key concepts involved. I close this Introduction with the Bohr–Einstein debate, which will shadow the argument of this book, just as it has shadowed the debate concerning quantum phenomena and QM for a century by now.

1.4 "Logically Possible Without Contradiction": The Bohr–Einstein Debate and the Nature of Quantum Theory

At stake in the Bohr–Einstein debate was not only the viability of quantum theory, QM or QFT, but also the nature of fundamental physics as a mathematical-experimental science, with mathematics, again, defining this conjunction, for Einstein as a form of conceptual realism. The conceptual realist imperative governed Einstein's thinking throughout his life, in particular, his work on general relativity, where the guidance of this imperative proved to be so successful. Bohr, by contrast, was ready to abandon this imperative even in his 1913 theory and found it to be no longer viable after the discovery of QM. He, briefly, returned to a realist view in his Como lecture of 1927, a view, however, quickly abandoned by him, following his exchange with Einstein at the Solvay Conference in Brussels about a month later. This discussion initiated their three decade-long debate, which ended only with Einstein's death in 1955.

It is a separate question how important this imperative was for Einstein's physics, as opposed to the mathematical and physical features that defined it. Thus, the guidance of this imperative notwithstanding, Einstein had not been able, after an unrelenting effort lasting 40 years (during which he also continued to ponder quantum foundations), to unify gravity and electromagnetism along the lines of a realist field theory of the type he aimed at, that of continuous fields. To do so was arguably impossible apart from taking QFT into account, which Einstein was reluctant to consider in view of his belief in the type of theory he pursued and his lack of confidence that QM or QFT could lead to such a theory (Einstein 1936, p. 378).

In any event, it was his conceptual realist imperative that grounded Einstein's discontent with QM, at least if the latter was understood on RWR lines, as it was by Einstein and by Bohr, with very different assessments of the RWR view itself. Einstein never accepted that QM or QFT offered a "useful point of departure for future development," as he stated as late as 1949, more than two decades after the introduction of QM and in the immediate wake of major breakthroughs in QED (related to its renormalization) (Einstein 1949, p. 83). Admittedly, these breakthroughs could have hardly affected Einstein's view. They did not convince many others either, such as Dirac, the founder of QED, who by the 1930s no longer saw it as ultimately viable either. These doubts in QFT, especially on the account of the manifested infinities of the formalism requiring renormalization, have never disappeared and are still around.

Einstein's discontent with QM is most widely known by his pronouncement that "God doesn't play dice" (e.g., Born 2005, p. 88). It is clear, however, that this discontent had more to do with the difficulty and perhaps impossibility of reconciling QM or QFT with his realist imperative. Einstein realized, just as Bohr did, that the generally probabilistic nature of quantum predictions is implied by the absence of realism, or by RWR-type interpretations, because of the impossibility, in the absence of realism, of attributing classical causality to the ultimate nature of the reality responsible for quantum phenomena. Einstein's arguably final pronouncement on the subject, undoubtedly with quantum (EPR type) correlations in mind, was this: "That the Lord should play dice, all right; but that He should gamble according to definite rules, this is beyond me" (reported in Wheeler and Zurek 1983, p. 8). It is sometimes said that quantum dice are loaded in view of such phenomena as quantum correlations (clearly on Einstein's mind here), discussed in Chap. 7. In fact, however, the language of dice is misleading. The workings of quantum probability or statistics are very different from those of classical physics, to which a throw of the dice belongs and which has nothing similar to quantum correlations. Accordingly, Einstein's statement confirms and even amplifies, rather than tempers, his discontent with QM.

Bohr characterized Einstein's position, expressed in Einstein's 1936 article "Physics and Reality" (a position that remained essentially the same thereafter), as follows:

> Einstein ... [argued] that the quantum-mechanical description is to be considered merely as a means for of accounting for the average behavior of a large number of atomic systems, and his belief that it should offer an exhaustive description of the individual phenomena [was] expressed in the following words: "To believe this is logically possible without contradiction; but it is so very contrary to my [Einstein's] scientific instinct that [he could not] forego the search for a more complete conception" [Einstein 1936, p. 375]. (Bohr 1987, v. 2, p. 61)

Einstein's point was given by him as an additional justification in his article. His statement, cited by Bohr, replied to the following rhetorical question: "Is there really any physicist who believes that we shall never get an inside view of these important alterations in the single system, in their structure and their [classically?] causal connections, and this regardless of the fact that these single happenings have been brought so close to us, thanks to the marvelous invention of the Wilson [cloud] chamber and the Geiger counter?" (Einstein 1936, p. 375). Einstein's question is not unreasonable. It is not easy to believe that it is possible to ever get an inside view, also literally in the sense of visualization [*Anschaulichkeit*], of the inner workings of individual quantum objects from QM as a probabilistic or statistical theory predicting the outcome of quantum experiments, without describing how they physically come about in terms of independent physical reality. There is an additional wrinkle, insofar as it is still possible to see QM, in Bayesian terms, as providing probabilistic predictions even in the case of individual quantum elements, and even in dealing with most elementary individual quantum objects, such as elementary particles. I shall consider this subject in Chap. 2. Either way, one is likely to need a different theory to have such a description, such an "inside view," and thus to satisfy Einstein's realist imperative, *if* quantum phenomena will ever allow one to do so, which is a big if. As I shall suggest presently, it was this "if" that ultimately defined the Bohr–Einstein debate.

That quantum-mechanical predictions were fully in accord with all available experimental evidence did not deter Einstein, in part, because several of his arguments, concerning EPR-type experiments for the incompleteness, or else nonlocality, of QM, were based on predictions with probability one, ideally possible in these experiments (Einstein et al. 1935). In fact, as discussed in Chap. 7, Einstein has continued to believe that QM could not actually predict all that was possible to establish as real in considering individual quantum system, unless one allows for Einstein-nonlocality, as an action at a distance. Also, even though, on this occasion Einstein appealed to observations of quantum phenomena in the Wilson chamber and the Geiger counter, Einstein's belief in the power of mathematized physical concepts, which by the time of his comments under discussion had reached its height, sometimes led him to diminish the significance of experimental evidence (van Dongen 2010, pp. 89–95). This evidence, he argued, could change. It could. But then it also could not. Or it could change so as further to confirm a given theory, as has happened in the case of QM or QFT thus far, including, recently, in LHC experiments, which, some hoped, would have challenged QFT or even proved it to be incorrect. These experiments still might do so. Even if a more radical change is required, however, a new theory may still be quantum or allow for RWR-type interpretations.

Bohr, by contrast, argued that "there could be no other way to deem a logically consistent mathematical formalism as inadequate than by demonstrating the departure of its consequences from experience or by proving that its predictions did not exhaust the possibilities of observation" (Bohr 1987, v. 2, p. 57). According to Bohr, "Einstein's argumentation could be directed to neither of these ends" (Bohr 1987, v. 2, p. 57). The argumentation here referred to was that concerning the so-called Einstein "photon box" experiment, but Bohr clearly had in mind Einstein's other arguments, in particular, those concerning EPR-type experiments. Bohr's contention just cited also shaped his response to Einstein's view that "the belief that [QM] should offer an exhaustive description of individual phenomena," while "logically possible without contradiction," was "so very contrary to [Einstein's] scientific instinct that [he could not] forego the search for a more complete conception." Bohr countered that a mere rejection of, rather than a counterargument to a given theory or argumentation, such as that of the RWR type by Bohr, and a desideratum for an alternative, "more complete conception" (as Einstein understood it, in accordance with his realist imperative) do not in themselves constitute a demonstration of logical or experimental deficiency of this theory or argumentation. Bohr said:

> Even if [Einstein's] attitude might seem well balanced in itself, it nevertheless implies a rejection of [rather than offers a counterargument to] the whole argumentation [essentially that of his ultimate RWR-interpretation] that, in quantum mechanics, we are not dealing with an arbitrary renunciation of a more detailed analysis of atomic phenomena but with a recognition that such an analysis is *in principle* excluded. (Bohr 1987, v. 2, p. 62)

This is one of Bohr's strongest expressions of his ultimate, strong RWR-type interpretation of quantum phenomena and QM. Bohr's statement may suggest, perhaps in accord with his view at least on this occasion, the impossibility of a realist alternative in general, rather than only the fact he adopts an RWR-type interpretation of quantum phenomena and QM, possible as more viable than realist alternative. Technically, the statement refers to Bohr's response to Einstein's outright rejection (in 1936) of this type of view rather than counterarguing it. While the phrase "aiming to show" allows one to read this statement as only referring to Bohr's interpretation, it is still possible to read this statement as saying that no interpretation of quantum phenomena or QM and no alternative theory of quantum phenomena will ever be able to circumvent this prohibition. From the present viewpoint, such statements by Bohr or others could, regardless of the author's intent, only represent a particular interpretation of quantum phenomena and QM or QFT. Here, they represent a strong RWR-type interpretation adopted by Bohr at this point, *in the present interpretation of his interpretation*, a qualification unavoidable for the same set of reasons.

Be it as it may on that score, the exchange returns me to the big "if," invoked above: *if* quantum phenomena will ever allow one to do so, that is, if they ever allow one to fulfill Einstein's desiderata for a fundamental theory of quantum phenomena or whatever future phenomena may replace or delimit them. The debate between Bohr and Einstein was not primarily about what QM could or could not do, on which point their disagreement, while not discountable, was not so crucial. Their debate was ultimately about whether nature could allow us to do better. While Einstein thought that it should and while it is possible that Bohr thought that it will not, the present philosophical position implies that it *may not*, which is not the same as it never will. Einstein was more certain, even if, firm as his conviction was, perhaps not entirely so, given that "should" is still not the same as "will." Admittedly, the RWR view is human, just as was that of Einstein. The difference is that, contrary to Einstein's or others' position advancing realism as imperative, the RWR view is not seen in this study as imperative, but only as practically justified, as things stand now.

Most have been and remain inclined to accept Einstein's view. The circumstance that our fundamental theories remain incomplete (as there is no quantum theory of gravity and even handling other forces of nature by means of QFT is still not finalized) has often been used to support the Einsteinian hope that physics will return to realism. Einstein prevailed in this part of his debate with Bohr, with most physicists and philosophers taking his side. Bohr, as noted in the Preface, realized this situation, in speaking "of skepticism [among the physicists themselves] about the necessity of going so far in renouncing customary demand as regards the explanation of natural phenomena" (Bohr 1987, v. 2, p. 63). He also realized, however, that physics itself is a different matter. While human, it does not necessarily obey human imperatives, which have been defeated by physics throughout its history, not the least by Einstein's own relativity theory. Bohr's argument concerning and his, by that point RWR type, interpretation of quantum phenomena and QM was and (as there has been no change in this regard since) has remained fully in accord with the experimental data available thus far and with the logical structure of QM (Bohr 1987, v. 2, p. 57). So is, I would argue, the present interpretation, different from that of Bohr as it may be as concerns its concept of a quantum object as an idealization applicable only at the time of measurement. Accordingly, while a realist theory of quantum phenomena is possible or might become necessary in view of new experimental or theoretical findings, it is equally possible the RWR view will remain viable in grounding our interpretations of QM and QFT, or (extended to the U-RWR view) other fundamental theories. Either bet is possible, but only as a bet, notwithstanding the certainty of some in taking either view, although it appears that most of those who are certain advocate realist views.

Whatever the future of fundamental physics may bring, the following strong, but I would argue, justified, claim could be made, at least if one adopts the RWR view. Bohr's thinking in his 1913 theory, especially Heisenberg's thinking leading him to his discovery of QM revolutionized the very practice of fundamental physics, both theoretical physics and, in effect, redefined the practice of experimental physics when dealing with quantum phenomena. Taking advantage of and bringing together two main meanings of the word "experiment" (as a test and as innovative creation), I would argue that the practice of quantum physics became the first practice of physics or science that is both, and jointly, *irreducibly* (structurally unavoidably) experimental and *irreducibly* mathematical.

First, it is irreducibly experimental as experimental physics in its conventional sense because it no longer consists, as in classical or relativistic physics, in tracking the independent behavior of the systems considered, but in *unavoidably* creating configurations of experimental technology containing traces of its interactions with nature that reflect the fact that what happens is *unavoidably* defined by what kinds of experiments we perform, by how we affect physical reality as a reality without realism. I emphasize "unavoidably" because, while the phenomena observed in classical physics or relativity may be affected by experimental technology and while we do stage experiments there, in principle, one can observe these phenomena without appreciably affecting their behavior (which also allows one to see these phenomena as ideally representing the corresponding objects) and thus follow what happens in any event. In quantum physics, the experiment is a creation of a new material and phenomenal configuration that enables one, by performing a measurement on the observable part of the instruments used, to define the probabilities for outcomes of future experiments, without it being possible to follow how these outcomes come about. No tracking or, in RWR-type interpretations, no assumption of classical-like continuity of events or of the continuous (or any other, for example, discrete) concept of the ultimate nature of reality that gives rise to these events is possible.

It is true that we do not really know either why the mathematical formalism and laws of classical physics or relativity work (as opposed to other possible mathematical representations and laws). We only know that these theories work within their proper scopes. But they work differently, including if one adopts the U-RWR view as ultimately underlying these theories. In classical physics and relativity, our predictions, deterministic or probabilistic, are defined by the idealized mathematical representations of the ultimate constituents of the reality considered and their behavior. While, in the U-RWR view, this idealization does not correspond to how nature ultimately works even within the proper scope of these theories, it is permissible for all practical purposes. This type of idealization is, however, no longer possible in RWR-type interpretations of quantum phenomena and QM or QFT. In these interpretations, the formalism of QM or QFT only enables probabilistic or statistical predictions of the numerical data found in quantum phenomena. (The fact alone that no other predictions of these data are possible on experimental grounds does not preclude realism.)

Accordingly, in the view assumed in this study, as the U-RWR view, the probabilistic relationships between a given theory and observed phenomena are always, in any physical theory, *human* because an assignment of probabilities is human. As stated, nature does not assign probabilities and makes no predictions. In classical physics, however, this assignment may, in the U-RWR view, be assumed to be *physically* grounded in the behavior of the systems considered because classical physics may be assumed to represent this behavior and relate this representation to the phenomena considered. In quantum theory, by contrast, at least in RWR-type interpretations, this assumption is no longer possible. These relationships are still ultimately grounded in our interactions with nature, but, in RWR-type interpretations, we don't know or even can't conceive of the way in which they are so grounded. Hence, we can no longer conceive either why these probabilistic relationships obtain and why the mathematics of quantum theory predicts them fully in accord with the data observed.

By the same token, the practice of theoretical physics is *irreducibly* mathematical because it no longer consists, as in classical physics or relativity, in offering an idealized mathematical representation of quantum objects and behavior, but in inventing abstract mathematical schemes enabling us to predict, probabilistically or statistically, the outcomes of quantum events. These schemes are unrelated to and thus are not helped by our general phenomenal intuition, as are those of classical mechanics or, in a more limited way, relativity. On the other hand, they are not limited by this intuition either. Indeed, one experiments with mathematics as well, and it is experimenting with abstract mathematics, divorced from our phenomenal intuition or even physical concepts that, with Heisenberg, became the way to new theories. As discussed in Chap. 8, this view of physics equally and even more radically applies, beginning with Dirac's work, in QED and then QFT, and experimental physics in high-energy quantum regimes, thus extending to high-energy physics the new epoch of the relationships between and "mutual stimulation of" mathematics and physics, which, according to Bohr, commenced with Heisenberg's discovery of QM (Bohr 1987, v. 1, p. 51). These relationships and the type of revolution in fundamental physics they brought about were not what Einstein wanted, even though he understood, more deeply than most, their radical nature. For Bohr, however, this revolution was what quantum physics needed, given his 1913 atomic theory, which built on Einstein's own and Planck's earlier thinking but which moved beyond this thinking and opened a new trajectory of thought leading to this revolution.

References

Bohr, N.: The Philosophical Writings of Niels Bohr, 3 vols. Ox Bow Press, Woodbridge (1987)

Bohr, N., Rosenfeld, L.: On the question of the measurability of electromagnetic field quantities. In: Wheeler, J.A., Zurek, W.H. (eds.) Quantum Theory and Measurement, 1983, pp. 479–522. Princeton University Press, Princeton, NJ (1933)

Bohr, N., Rosenfeld, L.: Field and charge measurements in quantum electrodynamics. In: Wheeler, J.A., Zurek, W.H. (eds.) Quantum Theory and Measurement, 1983, pp. 523–534. Princeton University Press, Princeton, NJ, USA (1950)

Born, M.: The Einstein-Born Letters, Trans. Born, I. Walker, New York, NY (2005)

De Bianchi, S., Wells, J.D.: Explanation and the dimensionality of space Kant's argument revisited. Synthese **192**(1), 287–303 (2015)

Einstein, A.: Physics and reality. J. Franklin Inst. **221**, 349–382 (1936)

Einstein, A.: Autobiographical Notes, Trans. Schilpp, P. A. Open Court, La Salle, IL (1949)

Einstein, A., Podolsky, B., Rosen, N.: Can quantum-mechanical description of physical reality be considered complete? In: Wheeler, J.A., Zurek, W.H. (eds.) Quantum Theory and Measurement, 1983, pp. 138–141. Princeton University Press, Princeton, NJ, USA (1935)

Hegel, G. W. F.: Hegel's Phenomenology of Spirit, Trans. Pinkard, T. Cambridge University Press, Cambridge (2019)

Heisenberg, W.: Physics and Philosophy: The Revolution in Modern Science. Harper & Row, New York, NY, USA (1962)

Kant, I.: Critique of Pure Reason, Trans. Guyer, P., Wood, A.W. Cambridge University Press, Cambridge (1997)

Kant, I.: Thoughts on the true estimation of living forces. In E. Watkins (ed.), Immanuel Kant: Natural Science, pp. 1–155. Cambridge University Press, Cambridge (2012)

Misner, C., Thorn, K. S, Zurek, W. H.: John Wheeler, relativity, and quantum information. Physics Today **62**(4), 40–46 (2009). https://doi.org/10.1063/1.3120095

Plotnitsky, A.: On the reasonable and unreasonable effectiveness of mathematics in classical and quantum physics. Found. Phys. **41**, 466–491 (2011). https://doi.org/10.1007/s10701-010-9442-2

Plotnitsky, A.: The Principles of Quantum Theory, from Planck's Quanta to the Higgs Boson: The Nature of Quantum Reality and the Spirit of Copenhagen. Springer/Nature, New York, NY (2016)

Riemann, B.: On the hypotheses that lie at the foundations of geometry. In: Pesic, P. (ed.) Beyond Geometry: Classic Papers from Riemann to Einstein, 2007, pp. 23–40. Dover, Mineola, NY, USA (1854)

Van Dongen, J.: Einstein's Unification. Cambridge University Press, Cambridge (2010)

Wheeler, J.A., Zurek, W. H. (eds.) Quantum Theory and Measurement. Princeton University Press, Princeton, NJ (1983)

Weyl, H.: Space Time Matter, Trans. Brose, H. L. Dover, Mineola, NY (1952)

Chapter 2
Fundamentals of the RWR View

But I can think whatever I like, as long as I do not contradict myself, i.e., as long as my concept is a possible thought, even if I cannot give any assurance whether or not there is a corresponding object somewhere within the sum total of all possibilities.
—Immanuel Kant, *Critique of Pure Reason* (1787) (Kant 1997, p. 115)

Abstract This chapter offers a philosophical outline of the RWR view and defines the key concepts considered in this book, both those specifically grounding the RWR view and more commonly used concepts, such as reality, causality, or determinism. I begin, in Sect. 2.2, with the concept of concept itself. Section 2.3 considers the concepts of a theory, model, and interpretation, and the relationships among them. Section 2.4 discusses the concepts of reality, realism, and reality without realism. Section 2.5 considers indeterminacy and probability in quantum theory. Section 2.6 addresses the question of idealization in RWR-type interpretations, in particular in the interpretation adopted in this study, in which the concept of a quantum object, while still defined as beyond conception, is only applicable at the time of measurement, rather than referring to something that exists independently.

Keywords Causality · Concepts · Measuring instruments · Quantum measurement · Quantum object · Reality · Realism · Reality without realism

2.1 Introduction

This chapter offers a general exposition of the RWR view of quantum phenomena and quantum theory, and defines the key concepts considered in this study, both those, such as reality without realism, specifically grounding the RWR view and more commonly used concepts, such as reality, causality, and determinism, in their definitions adopted here, because they can be defined otherwise. Most of these concepts are complex, and their full-fledged structure will emerge as this study proceeds. Some of them, such as quantum nonlocality, Bohr's concepts of complementarity

A. Plotnitsky, *Reality Without Realism*,
https://doi.org/10.1007/978-3-030-84578-0_2

and phenomenon, or elementary particles, will only be briefly sketched here and properly delineated in later chapters.

I begin, in Sect. 2.2, with the concept of concept. Section 2.3 considers the concepts of theory, model, and interpretation, and the relationships among them. Section 2.4 outlines the concepts of reality, realism, and reality without realism. Section 2.5 addresses randomness, indeterminacy, and probability. Section 2.6 discusses the question of idealization in RWR-type interpretations, in particular that adopted in this study, in which the concept of a quantum object, while still defined as beyond conception, is only applicable at the time of measurement, rather than referring to something that exists independently.

2.2 Concepts

The *concept* of concept is rarely adequately considered in the physical or philosophical literature, in part perhaps because the role of concepts is not always sufficiently appreciated there. If, however, as Franck Wilczek, a leading elementary particle theorist and a Nobel Prize laureate, argues, "the primary goal of fundamental physics is to discover profound concepts that illuminate our understanding of nature," then creative thinking in fundamental physics is defined and fundamental physics is advanced primarily by the invention of concepts (Wilczek 2005, p. 239). But what is a physical concept, and what is a concept in the first place? Wilczek does not explain it, taking it for granted or assuming some general sense of it shared by his readers. Given that his article was published in *Nature*, one might expect that these readers are sufficiently informed to understand the concepts that Wilczek considers, such as that of an elementary particle, associated with that of a symmetry group (an association explained later in this study). Thus, these concepts have mathematical components, the presence of which has defined the concepts of all modern theoretical physics. A physical theory might then be seen, as it will be here, as an organized assemblage of concepts. As such, it would then relate to experimentally observed physical phenomena in terms of logical propositions, thus confirming the character of modern physics as a mathematical-experimental science.

It is, as noted in the Introduction, the logical-propositional structure of physical theories, in conjunction with their mathematical formalism, that tends to dominate the philosophy of physics, in particular the analytic philosophy of physics. This structure is of course indispensable: no physical theory and no philosophical argument concerning theoretical physics can bypass it. I would, nevertheless, contend that it is far from sufficient for understanding physical theories, especially in their creative dimensions, where the discovery of new concepts plays a key and even decisive role. This is the case that I would like to make in this study. I begin this section by briefly considering Émile Borel's 1907 argument concerning the limits of the role of the logical aspects of mathematics, an argument also applicable to theoretical physics (Borel 1907).

Borel, a major figure at the time, who made key contributions to measure theory and probability theory (such as the concept of a Borel set and the thought experiment known as "the infinite monkey theorem"), questioned the logicist philosophy of mathematics, which theorized mathematics as a form of logic and which, championed by, among others, Bertrand Russell, was in vogue at the time. Borel was not alone in questioning logicism. A similar critique of it was, for example, undertaken about the same time by Henry Poincaré, whose mathematical work was closely connected to contemporary physics and who also made important contributions to physics itself (Poincaré 1982). Borel gave several examples of the failure of logic to capture the essence of mathematical thought, such as the fact that the formula expressing the invariance of the cross-ratio of four points on a line under a perspectivity is easy to find, but it took Michel Chasles to see in it the key to projective geometry; or the case of the polynomial identity (the icosahedral equation) that could only have been found to be valuable if discovered by a nonmechanical route, namely through Felix Klein's unification of Galois theory with the theory of the symmetries of the icosahedron. Borel's view was that a truly fertile invention in mathematics consists of the discovery of a new point of view from which to classify and interpret the facts, followed by a search for the necessary proofs by plausible reasoning (later considered by Pólya [1958]), and only then bringing logic in. The same argument could be made in physics, for example, in such cases as Heisenberg's invention of matrix variables, Schrödinger's invention of the complex wave function, and Dirac's invention of spinors, or of course, Einstein's concepts of general relativity, based on Riemannian geometry, to stay for the moment with mathematical concepts. Bohr's concept of quantum jumps provides an even more dramatic example, although in this case mathematics did not play as significant a role as it did in the other cases just mentioned. The structure of these concepts was and had to be logical, but their invention was not defined by logic, which played only an auxiliary role in this invention.

Borel's argument has not lost its importance in assessing the philosophy of mathematics or, as I argue here, the philosophy of physics. According to Jeremy Gray, "Borel's criticisms point quite clearly toward a problem that has not gone away in philosophers' treatment of mathematics: a tendency to reduce it to some essence that not only deprives it of purpose but is false to mathematical practice. The logical enterprise, even if it had succeeded, would only have been an account of part of mathematics, its deductive skeleton" (Gray 2008, pp. 202–203). This is, I would contend, equally true about much of the analytic philosophy of physics, indispensable as the logical-propositional aspects of theoretical physics may be in turn.

It is not my aim (any more than it was that of Borel) to diminish the role of logic or, one might add, the calculational aspects of mathematics or physics, or to take anything away from the contribution of the analytic philosophy of physics to our understanding of physics, including quantum theory. Several studies in this field were helpful to my argument in this book. Instead, I would like to give a proper emphasis to the role of concepts, especially, again, in creative thinking in theoretical physics, but not only there, because working with already established concepts is indispensable in all practice of theoretical (or experimental) physics. Besides, analytic philosophers, too, sometimes give concepts their due, beginning with one of the founders of analytic philosophy, sometimes even seen as *the* founder of it, Gottlob Frege, who gave priority to the role of concepts in mathematics and philosophy. Frege did not deal with physics, but his work shows the difficulty of making unqualified claims here, as opposed to indicating one or another degree of emphasis. Finally, the unconditional opposition between the logical-propositional and conceptual structures, or between calculations and concepts, is not so easy and is ultimately impossible to maintain, even in considering figures, such as Riemann, who gave a strong priority to thinking in terms of concepts vis-à-vis calculations. Riemann was perfectly capable of difficult calculations. Logic and calculations do involve concepts and lead to new concepts. Calculations can also function as statements, sometimes even as philosophical statements, thus connected to concepts. It is the balance of these aspects of mathematical and scientific thinking that I would like to emphasize, while giving concepts their due, as against a degree of neglect of their role, beginning with not considering the *concept* of concept.[1]

The centrality of concepts in philosophy was emphasized most by the post-Kantians, especially Hegel, then in mathematics by Riemann and Frege (both of whom might have been influenced by Hegel), and in physics by Einstein. Riemann, a major inspiration for Einstein's general relativity, likely also concerns the essential role of concepts, observed in his 1854 Habilitation lecture, discussed in the Introduction:

> From Euclid to Lagrange this darkness [in our understanding of geometry] has been dispelled neither by the mathematicians nor the philosophers who have concerned themselves with it. The reason [*Grund*] for this is undoubtedly because the general concept of multiply extended magnitudes, which includes spatial magnitudes, remains completely unexplored. I have therefore first set myself *the task of constructing the concept* of a multiply extended magnitude from general notions of magnitude. (Riemann 1854, p. 24; emphasis added)

This concept is his concept of manifold [*Mannigfaltigkeit*], which enabled Riemann to establish new foundations for geometry, and which became central to modern geometry and topology, and their use in physics (Plotnitsky 2017).[2] As discussed in Chap. 6, part of the genealogy of Bohr's concept of complementarity was one of Riemann's new concepts as well, that of a Riemann surface, which, introduced by Riemann in 1851, was also part of the genealogy of the concept of manifold. A Riemann surface is a manifold, as Riemann undoubtedly realized by 1854, even though he did not originally define it as a manifold. Einstein saw the invention of new

[1] There are still other dimensions of physical or mathematical thinking, for example, a narrative dimension, the role of which, as a constitutive rather than (more easily recognized) auxiliary role, in scientific or mathematical thinking, has received a considerable amount of attention during recent decades. See, for example, a representative collection of essays, by both scholars in the humanities and mathematicians and physicists, on the role of narrative in mathematics and physics, a collection that contains extensive further references (e.g., Doxiadis and Mazur 2012). (For a full disclosure, the present author was among the contributors.) Calculations, too, involve narrative dimensions: they tell stories. My main emphasis in this study, however, will remain on concepts as the primary creative technology of thought in physics, mathematics, and philosophy.

[2] Riemann's own last work, dealing with "the mechanism of the ear" is a remarkable earlier example of the use of the concept of manifold in physics (Riemann 1866).

concepts as defining foundational thinking in physics, and as noted earlier, he argued that only through "a free conceptual construction" one could approach the nature of physical reality, which view defined his realism as conceptual realism (Einstein 1949, p. 47).

The concept of concept that I adopt here, while in accord with the views of these figures, follows more expressly that of Deleuze and Guattari (1994). Their thinking was inspired, along with that of the post-Kantians, such as Hegel, by Riemann, including by his concept of manifold (Plotnitsky 2017). Deleuze and Guattari applied their concept of concept primarily to philosophical concepts, in a juxtaposition to scientific or mathematical concepts (Deleuze and Guattari 1994, pp. 24–25). However, as I have argued previously (Plotnitsky 2017), this concept of concept is equally applicable to mathematical and scientific concepts, as concerns both the structure of concepts and their capacity to pose new problems, which are the two defining characteristics of concepts for Deleuze and Guattari. Deleuze and Guattari were ultimately unable to sustain their juxtaposition between philosophical and mathematical or scientific concepts either. Accordingly, concepts may be seen, as they will be here, as primary vehicles of creative thought in all theoretical fields. Riemann's concept of manifold; Einstein's concept of gravitation based on Riemann's concept of manifold; Bohr's concept of quantum jumps; Heisenberg's concept of new quantum variables, as matrix variables; Schrödinger's concept of the wave function; Dirac's concepts, such as spinors, that established QED; or Bohr's concept of complementarity are all concepts in Deleuze and Guattari's sense. So are, of course, earlier physical concepts, those of classical mechanics, thermodynamics, or electromagnetism, in particular that of an electromagnetic field. Each is defined by a complex structure of their elements and each posed a new problem or set of problems and opened a space for creating new concepts, a space that we still explore now. What, then, is a concept, according to Deleuze and Guattari?

First of all, a concept is not merely a generalization from particulars (which is commonly assumed to define concepts) or a general or abstract idea, although a concept may contain such generalizations or ideas, specifically abstract mathematical ideas in physics or in mathematics itself, where these ideas may become concepts in the present sense. A concept is a multicomponent entity, defined by the organization, *composition,* of its components, and some of these components may be concepts in turn. In this respect, a concept is akin to a work of art, a parallel that reflects the creative nature of concepts. This is, of course, only a general definition. What is crucial is how the compositional structure is specifically instantiated in a given concept, which is defined by both the nature of each component and its composition, and how they relate to each other in the structure of the concept. All key concepts considered in this study will be discussed through the specificity of their compositional structure. This structure is special and even unique in the case of some concepts of quantum theory in the strong RWR view, such as Bohr's concept of phenomenon, insofar as these concepts are defined by the combination of classical physical concepts, on the one hand, and on the other, the ultimate constitution of the reality responsible for quantum phenomena, placed beyond conception. The present definition of concept allows something that is beyond conception, such as a reality without realism, to be considered a concept or placed within a concept, because it always involves connections to something that can be conceived or known, from which the existence of such a reality is inferred.

Simple, single-component concepts are rare, if ultimately possible at all, as opposed to appearing as such because their multicomponent structure is provisionally disregarded or cut-off. In practice, there is always a cut-off in delineating a concept, which results from assuming some of the components of this concept to be primitive entities whose structure is not specified. These primitive concepts could, however, be specified by an alternative delineation, which would lead to a new overall concept, containing a new set of primitive (unspecified) components. The history of a given concept—and every concept, however innovative, has a history—is a history of such successive specifications and changes in previous specifications. Concepts are never created out of nothing, as proverbially, Athena from the head of Zeus, fully grown and armed. They are constructed and composed of elements borrowed from earlier concepts, sometimes used in very different lines of development or even different fields. In this sense, concepts are also akin to any technology, which always emerges from previous technologies, sometimes by combining previously heterogenous elements, like the phone and the computer in the cellphone. When new concepts become to be used, either as they are or as further modified, they acquire their history. Hence, concepts also play an important role in the regular practice in the field in which they are created or a field which adopted concepts from other fields.

Consider, as an example, the concept of motion, first, as used in daily life: it involves various components, such as a change of place, speed, acceleration, moving bodies, etc., some of which are concepts in turn. These components belong to our phenomenal intuition and are not defined rigorously, especially mathematically, but are still parts of the concept of motion defined by the organization of these components. One can then view, as both Bohr and Heisenberg did, classical mechanics as a physical and mathematical refinement of daily concepts, including that of motion, by means of such mathematically defined concepts as coordinates, momentum, angular momentum, energy, and so forth. The history of the concept of motion in classical physics extends much further, even to the pre-Socratics, but certainly to Plato and, especially, Aristotle, whose *Physics* is central to this history, although the ancient Greeks never mathematized this concept in the way that modern physics did with Galileo. On the other hand, modern developments such as electromagnetism and then relativity introduced new concepts of motion, while QM, at least in RWR-type interpretations, abandoned this concept as applicable to quantum objects or the ultimate constitution of reality responsible for quantum phenomena.

In quantum theory, in RWR-type interpretations, classical physical concepts are no longer applicable to this constitution. They still apply to quantum phenomena, defined as the effects of the interaction between this reality and measuring instruments, which, in the present view, also defines quantum objects as an idealization applicable at the time of measurement. Thus, the concepts of quantum theory still have their history in classical physics, both physically (in considering measuring instruments and quantum phenomena) and mathematically, for example, by adopting the concept of the Hamiltonian, while changing the variables from those of the functions of real variables to operator variables in Hilbert spaces over $\mathbb{C}$. The standard conservation laws (those of energy, momentum, and angular momentum) are retained as well, although new conservation laws are added, such as the conservation of the baryon or lepton number. According to Heisenberg:

> The concepts of velocity, energy, etc., have been developed from simple experiments with common objects, in which the mechanical behavior of macroscopic bodies can be described by use of such words. The same concepts have then been carried over to the electron, since in certain fundamental experiments electrons show a mechanical behavior like that of the objects of common experience [or classical mechanics]. Since it is known, however, that this similarity exists only in a certain limited region of phenomena, the applicability of the corpuscular theory must be limited in the corresponding way. …
>
> As a matter of fact, it is experimentally certain only that light [too] sometimes behaves as if it possessed some of the attributes of a particle [as reflected in the uncertainty relations], but there is no experiment which proves that it possesses all the properties of a particle; similar statements hold for matter [e.g., electrons] and wave motion. The solution of the difficulty is that the two mental pictures [derived from classical physics] which experiments lead us to form—the one of particles, the other of waves—are both incomplete and have only the validity of analogies which are accurate only in limited cases. … Light and matter are both single entities, and the apparent duality arises in the limitation of our language. (Heisenberg 1930, pp. 13, 10)

In the RWR view, quantum objects and behavior are beyond any representation or, in the strong RWR view, beyond conception, including a mathematical one, which can be sufficiently, even if not completely, independent of ordinary language or concepts, as it is in the case of the mathematical formalism of QM and QFT. (In the present view, again, quantum objects, in RWR-type entities, are idealizations applicable at the time of measurement.) Heisenberg, on the same occasion, stresses the independence of mathematics from ordinary language and concepts, which he, argues, helped his invention of QM (Heisenberg 1930, p. 11). This independence eventually led him to believe, on Platonist lines, that an abstract mathematical scheme could represent the ultimate nature of physical reality, apart from physical concepts, at least as we ordinarily understand them (e.g., Heisenberg 1962, pp. 145, 167–186). Bohr, in contrast, rejected this possibility, especially in his ultimate, strong RWR-type interpretation.

In the framework of QM, in most interpretations, wave–particle duality is replaced by wave–particle complementarity. Using this concept, however, requires qualifications. As explained in Chap. 6, wave–particle complementarity does not play a significant, if any, role in Bohr's own interpretation (in any of its versions). The concept of complementarity underwent an evolution in Bohr's thinking, concomitant with the evolution of his interpretation of quantum phenomena and QM, before it reached its final form, defined in terms of complementary measurements or phenomena in Bohr's sense. As indicated in the Preface, it refers to measurements or phenomena that are mutually exclusive, yet equally necessary for a comprehensive account of the situation that obtains in quantum physics. In 1937, when his ultimate interpretation was introduced, Bohr said: "In the last resort an artificial word like 'complementarity' which does not belong to our daily concepts serves only briefly to remind us of the epistemological situation [found in quantum physics], which at least in physics is of an entirely novel character" (Bohr 1937, p. 87). The structure of the concept, which does more than merely serve as such a reminder, will be discussed in Chap. 6. For the moment, my main point is that complementarity is a new physical concept with several interrelated components, a concept that, again, "does not belong to our daily concepts," and which must, accordingly, be understood, *as a physical concept*, in the specific sense Bohr gives it. As most innovative concepts, complementarity, when introduced, was not defined by generalization from available entities: it was something new, although it, too, had its history in physics and beyond (Plotnitsky 2012, pp. 173–179, 2016, pp. 107–119). It was then instantiated by multiple specific complementary configurations, say, those of the position or the momentum measurement, which are always mutually exclusive at any given moment of time.

In his *Physics and Philosophy* (Heisenberg 1962), Heisenberg used the role of concepts in explaining the difference between positivism and "the Copenhagen interpretation of quantum theory," in his version of it. In his emphasis on concepts, Heisenberg followed Einstein, with whom he had exchanges concerning the subject (Heisenberg 1962, pp. 45–46, 1989, p. 30; Plotnitsky 2016, pp. 43–44). The phrase "the elements of reality" is borrowed from Einstein, who used it on several occasions, most famously in EPR's paper (Einstein et al. 1935, p. 138). Heisenberg says: "The Copenhagen interpretation of quantum theory is in no way positivistic. For, whereas positivism is based on the sensual perceptions of the observer as the elements of reality, the Copenhagen interpretation regards things and processes which are describable in terms of classical concepts, i.e., the actual, as the foundations of any physical interpretation" (Heisenberg 1962, p. 145). "The actual" refers to what is observed in measuring instruments and is described by classical physical concepts. Heisenberg's main point is that in order to be meaningfully used in quantum theory, "the idealization of observation" (Bohr 1987, v. 1, p. 55), defining quantum phenomena uses classical physical concepts, just as this idealization does in the case of observed phenomena in classical physics. There, however, it is, for all practical purposes, sufficient to assume that the physical objects considered, at least individual or sufficiently simple, are in principle represented by the corresponding observed classical phenomena. In quantum physics, in contrast, one assumes a new stratum of physical reality that is different from the *observable* parts of measuring instruments and that cannot be observed, in the RWR view, represented, or even conceived of, even in the case of the simplest possible systems considered. Measuring instruments, too, have an unobservable, quantum stratum, which enables their interaction with the part of this stratum assumed to exist independently, without allowing one to represent or even conceive of it, as would be the case in a realist understanding. Accordingly, in any interpretation, quantum phenomena, while representing classical objects associated with measuring instruments, are irreducibly different from quantum objects. In the present interpretation, moreover, while an RWR-type entity, the concept of a quantum object is an idealization only applicable at the time of measurement rather than to somethings that is assumed to exist independently, as against the ultimate, RWR-type reality responsible for quantum phenomena.

"The Copenhagen interpretation" discussed by Heisenberg is different from that of Bohr, for example, by introducing the concept of "potentiality" or "*potentia*" (Heisenberg 1962, p. 44), something quite different from "possibility" (a concept central to Bohr), or the assumption that mathematics could in principle be able to represent the ultimate structure of matter, a belief not shared by Bohr. Bohr would have, of course, agreed that his understanding of the quantum–mechanical situation was "in no way positivistic," in part for the reasons stated by Heisenberg. His assumption of the *independent* stratum of physical reality, in his understanding (as opposed to the present one) as quantum objects, ultimately responsible for quantum phenomena was essential for Bohr, even though this stratum could not be observed as such, but only inferred from its effects on measuring instruments. As noted, nobody has ever seen a moving electron or photon. It is only possible to observe traces of their interactions with measuring instruments, traces that make it difficult and, in the RWR view, impossible to reconstitute the ultimate nature of the reality responsible for them, whether one sees this reality in terms of quantum objects or only assumes, as here, quantum objects to be an idealization applicable at the time of measurement.

As discussed in the Introduction, the individual quantum objects that may appear to "have been brought so close to us" by our measuring instruments, such as "the marvelous invention of the Wilson [cloud] chamber and the Geiger counter," made Einstein question the RWR-type view of the ultimate constitution of the reality responsible for quantum phenomena (Einstein 1936, p. 375). Heisenberg, on the other hand, was led to his discovery of QM and then the uncertainty relations by realizing that one only deals with traces of the interactions between quantum objects and measuring instruments and not with quantum objects themselves (Heisenberg 1925, 1927, 1967). In his thinking leading to his discovery of the uncertainty relations, Heisenberg considered the visible and apparent continuous trajectory of an electron in a cloud chamber (Heisenberg 1967). He realized that the position of the electron was only known and the trajectory appears continuous because the water droplets, consisting of millions of atoms and thus much larger than the electron (which must be idealized as a point-like), condensed into this apparent continuity around the discrete events, the traces of which were spreading through water. If one could zoom in on what actually happens, one would see this discreteness of the underlying phenomena and, as it were, "quantum jumps" from one to another, and realize that the position and the momentum of the electron could not be known exactly or defined simultaneously. In both Bohr's and the present view, such variables could only be attributed to what is observed, classically, in the cloud chamber and not to the electron itself. It is clear that the continuity of quantum phenomena can only be apparent, while being underlain by the discreteness and uniqueness of each, in accord with the QD and QI postulates.

We do, of course, deal with indirect evidence in classical physics and relativity, from that of the movement of molecules (considered classical) to that of the Big Bang in the expansion of the universe, but in these cases, this reconstitution is possible, at least ideally and in principle, and effective within the scope of these theories. It is this situation that compels one, while adopting the RWR view, to introduce the concept of a quantum object, and, in the present interpretation, also the independent ultimate constitution of the reality responsible for quantum phenomena, while assuming quantum objects to be an idealization applicable only at the time of measurement. It is not us but nature that, in its interaction with our experimental technology (beginning with our bodies), is responsible for these effects, even though we cannot know or even conceive how they come about. While our bodies and our technology are natural, too, they are not sufficient to produce these effects. These effects require something in the constitution of nature apart from us and our technology, or what compels us to speak of this constitution, assuming that the term "constitution" (or "nature") applies.

I close this section by considering Bohr's argument that the quantum–mechanical situation does not require new physical concepts and could be adequately handled, in part by using complementarity, by means of already available, "old" empirical concepts, in effect, those of classical physics. This argument emerged in response to the contention of Einstein and Schrödinger that the quantum–mechanical situation needed new physical concepts to resolve its difficulties and paradoxes (e.g., Schrödinger to Bohr 5 May 1928; Bohr to Schrödinger 23 May 1928; Einstein to Schrödinger, 31 May 1928, Bohr 1972–1996, vol. 6, pp. 47–48). According to Bohr, writing to Schrödinger: "I am not quite in agreement with your emphasis on the necessity to develop 'new' concepts. We have not only, as far as I can see, no basis for such a new-fashioning so far, but *the 'old' empirical concepts* appear to me to be inseparably linked to the foundation of the human means of visualization" (Bohr to Schrödinger 23 May 1928, Bohr 1972–1996, vol. 6, p. [48]; emphasis added). "Visualization" also has a broader meaning here and elsewhere in Bohr's writings, in accord with German *Anschaulichkeit,* sometimes expressly used by Bohr as referring to our intuitive phenomenal comprehension in general. In other words, these old empirical concepts are linked to and refine our general phenomenal concepts. Bohr, then, sees no basis for the necessity of "fashioning" new concepts in quantum theory. This argument might appear to be in conflict with the preceding discussion, which emphasizes the role of new concepts in quantum theory. Were not Heisenberg's matrix variables and Schrödinger's wave function new concepts? Was not Bohr's complementarity a new concept? This conflict, however, is only apparent.

Einstein and Schrödinger thought that the new physical concepts might enable one to offer a physical representation (idealized and suitably mathematized) of the behavior of quantum objects in the way classical physics was able to do by means of its concepts (and at various points by introducing new concepts, as in electromagnetism), or relativity was able to do by means of its new physical concepts. Such concepts would, then, restore realism, which was conceptual in nature for both Einstein and Schrödinger, to quantum theory, and would allow one to avoid the paradoxes, or what so appear under realist assumptions, of QM. This view was in accord with Schrödinger's initial program for his wave mechanics, which by this point (in 1928) had run into difficulties, leading to both Schrödinger's own and especially Einstein's disenchantment with it. Although, as noted, it is possible to think, as Heisenberg did later, that such new representational concepts could be purely *mathematical,*

Einstein and Schrödinger were clearly thinking of new *physical* concepts, suitably mathematized, in the way those of classical physics or relativity are.

In contrast, Bohr's appeal to the use of the "'old' empirical concepts" of classical physics represents a very different, in effect RWR, view. Although the exchange in question occurred before Bohr's interpretation reached its strong, RWR-type version, the weak RWR view, held by Bohr at the time, was sufficient for his argument.[3] In Bohr's view, such old classical concepts are applied only to the behavior of observable parts of measuring instruments or quantum phenomena, registered in these instruments, but not to quantum objects. The latter were, at the time of this exchange, placed by his interpretation beyond a description or representation by means of any physical concepts, even if not yet beyond conception. While complementarity was a new concept, it only involved representational components, classical in nature, at the level of the observed phenomena, but not as applicable to quantum objects. Bohr, undoubtedly with the exchange under discussion in mind, stated this view in his "Introductory Survey" (1929) to his 1931 *Atomic Theoryand the Description of Nature* (Bohr 1987, v. 1):

> [A]ccording to the view of this author, it would be a misconception to believe that the difficulties of atomic theory may be evaded by eventually replacing the concept of classical physics by new conceptual forms. Indeed, … the recognition of the limitation of our forms of perception by no means implies that we can dispense with our customary ideas and their direct verbal expression when reducing our sense of impressions to order. No more is it likely that the fundamental concept of the classical theories will ever become superfluous for the description of physical experience. The recognition of the indivisibility of the quantum of action [h], and the determination of its magnitude, not only depend on an analysis of measurements based on classical concepts, but it continues to be the application of these concepts alone that makes it possible to relate the symbolism of the quantum theory to the data of experience. At the same time, however, we must bear in mind that the possibility of an *unambiguous* use of these fundamental concepts solely depends upon the self-consistency of the classical theories from which they are derived and therefore, the limits imposed upon the application of these concepts are naturally determined by the extent to which we may, in our account of the phenomena, disregard the element which is foreign to classical theories and symbolized by the quantum of action. (Bohr 1987, v. 1, p. 12)

Accordingly, Bohr's argument, defining his response to Einstein's and Schrödinger's appeal to the necessity of new physical concepts, was not that "old" empirical concepts, such as those of classical physics that were sufficient to offer a representation of the behavior of quantum objects. This argument was that *no concepts*, old or new, could do so, while old concepts, specifically those of classical physics, remained essential, although not sufficient (e.g., because complementarity, not found in classical physics, was necessary), for handling quantum phenomena observed in measuring instruments and to the observable parts of these instruments themselves. These instruments, again, also have quantum parts through which they interact with quantum objects. I shall discuss this point and Bohr's overall argument concerning the

---

[3] As noted, Bohr's Como lecture of 1927 (Bohr 1987, v. 1, pp. 52–91), which introduced complementarity and Bohr's first version of his interpretation, reinstated an instance of realism to his view. Bohr, however, abandoned this argument and returned to the RWR view, at least in its weak form, by the time of the exchange with Einstein and Schrödinger under discussion.

role of classical physical concepts in considering quantum phenomena and measuring instruments in Sect. 2.4. It is clear, however, that there is no conflict between Bohr's argument and the possibility of new concepts in quantum theory. That included Bohr's own concepts, such as complementarity and phenomenon, once one properly understood their structure, as defined by the RWR view, combining "old" classical concepts and that which is beyond concepts. There is only a difference between the two views of the possible functioning of physical concepts in quantum theory, the realist view of Einstein and Schrödinger and the RWR view of Bohr.

Thus, classical *physical* concepts are only part of the overall conceptual structure of Bohr's concept of phenomenon and then of his concept of complementarity, defined by the complementary character of some quantum phenomena. Classical physical concepts are used to describe the physics of the observable parts of measuring instruments, parts assumed to behave classically. In this respect, old classical physical concepts are indispensable. On the other hand, because no concepts of any kind can apply to the ultimate constitution of the reality responsible for quantum phenomena, the need for new concepts to represent this reality would be meaningless in Bohr's view.

Both Bohr and Heisenberg argued that classical physical concepts are the products of a suitably mathematized refinement of our general phenomenal intuition (*Anschaulichkeit*), conceptuality, language, and so forth (e.g., Bohr 1987, v. 2, pp. 68–69; Heisenberg 1930, pp. 11, 64–65, 1962, pp. 56, 91–92). These aspects of our thinking are determined by the evolutionary biological and neurological constitution of our bodies. As such, they have emerged in our interaction with the world of objects consisting of billions of atoms and, thus, on a scale very different from those considered in quantum theory (or at the opposite end of the available physical scales from those at the scale of the Universe). In *Physics and Philosophy*, Heisenberg addressed the key paradox (or what so appears) at the heart of "the Copenhagen interpretation": "[The Copenhagen interpretation] starts from the fact that we describe our experiments in the terms of classical physics and at the same time from the knowledge that these concepts do not fit nature accurately. The tension between these two starting points is the root of the statistical character of quantum theory" (Heisenberg 1962, p. 56). As noted earlier, "the Copenhagen interpretation" present in the book was a mixture of Bohr's and his own, in some respects, different views. In particular, again, unlike Bohr, Heisenberg at this stage believed that purely mathematical concepts might be able to represent the ultimate constitution of nature, apart from physical concepts (in contrast to Einstein's and Schrödinger's search for new physical concepts to assume this role). On the point in question at the moment, however, Heisenberg's view was fully in accord with that of Bohr. Heisenberg says the following Bohr's argument is just considered:

> Therefore, it has sometimes been suggested that one should depart from the classical concepts altogether and that a radical change in the concepts used for describing the experiments might possibly lead back to a nonstat[ist]ical [sic!], completely objective description of nature. . . . This suggestion, however, rests upon a misunderstanding. The concepts of classical physics are just a refinement of the concepts of daily life and are an essential part of the language which forms the basis of all natural science. Our actual situation in science is such that we

> do use the classical concepts for the description of the experiments, and it was the problem of quantum theory to find theoretical interpretations of the experiments on this basis. There is no use in discussing what could be done if we were other beings than we are. At this point we have to realize, as von Weizsäcker has put it, that "Nature is earlier than man, but man is earlier than natural science." The first part of the sentence justifies classical physics, with its ideal of complete objectivity. The second part tells us why we cannot escape the paradox of quantum theory, namely, the necessity of using classical concepts. (Heisenberg 1962, p. 56)

In other words, classical concepts reflect the essential workings of our biological and specifically neurological machinery born with our evolutionary emergence as human animals. Our thinking in general, as the product of this machinery, is classical or classical-like in that it is consistent with and leads to the concept of classical physics. Any concepts we actually form derive from and apply only to observed phenomena, and, as observed phenomena, quantum phenomena are physically classical. They are different from classical phenomena because the data observed in them prevents us from describing how they come about by classical physics (which incapacity led to quantum theory) or in the RWR view, by any physical theory. Quantum theory, however, in particular, QM or QFT, can probabilistically predict these data without describing or representing how they come about, or at least it allows for this, RWR, type of interpretation.

The case is somewhat but, as discussed in Sect. 2.4, not entirely different as concerns new mathematical concepts, which, as Heisenberg noted, are free from limitations of daily language and concepts, or even limitations of classical or other physical concepts that we can form as representing the behavior, such as motion, of physical objects. This freedom, Heisenberg argued (at the time still adopting the RWR view, as opposed to his later works just cited), enabled his creation of QM (Heisenberg 1930, p. 11). The creation of new mathematical concepts, such as Heisenberg's matrix variables, Schrödinger's wave function, or Dirac's spinors, was essential to QM or QFT, which would not have been possible without them. Nor would classical physics and relativity have been possible apart from their mathematical concepts. These concepts, however, have a representational function in these theories, while the quantum–mechanical concepts just listed do not, at least in the RWR view.

Are, then, new quantum–theoretical *physical* concepts possible in the RWR view? And what is the status of Bohr's concepts such as phenomenon and complementarity, which appear to be necessary, along with mathematical concepts, but which are not mathematical? Yes, such concepts are possible, and Bohr's concepts of phenomena and complementarity are among them. The structure of such concepts is different, however. Bohr's concept of phenomenon has representational classical components, yet entails that the ultimate nature of reality responsible for any phenomena is beyond representation or even conception, which defines its RWR component. This is in accord with the two-component structure of measuring instruments, and hence of the *concept* of measuring instruments, consisting in their classical describable observable parts or their quantum strata through which they interact with quantum objects or, in the present view, the ultimate, RWR-type physical reality responsible for both quantum phenomena and quantum objects (with the latter still idealized as entities of RWR type).

In the case of complementarity, which, in Bohr's ultimate version of the concept, is that of phenomena in his sense, this architecture still applies, thus making it an RWR concept as well. But complementarity contains other features, which define it as a concept. As indicated in the Preface and as will be discussed in detail in Chap. 6, the fact that complementary quantum phenomena, say, that defined by a position measurement and that defined by a momentum measurement (in correspondence with the uncertainty relations), are mutually exclusive implies that, while each phenomenon could be described by means of classical physics, their emergence and their complementary nature cannot be so described. This in turn implies that they cannot be seen as parts combinable into a whole, a single entity, existing at the same moment in time and the same location in space. Either phenomenon is the only whole that is available at the time when it is established by a measurement. We can, by the definition of complementarity, always establish the alternative complementary phenomenon at the same point in time. But then, the first phenomenon is in principle excluded at this point in time. In classical physics and relativity, in contrast, it is always, in principle, possible to establish both quantities simultaneously and to represent their emergence by a mathematized classically causal process connecting them. Indeed, this emergence is defined by the fact that both quantities could be established at any point in time, thus allowing for classical causality. In Bohr's interpretation of the uncertainty relations, these quantities cannot even be defined simultaneously, even though each one of them can be determined exactly by the corresponding measurement and hence defined at any point. Accordingly, in this interpretation, the physical meaning of the uncertainty relations is complementary of two *exact* measurements—their mutual exclusivity at any given point in time and yet the possibility, by a conscious decision, to measure either one or the other exactly at any given point in time. (Their exact determination is of course crucial here, because both variables can be simultaneously defined inexactly.)

The overall conceptual structure of QM or QFT, in RWR-type interpretations, is then essentially defined by the abstract mathematical formalism (thus far, over $\mathbb{C}$) and the physical concepts applied to observed phenomena. Some of the latter concepts are just classical concepts. Others, however, as defined by the RWR view, have the double structure consisting, on the one hand, of classical concepts, describing measuring instruments, observed phenomena, and the data found in them, and on the other, the ultimate nature of physical reality responsible for quantum phenomena and quantum objects, which are RWR-type idealizations that may be applicable to something that exists independently or, as in the present view, only at the time of observation. The relationships between the formalism and the observed phenomena are defined by probabilistic or statistical predictions, which also requires Born's or analogous rules (explained later in this study), added to the formalism, rather than derived from it. The formalism, thus, has no conceptual connections at all to the physical concepts, of either type, just described. Accordingly, why the formalism and Born's rule work, and work jointly, is beyond representation or conception as well. It is mysterious. But if so, it is not because there is some mystical agency in charge of this situation, as in so-called mystical or negative theology, which presupposed such an agency, while denying that any humanly conceivable properties could be assigned to it. This mystery is, in Bohr's words, free from any "mysticism incompatible with the true spirit of science," although our scientific thinking in quantum physics has plenty of human spirit that drives it (Bohr 1937, p. 83, 1987, v. 2, p. 63). It is mysterious without being mystical, a mystery without mysticism.

2.3 Theories, Models, and Interpretations

By a theory I understand an organized assemblage of concepts, as concepts are defined in the preceding section. Just as concepts do, every theory has its history in preceding concepts and theories, and in its own historical development which can change it by modifying its concepts or the relationships among them. A viable physical theory relates (usually in an idealized way) by means of logically consistent and experimentally verifiable propositions, to the multiplicity of phenomena or, via phenomena, objects or other forms of physical reality considered by this theory. This relationship, in modern physics by means of mathematical models (defined below), might be representational and derive its predictive capacity from this representation, as in classical physics or relativity. Doing so was assumed to be possible, at least in principle, in the case of all classical objects, even when they could not be observed, as in the case of atoms or molecules in the kinetic theory of gases. Certain classical theories, such as classical statistical physics or chaos theory, introduce additional complexities into this representational scheme that require qualifications, but, as explained below, these complexities are not fundamental in nature. These assumptions are, as discussed in the Preface, permissible, for all practical purposes, in classical physics and relativity, even in the U-RWR view. This relationship may also be strictly predictive, possibly only probabilistically or statistically predictive, as in the case of quantum phenomena and QM or QFT, in any interpretation. RWR-type interpretations do not offer and even preclude a representation or, in the case of strong RWR-type interpretations, forming a conception of the ultimate nature of reality responsible for quantum phenomena. This, again, does not mean that a realist interpretation of quantum phenomena and QM or QFT is impossible. There is, as I said, no shortage of such interpretations, which are likely to continue to proliferate, although this proliferation also reflects a lack of consensus concerning their relative validity. There are considerably fewer nonrealist interpretations, *in the present definition of the term*, as the assumption of the possibility of a representation or at least a conception of the reality considered. I qualify because some see merely the assumption of the *existence* of physical reality as a form of realism. I shall comment on such views in the next section.

Quantum phenomena, and thus quantum physics, were initially defined by the fact that in considering them, Planck's constant, *h*, must be taken into account. Doing so allowed one, as it still does, to use classical physics in *describing* quantum phenomena, as observed in measuring instruments. Classical physics, however, could not predict these phenomena. This incapacity led to the assumption that there must exist entities in nature the behavior of which could not be described by classical physics, for otherwise classical physics would be able to predict them. These entities are now understood or idealized as quantum objects, and in the present interpretation, this idealization is only applicable at the time of measurement. It is true that quantum objects, such as electrons, were discovered within the framework of classical physics, and were expected to behave classically. With the introduction of the photon, a relativistic quantum object by definition (and hence requiring QED for properly handling it), and then quantum theory meeting these expectations became increasingly difficult. In the RWR view, *h*, is only associated with what is observed in measuring instruments, because placing the ultimate constitution of the reality responsible for quantum phenomena beyond representation or conception obviously precludes associating any numerical constant with it.

While, however, measuring any quantum phenomenon known thus far involves *h*, its role may not be sufficient to fully distinguish quantum phenomena from classical ones, and their specificity as quantum appears to be defined by a broader set of features, some of which are not expressly linked to *h*. Some of them are exhibited by classical phenomena or found in theories or "toy" models different from those of QM.[4] Accordingly, one now has a more complex sense of quantum phenomena. This book, for example, considers, *within the strong RWR view*, the following features of quantum phenomena, all, it appears, necessary (even if some of them are correlative) to define them vs. classical phenomena—(1) the role of *h*, (2) the irreducible role of measuring instruments in defining quantum phenomena, (3) discreteness, (4) individuality, (5) complementarity, (6) entanglement, (7) quantum nonlocality, and (8) the irreducibly probabilistic or statistical nature of quantum predictions. It might be preferable to have fewer such features and derive the rest from them, perhaps only one such feature.

I am not claiming that it is in principle impossible to do so. For example, if one adopts the RWR view, it is tempting to argue, following Bohr, that, if there is any single feature distinguishing classical and quantum phenomena, it is (2). Even if Bohr thought so, however, one might prefer to err on the side of caution, even if assuming this feature to be unavoidable. Thus, although an entanglement is part of the interaction between quantum objects and measuring instruments, and although these interactions relate to quantum nonlocality, neither entanglement nor quantum nonlocality appears to be derivable from these interactions. This difficulty is manifested in the EPR-type experiments, although, as Bohr argued, the irreducible role of measuring instruments remains crucial there as well (Bohr 1935). What appears to distinguish quantum and classical phenomena the most is the structure of information, physically classical but in its organization beyond the capacity of classical physics (or relativity) to predict it. As Bohr said in one of his final works in 1958 (perhaps by then with Clause Shannon's information theory in mind): "The quantal features of the phenomenon are revealed in the information about the atomic objects derived from the observations" (Bohr 1987, v. 3, p. 4). This structure, however, involves other features just mentioned. Accordingly, it is more prudent to consider all features listed above in their interactions, and these features may still not exhaust the

[4] See, for example, (Spekkens 2007, 2016). I shall discuss this subject in Chap. 9.

nature of quantum phenomena. These considerations confirm the view of "quantum phenomenon" as a complex, multicomponent concept, still open to modifications in view of possible new features added.

By quantum theory I refer to a set of conceptual schemes accounting for quantum phenomena, comprising the three standard versions of quantum theory, the only quantum theories considered in this study. (As noted from the outset, alternative theories of quantum phenomena, such as Bohmian mechanics or spontaneous collapse theories, will only be mentioned in passing.) These three versions were discovered in quick succession between 1925 and 1927, following a quarter of a century of the old quantum theory, ushered in by Planck's discovery of his black-body radiation law, quantum in character, in 1900. All these theories are probabilistic or statistical in nature. The first is quantum mechanics for continuous variables in infinite-dimensional Hilbert spaces (QM), the second is quantum mechanics or theory for discrete variables in finite-dimensional Hilbert spaces (QTFD), to be discussed as such in Chap. 9, but until then subsumed under the heading of QM, and the third is quantum field theory in Hilbert spaces that are tensor products of finite and infinite-dimensional Hilbert spaces (QFT), initially introduced in the form of quantum electrodynamics (QED). (There are nonrelativistic quantum field theories, which contain important differences from QM, but they will be mostly put aside here.) QFT, in turn, consists of several theories, now comprising the standard model of elementary particle physics: QED, the quantum field theory of weak forces, and the quantum field theory of strong forces, known as quantum chromodynamics (QCD). The first two are unified in the electroweak theory. The unification of all three, sometimes referred to as the grand unification (defined by a single symmetry group), has not been achieved. Nor is it always seen as imperative, insofar as the standard model in its present form correctly predicts all known quantum phenomena within its scope. In contrast, QFT and general relativity, the current standard theory of gravity, are inconsistent with each other, which, it is worth keeping in mind, does not mean that the phenomena themselves considered by these theories are inconsistent. Resolving this inconsistency, whether by means of one or another form of quantum gravity theory or otherwise, is one the greatest outstanding problems of fundamental physics. It motivated string theory and related approaches, such as brane theory or AdS/CFT (anti de Sitter [space]/conformal field theory), and more recently, the so-called ER = EPR (relating the Einstein-Rosen wormhole bridge of general relativity and quantum entanglement) of Leonard Susskind and Juan Maldacena, or alternative approaches, such as loop quantum gravity or those based in quantum information theory. While many of these approaches are impressive mathematically (and have contributed to contemporary mathematics), all of the proposals that have emerged from them thus far remain hypothetical. In any event, these developments are beyond the scope of this study, which only addresses QM and QFT. In most of this chapter, I shall primarily discuss QM (in either finite or infinite dimensions), although most of my claims apply to QFT.

The history of a theory is accompanied by the history of its interpretations, most generally, defined by concepts added to a theory (which, in the present definition, is already an assemblage of concepts), such as those that establish how the theory refers to the phenomena it considers. This distinction between a theory and its interpretation is not unconditional, and it may be subsumed by the present concept of a theory as an assemblage of concepts, on which I shall comment presently. But it is common and often useful. The history of QM has been shaped by a seemingly uncontainable proliferation of, sometimes conflicting, interpretations. It is not possible to survey these interpretations here. The literature dealing with the subject is immense. Even each rubric, on by now a long list (e.g., the Copenhagen, the many-worlds, consistent-histories, modal, relational, pragmatist, and so forth), contains different versions. The situation is somewhat more manageable in QFT, but interpretations multiply there as well, for example, particle versus field interpretations. Standard reference sources would summarize the most prominent rubrics, but are unlikely to adequately establish differences between interpretations within each rubric. This can only be done by properly considering each such interpretation, in the way I attempt to do here in considering Bohr's interpretation, which, too, has several versions.[5]

As I am primarily concerned with RWR-type interpretations of QM or QFT, the general subject of interpretation of either theory or, more broadly, of interpretation of physical theories will only be addressed in a limited fashion. But they cannot be avoided either, in part because of the essential difference between realist and RWR-type interpretations. Consider Michael Redhead's claim that an interpretation of QM "is simply some account of the nature of the external world and/or our epistemological relation to it that serves to *explain* how it is that the statistical regularities predicted by the formalism ... come out the way they do" (Redhead 1989, p. 44). A great deal depends on this "and/or." While RWR-type interpretations do say something on the subject of our epistemological relation to the world, via our thought and experimental technology, they do not "serve to explain," or even allow one to conceive, of "how it is that the statistical regularities predicted by the formalism ... come out the way they do," central to Redhead's ideal of interpretation.[6]

[5] A few recent monographs might be mentioned here as, in various ways, comparable to the present book (Timpson 2016; Svozil 2018; Healey 2019; Wilson 2020). Christopher Timpson's book addresses related epistemological perspectives in the context of quantum information theory and quantum Bayesianism, QBism, without, however, relating it to the work of founding figures, specifically Heisenberg's invention of QM, as the present book does. Karl Svozil's book, more technical in nature, also brings up related philosophical arguments and offers several deep insights. There are affinities between Richard Healey's departure from realism (and toward pragmatism) in his book and the present book, which is, however, more radically nonrealist, especially by virtue of the strong RWR view it assumes. Neither Svozil's nor Healey's book addresses the work of the founding figures extensively discussed in the present study. Healey comments on a single reported statement by Bohr, and critically (and in my view, rightly) on Dirac's view of quantum states (Healey 2019, pp. 253–254, 38).

[6] This interpretation was perceptively discussed by Laura Ruetsche, who cited the passage and Redhead's elaboration on it (Ruetsche 2011, p. 4). Ruetsche's book, which focuses on QFT, offers a helpful treatment of the question of interpretation of QM as well, in both cases from a realist perspective that appears to be Aristotelian in flavor. I find this perspective appealing, in part because it enables the book's critique of Redhead's "pristine" ideal, and related, such as "imperialist" and "universalist," interpretive ideals, which represent more rigid versions of realism. It is of some interest that neither Bohr's nor other views in the spirit of Copenhagen are mentioned by Ruetsche, while they could have provided helpful angles on several subjects she considered. I might note two:

Although often implicit, an interpretation is essential and ultimately unavoidable in any workable physical theory, first of all, by virtue of establishing the relationships between it and the phenomena or objects it considers. In modern physics, from Galileo on, this is customarily done by means of mathematical models. I define a *mathematical* model *in physics* as a mathematical structure or a set of mathematical structures that enables such relationships. (The concept of a model in mathematical logic is a separate subject.) As that of theory and other major concepts discussed here, the concept of a mathematical model or model, in the first place, has a long history, which is also a history of diverse definitions, and literature on the subject is extensive. It is not my aim to discuss the subject as such or engage with this literature, which would be difficult to do in a contained manner and is unnecessary for my aims here. The present concept of a mathematical model, while relatively open, is sufficient to accommodate those models that I shall consider and their functioning.[7] Mathematical models used in physics may be *more* geometrical, as in general relativity, or *more* algebraic, as in QM and QFT. I qualify because these geometrical models contain algebraic elements, and these algebraic models, specifically those used in QM and QFT, have their own geometry, discussed in Chap. 4.

The relationships between a model and the objects or phenomena considered by it may be representational. In this case the elements of a model and relations among them would map, in an idealized way, the relevant properties of reality and the relations among them, and relate the theory to reality by means of this mathematical representation. The predictive capacity of the theory would derive from this representation. The mathematical models used in classical mechanics or relativity are examples of such models. Models may also be strictly predictive, without being representational, as are the mathematical models used in QM or QFT, in RWR-type interpretations, the predictions of which are probabilistic or statistical, even in the case of elementary individual quantum objects and behavior. An interpretation of a given theory is, thus, always an interpretation of how the mathematical model or models used by it relate to the phenomena or objects considered. A theory may, however, involve other interpretive aspects, defined by either its own concepts or by some additional concepts. For example, the basis for Bohr's and, following Bohr, the present interpretation, is the irreducible role of measuring instruments in the constitution of quantum phenomena. The observable parts of measuring instruments are described by means of classical physics, while these instruments also have quantum parts, through which they interact with quantum objects, or, again, RWR-type reality via quantum objects, an interaction "irreversibly amplified" to what is observed in measuring instruments (Bohr 1987, v. 2, p. 51, v. 3, p. 3). The structure of

(1) "the measurement problem," only addressed by Ruetsche under the realist assumption that the independent behavior of a quantum system is described by its "Schrödinger evolution, [which is] deterministic, continuous, and reversible" (Ruetsche 2011, pp. 343–344), a view challenged below (notes 14 and 24) and in Chaps. 4, 5 and 6 of this study; and (2) the nature of elementary particles, a subject extensively considered by Ruetsche (in Chaps. 8–11 of her book) and discussed in Chap. 8 of this study.

[7] A discussion of modeling along the lines of the analytic philosophy of science is given in (Frigg and Hartmann 2012) and (Frigg 2014), without, however, considering RWR-type views.

quantum measurement is entirely independent of the mathematical structure of QM, which (with Born's rule added) can, however, correctly predict the data observed in measuring instruments. As an idealization, the RWR view, weak or strong, of the ultimate constitution of the reality responsible for quantum phenomena is, too, an interpretative assumption, as is the three-partite stratification of physical reality adopted in this study as defining quantum physics:

> (Q-RWR): the ultimate constitution of the reality responsible for quantum objects and quantum phenomena → quantum objects (defined, as RWR-type concepts or idealizations, at the time of measurement) → measuring instruments (observed as classical physical objects) → quantum phenomena (defined by what is observed, as classical objects, in measuring instruments).

It is possible and is more rigorous, given the present definition of a theory, to see a different interpretation of the mathematical model defined by a given theory as forming a different theory, because each interpretation may involve concepts not be shared by others. This is the case, for example, in different interpretations in the spirit of Copenhagen, not all of which are of the RWR-type, and some are different RWR-type interpretations. What is shared is the mathematical model used, at least in terms of the essential equivalence or mutual translatability of its different versions. For simplicity, however, I shall speak of the corresponding interpretation of the theory itself containing a given mathematical model, specifically, of one or another interpretation of QM or QFT.

A historical example of this situation is the introduction of QM. Initially, Heisenberg's and Schrödinger's *versions of the formalism* of QM appeared as two different mathematical models, giving the same predictions. They were also accompanied by *two different theories*, designated as quantum mechanics and wave mechanics, respectively. The first was more algebraic and strictly probabilistically predictive rather than representational, and the second was more geometrical and representational (at least in aim), by virtue of conceiving of quantum-level reality as a continuous wave-like process. Each theory was given several interpretations at the time, although interpreting either posed major difficulties, and these interpretations were initially mostly implicit. These two models were then proven to be mathematically equivalent, with one proof offered by Schrödinger himself. There is some debate concerning how rigorous some of these proofs were, with a consensus that the ultimate demonstration was given by von Neumann in the 1930s. In practice, however, physicists started to use both formalisms interchangeably in 1926.

In any event, this equivalence allowed one to unify the mathematical model of QM, ultimately, with von Neumann, in terms of its Hilbert-space formalism, with some yet more abstract versions, still mathematically essentially equivalent, added later. In contrast, the two theories—quantum mechanics (underlying matrix mechanics) and wave mechanics—which were based on two different sets of concepts, remained different. I would argue that they were primarily different as *theories* rather than as interpretations, given how different their physics were, although it is possible to see them as two interpretations of the same mathematical model. Schrödinger's theory, based on the idea of a wave-like ultimate constitution of the reality responsible for quantum phenomena, had receded by the late 1920s, in part because Schrödinger's wave function received a probabilistic interpretation. The theory has not been entirely abandoned, and Schrödinger returned to it in the 1950s (Schrödinger 1995). Its impact, however, has been limited. The probabilistic interpretation of the wave function has remained dominant even in realist interpretations.

2.4 Reality, Realism, and Reality Without Realism

The concept of reality without realism, RWR, is grounded in more general concepts of reality and existence, assumed in this study to be primitive concepts and not given analytical definitions. These concepts are, however, in accord with most, even if not all (which would be impossible), available concepts of reality and existence in realism and nonrealism alike.[8] By "reality" I refer to that which is assumed to exist, without making any claims concerning the *character* of this existence. The absence of such claims, which define realism, allows one to place this character beyond representation or even conception, as, when it comes to the ultimate nature of reality, is the case in either the Q-RWR view in considering quantum phenomena, or the U-RWR view in considering all physical phenomena. I understand existence as a capacity to have effects on the world with which we interact. The very assumption that something is real, including the RWR-type, is made on the basis of such effects. Following L. Wittgenstein, I understand "the world" as "everything that is the case," in particular, the world of events (Wittgenstein 1924, p. 1). Quantum events are observed as phenomena defined by measuring instruments or their equivalents in nature, which, in the Q-RWR view, are the only quantum events present. In the U-RWR view, there are no other physical events in general either, assuming, as I do here, that our bodies are observational technology as well. On the other hand, while unobservable as such, the ultimate constitution of nature in the U-RWR view or the ultimate constitution of the reality responsible for quantum phenomena in the Q-RWR view is assumed to be "the case" as well and, hence, is part of the world. This constitution never appears as such and hence is not an event, but it is responsible for certain events, as its effects, from which its unthinkable character is inferred.

To ascertain observable effects of physical reality entails a representation of these effects but not necessarily a representation or even a conception of how they come about, which may not be possible and is not in the RWR view. The concept of an

[8] These concepts could be defined in a great variety of ways, and they have been discussed and debated from the pre-Socratics on. During recent decades, this debate has acquired a new vigor and a set of revisionary perspectives, in the work of Thomas Kuhn, Imre Lakatos, Paul Feyerabend, and their followers in the so-called constructivist studies of science. The literature on these subjects (both more traditional and more revisionist) is massive. I provide a few representative references. For analytic-philosophical approaches, see (Pincock 2012) and (Van Fraassen 2008); for more restrained post-Kuhnian approaches, see (Cartwright 1983) and (Hacking 1983), and in the context of the relationships between classical and quantum physics (Cartwright 1999, pp. 177–233); and for more radically constructivist treatments, see (Galison 1997; Latour 1999; Hacking 2000).

effect (in the absence of classical causality) is crucial here, and it is not surprising that the appeal to effects (in the absence of classical causality) becomes persistent in Bohr's writings (e.g., Bohr 1987, v. 1, p. 92, v. 2, pp. 40, 46–47). Consider Poincaré's revealing remark: "[B]eyond doubt a reality completely independent of the mind which conceives it, sees or feels it, is an impossibility. A world as exterior as that, even if it existed, would for us be forever inaccessible" (Poincaré 1982, p. 209). The first sentence suggests something akin to Bishop Berkeley's view (which was, however, rejected by Poincaré), unless Poincaré merely means that *conceiving* of the character of such a world is an impossibility, while still assuming that some form of the world does exist, as the second sentence suggests. This sentence would, then, be in accord with the strong RWR view, except that this world can have effects that are accessible to us. The inaccessibility of the ultimate character of the world is an assumption based on the totality of such effects, such as those manifested in quantum phenomena in the Q-RWR view, which may then be extended, as it is in this study, to the U-RWR view. This possibility does not appear to have been contemplated by Poincaré. While aware of then (in 1905) emerging quantum theory, Poincaré, who died in 1912, made his comment before Bohr's 1913 theory, which inaugurated the RWR view. The main context of his remark was more general, including his questioning of Kant's concept of things-in-themselves.

While, as I argue here, the RWR view changes the nature of thought and knowledge, it does not in any way impede them. Instead, it opens new ways for them. The assumption that this reality is beyond thought allows this reality to be immensely, *unimaginably* (literally speaking!) rich. It is likely that, limited by our evolutionary biological and neurological nature, we can imagine very little, as against what lies beyond our thought, of how nature is, if the word "to be" applies to this beyond our thought of nature, or possibly to thought itself. Over a hundred years ago, Henry Lebesgue, one of the founders of modern integration and measure theory, observed (while commenting on the paradoxes of set theory, shaking the foundations of mathematics then) that the fact that we cannot imagine or mathematically define objects, such as "sets," that are neither finite nor infinite, does not mean that such objects do not exist (Dauben 1990, p. 258). Lebesgue did not specify in what type of domain, material or mental, they might exist. The statement is, however, a profound reflection of the limit of our thought concerning the nature of reality, material or mental. Similarly, the fact that we cannot conceive of entities that are neither continuous nor discontinuous does not mean that such entities do not exist in nature. The concept of reality without realism is a response to this type of difficulty brought about by quantum phenomena, beginning with the dilemma of the continuous versus discontinuous character of the ultimate constitution of the reality responsible for quantum phenomena. In RWR-type interpretations, this constitution is neither any more than it is anything else that thought can reach.

Although there is little direct evidence to support this claim, it is conceivable that Bohr's thinking was influenced by these developments in post-Cantorian foundations of mathematics. Bohr's brother, Harald Bohr, with whom Bohr talked on a daily basis, was a prominent mathematician in the field of functional analysis and was well familiar with these developments. He spent some time in Göttingen when David Hilbert, whom Harald Bohr knew well, was the reigning figure there. Hilbert made momentous contributions to both functional analysis and the foundations of mathematics. He also had abiding interests and involvement in many areas of physics, including both relativity (where he made a major mathematical contribution in formulating the field equations of general relativity by variational methods) and quantum theory, in this case, more indirectly, through his work in functional analysis. This work led to the concept of Hilbert space (defined by von Neumann), crucial in QM and QFT. Hilbert was also at (Niels) Bohr's Wolfskehl lectures on the history of atomic theory following his 1913 theory in Göttingen in June 1922, later called the "Bohr-Festival," where Bohr first met Heisenberg. This is a rich network of historical and conceptual interconnections, in which the possibility of the RWR view can be detected at several junctures, as Lebesgue's statement suggests. Bohr's 1913 theory was, however, a bold act of making the RWR view a defining part of a physical theory.

It follows from the definition of reality that a given theory or interpretation might assume different levels and different types of *idealizations* of reality, some allowing for a representation or conception and others not (keeping in mind that "idealization" refers here to a workable conception of something or, in the strong RWR view, a lack of such a conception, thus, possibly, something different from what it idealizes, rather than to any form of approximation of something). As stated from the outset of this study, the present interpretation of quantum phenomena and QM or QFT, assumes three fundamental idealizations. The behavior of the observable parts of measuring instruments in defining quantum phenomena is, thus, idealized as representable. It is, to return to Bohr's language, "the idealization of observation" (Bohr 1987, v. 1, p. 55). In contrast, the RWR-type reality ultimately responsible for these phenomena is idealized as that which cannot be represented or even conceived of. The third idealization, which stratifies the second, is that of quantum objects. The reason for assuming this additional idealization is as follows. On the one hand, in contrast to classical physics or relativity, in quantum physics, in each experimental arrangement one must, as Bohr argued, always discriminate "between those parts of the physical system considered which are to be treated as measuring instruments and those which constitute the objects under investigation" (Bohr 1935, p. 701). On the other hand, these two parts themselves are not uniquely defined, the situation sometimes expressed as the arbitrariness of the "cut," discussed later in this chapter. Accordingly, it is how we set up an experiment that defines what is the quantum object in this experiment, while its quantum nature, including as an RWR-type entity, is defined by the ultimate, RWR-type reality, whose existence is independent of any experiment. This conception applies to all quantum objects, from elementary particles to macroscopic quantum objects, such as Bose–Einstein condensates or Josephson's junctions (although their quantum nature is defined by their microscopic constitution). Macroscopic quantum objects, too, can only be established as quantum objects by means of observations made in measuring instruments. This tripartite idealization is a key feature of the present interpretation of quantum phenomena. It is not found in Bohr's interpretation, although it might be seen as a consequence of his argument, a consequence, however, not expressly derived by him.

Realist or ontological thinking in physics is manifested in the corresponding theories, which are commonly representational in character.[9] Such theories aim to represent the reality they consider, usually by mathematized models based on suitably idealizing this reality. As noted, it is possible to aim, including in quantum theory, for a strictly mathematical representation of this reality apart from physical concepts, at least as they are customarily understood, as in classical physics or relativity. It is also possible to assume an independent architecture, structure, of the reality considered, without assuming that it is either (A) not possible to adequately represent this architecture by means of a physical theory, or (B) even to form a rigorously specified concept of this architecture, either at a given moment in history or even ever. Under (A), a theory that is merely predictive could be accepted for lack of a realist alternative, but usually with the hope that a proper representational theory will eventually be developed. This was Einstein's attitude toward QM or QFT, which he expected to be eventually replaced by a realist theory. Even under (B), however, this architecture is customarily conceived on the model of classical physics (to which relativity philosophically, although not physically, conforms), while leaving the determination of its specific physical or mathematical form to the future.

What, then, grounds realism most fundamentally is the assumption that the ultimate constitution of reality possesses properties and the relationships between them, or, as in (ontic) structural realism (Ladyman 2016), just a structure, in particular, a mathematical structure. This constitution, then, may either be ideally represented and hence, known by a theory or be unrepresented or unknown, or even unrepresentable or unknowable, but still conceivable, usually with a hope that it will be eventually represented and known. When this constitution is assumed to be unknowable, but still conceivable, realism, in affinity with Kant's philosophy, borders on the weak RWR view, except that the latter does not, in general, imply such a hope, and more significantly, does not assume the existence of such properties or the (the structure of the) relationships between them, or just of a structure, on lines of ontic structural realism. The concept of a structure apart from elements constituting it entails further complexities, which I put aside, given that this concept is not applicable in the RWR view.[10]

---

[9] Although the terms "realist" and "ontological" sometimes designate more diverging concepts, they are commonly close in their meaning and will be used, as adjectives, interchangeably here. I shall adopt "realism," as a noun, as a more general term and refer by an "ontology," more specifically, to a given representation or conception of the reality considered by a given theory. Another, relatively common, term for realist theories is "ontic," in part used because "ontology" has other meanings, especially in post-Heideggerian philosophy.

[10] One could, in principle, also see the assumption of the existence or reality of something to which a theory can relate without representing it as a form of realism, even metaphysical realism, insofar as this assumption is independent of a physical theory. This use of the term realism is sometimes found in advocating interpretations of QM, for example that are nonrealist in the present sense, for example, in (Cabello 2017; Werner 2014), or along the lines of quantum Bayesianism (QBism) in (e.g., Fuchs 2016; Fuchs et al. 2014). It is of some interest that, while Fuchs, in building his argument on Wheeler's views, speaks of "participatory *realism*" (Fuchs 2016), Wheeler himself, whose views are closer to those of Bohr, only speaks of a "participatory *universe*," which is not the same (Wheeler 1981, 1983, p. 194, 1990, p. 5). Fuchs, perhaps unintentionally, reveals this

Thus, classical mechanics (used in dealing with individual objects and small systems, apart from chaotic ones), classical statistical mechanics (used in dealing, statistically, with large classical systems), or chaos theory (used in dealing with classical systems that exhibit a highly nonlinear behavior) are realist theories. While classical statistical mechanics does not represent the behavior of the systems considered because their great mechanical complexity prevents such a representation, it assumes that the individual constituents of these systems are represented by classical mechanics. In chaos theory, which, too, deals with systems consisting of large numbers of atoms, one assumes a mathematical representation of the (classically causal) behavior of these systems.

Our phenomenal experience can only serve us partially in relativity. This is because, while we can give the relativistic behavior of photons a concept and represent it mathematically, which makes relativity a realist and classically causal and, in fact, deterministic theory, we have no means of phenomenally visualizing this behavior, or the behavior represented by Einstein's velocity-addition formula for collinear motion $s = \frac{v+u}{1+(vu/c)^2}$. Thus, when the velocity is close to c (or is c), the relativistic concept of motion is no longer a mathematical refinement of our phenomenal sense and the corresponding ordinary concept of motion in the way the classical concept of motion is. Relativity was the first physical theory that defeated our ability to form a phenomenal conception of elemental individual physical behavior, and it was a radical change in the history of physics. Photons, which only exist in motion with a velocity equal to c in a vacuum, represent the limit case. If one could put a clock on a photon, it would stand still, and a photon would be in all locations in space at once, which of course only means that no frame of reference, defined by rods and clocks, could be associated with a photon. Ultimately, photons are relativistic quantum objects treated by QED. Nevertheless, relativity still offers a mathematized conceptual representation of the behavior of individual systems. This behavior could, moreover, be treated classically causally and indeed deterministically, although because all physical influences are limited by c, relativity imposes new limits, considered below, on causal relationships between events.

All theories just mentioned are, again, based on the idea that we can observe the phenomena considered without disturbing them sufficiently to affect them. As

---

difference: "it also seemed a worthy tribute to John Wheeler, as I thought he captured the appropriate sentiment with his phrase "participatory universe": "it's as full-blown a notion of reality as anyone could want, recognizing only that the users of quantum mechanics have their part of it too. That's not less reality, that's more" (Fuchs 2016, p. 3). This may be, but this full-blown notion of reality is still not the same as "realism," unless Fuchs only means by realism the assumption of the existence of reality, including that we create and observe in experiments, which would, in effect, make his view nonrealist in the present definition. In any event, I would argue that the present definition of realism or ontology is more in accord with most, even if, again, not all, uses of the term in physics or philosophy. Besides, none of the authors just mentioned adopts the strong RWR view, although QBism may be seen as consistent with this view or the U-RWR view. As in the U-RWR view, in the QBist view, rather than describing the ultimate nature of physical reality, our physical theories only allow us to experience observed phenomena and make predictions concerning them. In some cases, such predictions are even possible with probability one, which, however, this still makes them only predictions and not reality.

a result, we can identify these phenomena with the corresponding objects in nature in their independent behavior and (ideally) represent and predict this behavior by using this representation for all practical purposes by these theories, keeping in mind that qualifications are necessary for classical statistical physics or chaos theory. This identification helps realism, but it does not guarantee it, even in the case of classical mechanics, where representational idealizations are more in accord with our phenomenal experience, as Kant already realized, even though, he ultimately granted Euclidean geometry and Newtonian physics the capacity to represent the ultimate nature of physical reality, at least with a practical justification. It has been also asked to what, if any, degree the mathematical architecture of relativity (further removed from our phenomenal intuition) corresponds to the architecture of nature, as opposed to serving as a mathematical model for correct predictions concerning relativistic phenomena (e.g., Butterfield and Isham 2001). In this case, these predictions are ideally exact, deterministic, as opposed to the probabilistic or statistical predictions of quantum theory, even in dealing with most elementary individual quantum phenomena, at least as things stand now. This is a fundamental difference, arising, one is compelled to argue, because of the impossibility of, in principle, controlling the physical interference of measuring instruments with the object under investigation, in any interpretation of quantum phenomena or QM or QFT. As, however, indicated in the Preface and discussed in detail in Chaps. 6 and 7, in the RWR view, even exact predictions, the predictions with probability one, which are also possible (with further qualifications) in the case of certain quantum phenomena, are still only predictions concerning possible events. They do not, in this view, imply that the theory enabling them represents the ultimate nature of physical reality responsible for these events, although this assumption is, even in the U-RWR view, possible for all practical purposes, in classical physics and relativity. It remains crucial, however, that such predictions require special circumstances, while in general, quantum predictions are not of this kind, even in considering the simplest possible systems, the behavior of which can always be predicted ideally exactly in classical physics and relativity.

The representation of individual physical objects, quantum objects, and behavior became partial in Bohr's 1913 atomic theory, which make it the inaugural event of the RWR view. The theory only provided representations, in terms of orbits, for the stationary states of electrons in atoms (in which electrons had constant energy levels), but not for the discrete transitions, "quantum jumps," between stationary states. This was an unprecedented and at the time nearly unimaginable step because this concept was incompatible with classical mechanics and electrodynamics alike. It was expected at the time that Bohr's theory was a temporary expedient that would no longer be necessary when a proper theory of quantum phenomena was developed. It was, however, this concept that became central for Heisenberg, who built on it by abandoning an orbital representation of stationary states as well. This led him to his discovery of QM, the first physical theory that allowed for an RWR-type interpretation, at least of the weak type, of QM as a whole, as opposed to only partially conforming to the RWR view, as Bohr's 1913 theory did. Bohr offered an RWR-type assessment of Heisenberg's theory after it was given its full-fledged form as matrix mechanics by Born and Jordan (Born and Jordan 1925) but before

Schrödinger's introduction, based on a realist approach, of his wave mechanics in 1926. (As discussed below, this history played a role in the development of Bohr's thinking.) Bohr said:

> In contrast to ordinary mechanics, *the new quantum mechanics does not deal with a space–time description of the motion of atomic particles.* It operates with manifolds of quantities [matrices] which replace the harmonic oscillating components of the motion and symbolize the possibilities of transitions between stationary states .... These quantities satisfy certain relations which take the place of the mechanical equations of motion and the quantization rules. (Bohr, 1987, v. 1, p. 48; emphasis added)

Following Heisenberg's own thinking at the time, this assessment was thus based on the RWR view, at least the weak RWR view, and the corresponding interpretation of QM, albeit only implicit at this point. By contrast, the first worked-out version of Bohr's interpretation, in his 1927 Como lecture, attempted to restore, ambivalently, an instance of realism and classical causality to QM, by assuming that the independent behavior of quantum objects could be represented, moreover, classically causally, by the formalism of QM (Bohr 1987, v. 1, 52–91). Bohr's Como argument reflected and was affected by several intervening developments, such as, on the RWR side, Heisenberg's discovery of the uncertainty relations, and on the realist side, Schrödinger's introduction of his wave mechanics (Plotnitsky 2009, pp. 179–218, 2012, pp. 41–58). The Como version of his interpretation was, however, quickly abandoned by Bohr, following his discussion with Einstein in October of 1927 at the Solvay conference in Brussels. This discussion initiated his path toward his ultimate, RWR-type interpretation, as apparent in Bohr's next article, "The Quantum of Action and the Description of Nature" (Bohr 1987, v. 1, pp. 92–101; Plotnitsky 2009, pp. 219–238, 2012, pp. 59–70).

Bohr's ultimate interpretation was first sketched in his 1937 article, "Causality and Complementarity" (Bohr 1937). He did not speak of a "reality without realism," but his view, as defined by the irreducible role of measuring instruments in the constitution of quantum phenomena, clearly amounted to the RWR view, now the strong RW view. He referred to "our not being any longer in a position to speak of the autonomous behavior of a physical object, due to the unavoidable interaction between the object and the measuring instrument," which by the same token, entails a "renunciation of the ideal of causality in atomic physics" (Bohr 1937, p. 87). Elsewhere, Bohr appeals to "a final renunciation of the *classical* ideal of causality" (Bohr 1935, p. 697; emphasis added). As noted in the Introduction, this and related statements by Bohr might have been making a stronger claim implying the incapacity of any theory to restore this position to us. In this study, again, all such statements, are assumed to be part of Bohr's interpretation, from 1937 on, as a strong RWR-type interpretation, placing quantum objects and behavior beyond conception, which is clearly the case in this article. For, if one is no "longer in a position to speak of the autonomous behavior of a physical object," this behavior must also be beyond conception, because, if one had such a conception, one would be able to say something about it.

There is still the question of whether our inability to do so only (A): characterizes the situation as things stand now, while allowing that quantum phenomena or whatever may replace them will no longer make this assumption and thus RWR-type interpretations viable, thus reverting to a realist view, or (B): reflects the *possibility* that this reality will never become available to thought. Logically, once (A) is the case, then (B) is possible, but is not certain. There does not appear to be any experimental data compelling one to prefer either. (A) and (B) are, however, different in defining how far our mind can, in principle, reach in understanding nature. This is the main reason to distinguish these views, although my argument equally applies to both. As discussed in Chap. 6, Bohr at least assumed (A), and some of his statements, especially those that make stronger than interpretive claims concerning our lack of access to the ultimate nature of reality responsible for quantum phenomena, suggest that he might have entertained (B). The qualification "as things stand now" applies, however, to (B) as well, even though it might appear otherwise given that this view precludes any conception of the ultimate constitution of the reality responsible for quantum phenomena not only now but also ever. It still applies because, as indicated in the Preface, a return to realism in quantum theory is possible, if quantum theory, as currently constituted, is replaced by an alternative theory that requires a realist interpretation. This might make the strong (or weak) RWR view obsolete, replaced by a more realist view, even for those who hold it with quantum theory in place in its present form.

In any event, in either (A) or (B) form, the RWR view requires a reconsideration of causality and possibly, a "renunciation of the classical ideal of causality," as Bohr said, although it is, as always, difficult to assume such a renunciation to be "final." As is clear from Bohr's argument in this article and elsewhere, this ideal is grounded in the concept of classical causality, sketched in the Preface. This concept is defined by the claim that the state, state X, of a physical system is definitively (rather than with any probability other than equal to one) determined, in accordance with a law, at all future moments of time once it is determined at a given moment of time, state A, and state A is determined definitively in accordance with the same law by any of the system's previous states. This assumption, thus, implies a concept of reality, which defines this law, thus making this concept of classical causality ontological. Accordingly, by precluding the possibility of applying such a concept to the ultimate constitution of the reality responsible for quantum phenomena, the RWR view, weak or strong, precludes the application of the concept of classical causality to this constitution and, thus, to the relationships between quantum phenomena or events. This makes these relationships irreducibly probabilistic even in considering the most elementary individual quantum objects, such as elementary particles, rather than only mechanically complex systems as in classical physics. In the present view, again, the concept of a quantum object only applies at the time of measurement.

Several qualifications of the definition of the concept of classical causality are in order, however. In particular, this definition need not imply that A is a cause of X, in accord, say, with Kant's definition of causality, commonly used since Kant, and in fact, this definition does not refer to a "cause." Bohr, while commonly speaking of causality (in this classical sense) does not appear to use "cause" either. On one occasion, in one of his last works, Bohr does say that "the classical description of the experimental arrangement and the irreversibility of the recordings concerning the atomic objects ensure a sequence of cause and effect conforming with elementary demands of causality," accompanied by "the irrevocable abandonment of the ideal of determinism" (Bohr 1987, v. 3, pp. 4–5). I shall discuss this passage and the concept of a cause that it uses in Chap. 6 (Sect. 6.5) in the context of quantum causality and complementarity as a generalization of causality (Bohr 1987, v. 2, p. 65). Given, however, the connections between events corresponding to these recordings are, in view of the lack of determinism (which would accompany it in classical mechanics of individual or simple systems), it is clear that these concepts of cause and causality are not in accord with "the classical ideal of causality" invoked by Bohr earlier, and the concept of classical causality grounding them.

Kant was a key figure in the modern history of the question of causality, although what he defines as the principle of causality had been used much earlier, beginning with Plato and Aristotle, or even the pre-Socratics. Thus, according to Plato, our "inquiry into nature" is a search for "the causes of each thing; why each thing comes into existence, why it goes out of existence, why it exists" (Plato 2005, *Phaedo*, A 6–10; cited in Falcon 2015). Aristotle adopted the same type of view as well (e.g., Falcon 2015). Kant defined, on this model, the principle of causality, as classical causality, as follows: if an event takes place, it has, at least in principle, a cause of which this event is an effect, defined by a given rule or law (Kant 1997, pp. 305, 308). It is commonly (although there are exceptions) assumed that the cause must be prior to, or at least simultaneous with, the effect, an assumption also known as the antecedence postulate. Special relativity would make the antecedence postulate required. The term causality is sometimes used in accordance with the requirements of special relativity, which restricts (classically) causal influences to those occurring in the backward (past) light cone of the event that is seen as an effect of this cause, while no event could influence, be a cause, of any event outside the forward (future) light cone of that event. In other words, no physical causes can propagate faster than the speed of light in a vacuum, *c*. This limit is a manifestation of a more general concept or principle, that of locality, discussed later in this study, which states that physical systems can only be physically influenced by their immediate environment. Relativity, however, only restricts classical causality by a relativistic antecedence postulate, rather than precludes it: relativity theory, special or general, is a classically causal and in fact deterministic theory.

Now, the fact that the physical state of a body at point t_1 determines, by Newton's law of gravity, the state of this body at any other point t_2 does not mean that the state at t_1 is the (classical) cause of the state at t_2. One might argue that the real (classical) physical cause for any determination, including that of the initial state, A, that defines any particular case considered, is (in our language) the gravitational field defined by the Sun and other physical objects in the Solar system, as encoded in Newton's law of gravity. In this view, a given state, A, of any single object can only be seen as a (classical) physical cause of its future states insofar as the whole configuration of bodies and forces involved, which determines the law of motion and hence causality, is seen as embodied in this state but is not necessarily conceptually specified (which may not be possible). Its existence is assumed. Newton bracketed the physical nature of and hence the causes of gravity and was (wisely) content with just defining a law of gravity in considering, as classically causal, the behavior of any given object under this law, and other laws of Newton's mechanics. All earlier history of the emergence and development of the solar system was bracketed as well. In general, if considered at all, the history of classical systems is usually bracketed and involves the suspension of more remote classical causes, let alone of the ultimate classical cause of this history. Assuming the ultimate cause (classical or probabilistic) of anything is a major difficulty, put aside here; and it is, in general, not necessary, as a more limited set of causes (which may be multiple) is usually sufficient, as a workable idealization.

This type of bracketing is workable on very large spatial and temporal scales, even that of the Universe itself, using Newton's theory of gravity or general relativity, *up to a point*. The situation changes once one gets closer to the Big Bang or in considering the Big Bang itself (assumed to be a complex process in which a great many things happen even if in a very short time, by our measure), because of the quantum aspects and possibly still other aspects of the Big Bang have to be considered. This leads to the well-known complexities that thus far have defeated all our efforts to deal with them and that may require physical concepts and mathematics beyond anything we have, or possibly can even imagine now. Some of these complexities may be insurmountable in principle for philosophical reasons, or even due to the basic ways in which our thinking and language work. For example, even if one assumes, as some do, that the Universe started (with the Big Bang or some pre-Bing-Bang process) from nothing, as a fluctuation of nothing, one could still ask: Who or what put this nothing there? This question is in the nature of our thought and language; and there are proposals to the effect that our Universe emerged, in the Big Bang (or what we assume to be the Big Bang), from an earlier Universe, say, as the collision of M-brains, or those of infinite cosmological cycles of Universes. Mathematics could be different in each such theory, but there is a kind of phenomenological and conceptual limit on what we can conceive. Leibniz's famous question, known as the main question of metaphysics, "Why there are things rather than nothing?" may be unanswerable because of this nature of our thinking and language. In considering questions concerning the origin of any given situation in physics this thinking and language cannot be circumvented regardless of how free from them is the mathematics involved. One might say that such questions are inapplicable in this case, which may be true, especially if one assumes the RWR view (the Q- or U-RWR view) of the early, pre-Big-Bang universe by virtue of its possible quantum nature (in the Q-RWR view of it). But that cannot stop others from asking these questions, including by denying the RWR view of the early Universe.

For the moment, a well-defined situation in physics usually entails the cut-off defined by the initial state and the actual or possible final state of the system considered, with either one or both determined by measurements we do or can perform. When at stake are predictions, this demarcation is defined by the initial measurement, performed by us, possibly using nature as part of our observational technology, and a possible future measurement that can verify the prediction made. Such predictions can only be made by us, even when they concern the behavior of objects in accordance with classical causality and, in the case of some classically causal systems, are deterministic. As stated from the outset, nature does not make measurements or predictions. In the present view, nature has no quantum objects, including elementary particles, either. Perhaps, as discussed in Chap. 8, it has quantum fields, which is to say, such a concept may reflect better the ultimate constitution of nature or part of this constitution responsible for quantum phenomena.

Because of the difficulties, such as those just mentioned (and there are still others), of applying the concept of (classical) cause even in classical physics and relativity, the viability of the idea of causality in modern physics has been challenged, beginning with Russell's 1913 essay, which had a significant influence on the subsequent history of this challenge (Russell 2013).[11] At the same time, the view of classical physics or relativity, or as discussed below, quantum theory as conforming to the concept of classical causality defined here (in principle, again, not requiring the idea of cause, but only the law connecting events, which would only imply a set of classical causes), has been nearly universally accepted, including by Russell. Thus, he says, appealing to the law of gravitation, a paradigmatic case:

> The law of gravitation will illustrate what occurs in any advanced science. In the motions of mutually gravitating bodies, there is nothing that can be called a cause, and nothing that can be called an effect; there is merely a formula. Certain differential equations can be found, which hold at every instant for every particle of the system, and which, given the configuration and velocities at one instant, or the configurations at two instants, render the configuration at any other earlier or later instant theoretically calculable. That is to say, the configuration at any instant is a function of that instant and the configurations at two given instants. This statement holds throughout physics, and not only in the special case of gravitation. But there is nothing that could be properly called "cause" and nothing that could be properly called "effect" in such a system. …
>
> There is no question of repetitions, of the "same" cause producing the "same" effect; it is not in any sameness of causes and effects that the constancy of scientific laws consists, but in sameness of relations. And even "sameness of relations" is too simple a phrase; "sameness of differential equations" is the only correct phrase. It is impossible to state this accurately in non-mathematical language; the nearest approach would be as follows: -- "There is a constant relation between the state of the universe at any instant and the rate of change in the rate at which any part of the universe is changing at that instant, and this relation is many-one, i.e. such that the rate of change in the rate of change is determinate when the state of the universe is given." If the "law of causality" is to be something actually discoverable in the practice of science, the above proposition has a better right to the name than any "law of causality" to be found in the books of philosophers. …
>
> The law makes no difference between past and future: the future "determines" the past in exactly the same sense in which the past "determines" the future. The word "determine," here, has a purely logical significance: a certain number of variables "determine" another variable if that other variable is a function of them. (Russell 2013, pp. 13–15)

While rejecting causes, this view is clearly fully in accord with the concept of classical causality as defined here. This way of understanding physical law may be and has been defined as "determinism," which, while ontological, allows one to avoid the concept of cause. I shall comment on this term and its (different) use in this study below. Russell's argument has its merits in its critique of unproblematized concepts of causality and in influencing the subsequent thinking concerning it in physics. On

[11] See (Ross and Spurrett 2007) for a reconsideration of Russell's argument from a contemporary perspective, allied with structural realism (Ladyman 2016).

the other hand, his view contains uncritical assumptions, first of all, by disregarding the role of an experiment, which even in classical physics or relativity, brings back temporality and the difference between the past and the future, as against Russell's assertion and similar assertions by others. Bohr's atomic theory, introduced in the same year, brings these considerations into play, although it took a while to realize them. As Bohr said later, "The unrestricted applicability of the [classically] causal mode of description to physical phenomena has hardly been seriously questioned until Planck's discovery of the quantum of action" (Bohr 1938, p. 94). Russell's critique of causality does not amount to such "serious" questioning either. Perhaps, one needed quantum theory for this type of questioning in physics.

Quantum phenomena, at least in the RWR view, expressly violate the principle of (classical) causality, because no determinable event (or quantum phenomenon) could be established as the classical cause of a given event (another quantum phenomenon). Only statistical correlations between events could be ascertained. As Heisenberg said in his uncertainty relation paper, "[quantum] physics ought to describe only the correlation of observations" (Heisenberg 1927, p. 83). But then, there is, in the RWR view, no relation of the type claimed by Russell and others either, although such claims are found in considering QM or QFT as well. In the RRW view, the differential equations of QM or QFT, such as Schrödinger's or Dirac's equation, only provide (with the help of Born's rule) probabilities of possible future experiments of the basis of previously performed experiments, which implies a (local) arrow of time.

It is possible to introduce alternative, specifically probabilistic, concepts of causality, applicable in QM, including in RWR-type interpretations, in which case classical causality, as defined here, does not apply, except as a limit case. Such concepts were proposed by several authors, the present author among them, under the heading of quantum causality, to be adopted in this study as well (e.g., Plotnitsky 2011b; 2016, pp. 203–207). I shall comment on these concepts and relate them to complementarity as a "generalization of causality," as Bohr saw it, in Chap. 6 (Bohr 1987, v. 2, p. 41). Briefly, quantum causality, as understood in this study, is defined as follows. An *actual* event of measurement (as, in the RWR view, there are no other actual events) that has happened determines which events *may or (in view of complementarity) may not* happen in the future and be predicted with *one probability or another*, which is not the same that any of these events *will* happen. This event, A, at time t_0, defines certain *possible*, but *only possible* future events, say, X, at time t_1. The temporal precedence of A and the corresponding (local) arrow of time is crucial to quantum causality. It follows, however, that quantum causality works without, as in the case of classical causality, definitively establishing, or rather (since it would already be definitively established in advance) being definitively connected to, any future state of the system considered. Only the temporal precedence of A vis-à-vis X is definitive. Quantum causality allows one to relate actual events in terms of statistical correlations, such as those of the EPR-type between them, which events, however, are specifically prepared by repeated initial measurements. By the quantum indefinitiveness postulate, no definitive relationship between any two actual events that have already occurred can be established even ideally, as they can always be in classical mechanics. There these events are defined by quantities that *preexist measurement* rather than are *established by measurement*, under the conditions of complementarity and the uncertainty relations, as they are in quantum physics in the RWR view. Hence, there is no probability (except equal to one) in classical mechanics of individual or simple systems, while there is always probability, in general not equal to one, in quantum physics, no matter how simple the system considered may be. *A*, as an event of measurement, is reminiscent of a classical "cause." This reminiscence may compel one to adopt this type of probabilistic connections between events for a general definition of causality with what is defined here as classical causality as the limit case (e.g., D'Ariano 2018; D'Ariano et al 2014). For the reasons explained in Chap. 6, I shall adopt the term probabilistic causality for this type of concept, with quantum causality as an instantiation of this concept in quantum physics. In the case of individual or simple (nonchaotic) classical or relativistic systems, this concept reduces to classical causality and, correlatively, determinism as well. Quantum causality may, thus, be seen as a *symbolic* version of classical causality, just as QM itself may be seen, as it was by Bohr, as symbolic mechanics because it uses the symbols of classical mechanics as part of a radically different, probabilistic, theory.

Some, beginning with P. S. Laplace, have used "determinism" or its avatars such as "deterministic causality," for classical causality, including, as noted above, on occasion, but not commonly, Bohr (1987, v. 3, p. 5). Russell's and related views may in fact be seen as forms of *ontological determinism*, defined by the mathematics of the theory, with all possible predictions with probability either one or zero. Both Laplace (one of the founders of probability theory) and Russell were of course well aware of the unavoidable use of probability in the case of some phenomena in classical physics. Both, however, and most others who use determinism along these lines assumed it necessary only for practical, epistemological reasons, due to our lack of knowledge concerning classically causal or deterministic relationships ultimately defining all events considered. In this study, I define "determinism" as an epistemological category referring to the possibility of predicting the outcomes of causal processes ideally exactly in accordance with laws that define them as causal. In classical mechanics, when dealing with individual objects or small systems (apart from chaotic ones), both concepts are correlatively applicable. On the other hand, classical statistical mechanics or chaos theory are classically causal but not deterministic in view of the complexity of the systems considered, which limit us to probabilistic or statistical predictions concerning their behavior.[12]

[12] It is possible to assume that the ultimate nature of reality is random or mixed, and, while classically causal conceptions of reality have been dominant, random ones have been around since the pre-Socratics, as in Democritus's and then Epicurus's and Lucretius's atomism, without, however, assuming, as in the RWR view, that the reality ultimately responsible for such random events is beyond conception, and hence may not be random any more than classically causal. There are also arguments for the possibility of classical causality and, in the first place, realism in the case of discrete events (e.g., [Sorkin 1991, p. 55; Smolin 2018, pp. 257–261; Rovelli 2016). Rovelli's article is especially relevant in the present context, because, while adopting a realist position in advocating a discrete ontology of the ultimate constitution of the reality considered in QM, it argues against a realist interpretation of the wave function, which may be expected given the continuous nature of the wave function. As discussed in the Introduction, already Riemann noted the discreteness of the ultimate constitution of reality, including space itself, will imply different physics (Riemann

In the case of quantum phenomena, deterministic predictions are, again, in general (apart from special idealized cases) not possible even in considering the most elementary quantum phenomena, such as those associated with elementary particles. This is because the repetition of identically prepared quantum experiments, in general, leads to different outcomes, and unlike in classical physics, this difference cannot be diminished beyond the limit defined by Planck's constant, h, by improving the capacity of our measuring instruments. This impossibility is manifested in the uncertainty relations, which would remain valid even if we had perfect instruments and which pertain to the data observed, rather than to any particular theory. Hence, the probabilistic or statistical character of quantum predictions must also be maintained by interpretations of QM or alternative theories of quantum phenomena that are classically causal. Such interpretations and theories are also, and in the first place, realist because classical causality implies a law governing it and thus a representation of the reality considered (defined by the behavior of quantum objects) in terms of this law.[13] RWR-type interpretations are, again, not classically causal because they preclude a representation or even conception of the ultimate constitution of the reality responsible for quantum phenomena.

1854, p. 33). The difficulty here is that of explaining the physical mechanisms by means of which classical causality can be established in a discrete set, although one could have mathematics for exact predictions of discrete events. It is not clear that it is possible to conceive of such a mechanism apart from assuming that connections between events are continuous, an assumption that grounds classical causality in classical physics or relativity. This problem is circumvented in the RWR view by precluding a representation, even a purely mathematical one, of how quantum phenomena, including correlations, come about, and thus precluding classical causality. Admittedly, this type of thinking cannot satisfy those who want realism, as all these authors do, Lee Smolin most expressly, even by advocating a "naïve realism," which would, as discussed earlier, not be acceptable to Einstein, Smolin's main inspiration (Smolin 2018).

[13] As noted, the meanings of these terms fluctuate in physical and philosophical literature, without a strong consensus concerning any of them. Thus, Schrödinger's or Dirac's equation is sometimes seen as "deterministic" or (classically) "causal" under the assumption that it describes, in a classically causal way, the independent behavior of quantum objects, with the recourse to probability only arising because of the interference of measurements into this behavior. This assumption is shared by both von Neumann's and Dirac's classic books (von Neumann 1932; Dirac 1958), and ambivalently, by Bohr in his Come lecture (Plotnitsky 2009, pp. 191–218). It poses difficulties, beginning with the fact that the variables involved are complex quantities in Hilbert spaces over $\mathbb{C}$, difficulties that, as discussed in Chap. 5, were confronted by Schrödinger because his wave-equation was dealing with waves in the configuration space rather than physical space. I shall put these difficulties aside because they do not affect my argument at the moment. In RWR-type interpretations, either equation only determines the mathematical state, the "quantum state," of the corresponding wave-function as a Hilbert-space vector at any future point once it is determined at a given point, mathematically. Indeed, more accurately, either equation determines this state for any given value of the parameter t in the equation, which values could be related to measurements at different points in time. The relationships between these measurements themselves are probabilistic. Accordingly, physically, each equation only determines, in Schrödinger's terms, an "expectation-catalog" concerning the outcomes of possible future experiments, to be observed in measuring instruments, without representing either how they come about or these outcomes themselves, represented by classical physics (Schrödinger 1935, p. 154). Hence, contrary to another common claim, neither equation can be seen as (physically) time-reversible, a point considered further in Chap. 6. See also note 24.

As stated from the outset of this study, the RWR view, while historically connected to Kant, moves beyond Kant's argument, defining his philosophical system, which distinguishes between "noumena" or "objects" as things in themselves, as they exist independently, and "phenomena" or appearances to our mind, including those that we assume to represent "objects" of nature or mind. This study is primarily concerned with nature or matter. Kant's "things-in-themselves" may also be mental, as can in fact be an RWR type reality as well, for example, in mathematics, as is suggested by Lebesgue's comment, cited above, concerning a possible existence of a set that is neither finite not infinite or, more generally, the development of foundations of mathematics following Cantor's set theory (Plotnitsky 2019, 2020). Phenomena are available to and noumena or objects are beyond representation or knowledge. While Kant, thus, positions objects beyond *knowledge*, he allows, at least with a practical justification, that one can *think*, form a conception of, objects, rather than, in the strong RWR view, placing them beyond thought altogether. As I shall explain presently, Kant's view does not reach the weak RWR view either and remains a form of realism. According to Kant:

> [E]ven if we cannot *cognize* these same objects as things in themselves, we at least must be able to *think* them as things in themselves ... To *cognize* an object, it is required that I be able to prove its possibility (whether by the testimony of experience from its actuality or *a priori* through reason). But I can *think* whatever I like, as long as I do not contradict myself, i.e., as long as my concept is a possible thought, even if I cannot give any assurance whether or not there is a corresponding object somewhere within the sum total of all possibilities. But in order to ascribe objective validity to such a concept (for the first sort of possibility [that of conceiving of it] was merely logical) something more [than logic] is required. This "more," however, need not be thought in the theoretical sources of cognition; it may also lie in practical ones. (Kant 1997, p. 115, p. 115 note)

A justification in "theoretical sources of cognition" refers to what is justified by a concept that is assumed to be *objectively true* concerning the object it represents and a justification in "the practical sources of cognition" to a concept that works in practice and is *objectively accepted* (has "objective validity"), even if it cannot be rigorously proven to correspond to the object in question. For simplicity, I shall refer to a theoretical vs. a practical justification. Both forms of justification are *objective* in the sense of being unambiguously definable and communicable.[14] Whether, according to Kant, such a proof is ever possible in dealing with things-in-themselves is a

[14] As its etymology indicates, the origin of the term "objective" is linked to the concept of an *object* (as something existing independently of us), as it was by Kant. This allows our claims concerning it, as opposed to those concerning phenomena, to be more general and, in principle, even universal. Such more general claims are possible even in the RWR view, *as practically justified*, in considering physical objects in classical physics and relativity, but not in considering quantum objects or the ultimate constitution of the reality that gives rise to both quantum objects and quantum phenomena at the time of measurement. Objectivity is only possible at the level quantum phenomena, because they are manifested in the *observable parts* of measuring instruments. Hence, as the observable parts of measuring instruments themselves, they can be described classically and identified with the corresponding physical objects (in this case, things observed in measuring instruments), in the U-RWR view still only with a practical justification.

complex question, which could be put aside here, given that my concern is RWR-type interpretations, in which such a proof is not possible. On the other hand, this study still assumes here that these interpretations are only practically justified. Kant's statement "I can *think* whatever I like" is also an expression of creative freedom of thought in approaching reality either with a theoretical justification, as Einstein aimed to do by means of "a free conceptual construction" (Einstein 1949, p. 47), or only with a practical justification.

In physics, too, more than merely logical consistency is required for establishing the architecture of our concepts and theories and the way in which they are related to experimental data. Newton's use of calculus in classical mechanics, Einstein's use of Riemannian geometry in general relativity, and Heisenberg's use of matrix calculus, and the ways (realist or RWR) these mathematical concepts relate to experiments are obvious examples of new mathematical and physical concepts decisive for physics. On the other hand, while Kant's argumentation, again, manifestly short of the strong RWR view, borders on the weak RWR view, it does not amount to this view. This is because, once a conception of some things-in-themselves is practically justified, a representation of them based on this conception becomes practically justified as well, even though it is impossible to speak of knowledge of things-in-themselves, which, in Kant, requires a theoretical justification. In contrast, in the RWR view what is practically justified is not a representation or conception of the ultimate nature of reality responsible for quantum phenomena, but the inapplicability of any such representation or conception, either as things stand now, even if such a conception is in principle possible, or ever. If such a conception becomes possible, even if only as practically justified, then a representation based on this conception becomes possible as well, at least as practically justified, in accord with Kant's argumentation. It may, however, not become possible. On the other hand, as stated, the RWR view may become obsolete even for those who hold it and be replaced by a realist alternative defined by an interpretation of QM or QFT or by a new theory of quantum phenomena.

What might, in the RWR view, be assumed to be *theoretically* justified are our experimental findings and events, defined by effects of the interactions between the ultimate constitution of the reality responsible for quantum phenomena and measuring instruments, with, in the present view, quantum objects as an idealization only applicable in these interactions. This assumption, however, still only holds as things stand now and is subject to interpretation, which amounts to only a practical justification after all. It is virtually never possible rigorously to speak of justifying any concept or theory for *all* practical purposes without qualifications, beginning with those concerning the assumptions that ground it, even if it is objective in the sense of being unambiguously definable and communicable, a sense of objectivity assumed by Bohr as well (e.g., Bohr 1987, v. 2, pp. 67–68). Experimental findings, too, involve assumptions shaping their formulations and communication, and may, as a result, be subject to interpretive qualifications and disagreements. It is even more difficult to assume a theoretical justification for a theory or a mathematical model that predict experimental findings. Theories can change or become obsolete, even in the absence of new experimental data that would contradict them.

I do, however, adopt the view that a theory, although not necessarily all assumptions grounding it, should be falsifiable.[15] Glossing Karl Popper's definition, a theory is falsified if it can be contradicted by a basic statement, always referring to a specific entity or specific location and time, as experimentally observed quantum phenomena would be, a statement that must correspond to a true observation in a successful falsification and to a hypothetical observation in a failed falsification (e.g., Popper 1983, p. xxii). In the RWR view, because nothing can be said or even thought about quantum objects or aspects of nature so idealized, all falsifiable concepts, statements, and theories can only concern quantum phenomena. On the other hand, the assumption, realist or of the RWR-type, of the existence of an independent physical reality, of the existence of matter, is not falsifiable. This fact may give support to those, such as Parmenides and Plato or, in modern times, Bishop Berkeley, who deny that independent material reality, matter, exists. While, as indicated in the Preface, they have a point insofar as the assumption of the existence of this reality or any conception of it, or placing it beyond conception, is a product of human thought, none of them considered the RWR view of matter as a possibility.

QM and QFT are falsifiable by the data observed in measuring instruments, which compels me to define either theory strictly in terms of its mathematical formalism. Because, however, in contrast to classical mechanics and relativity, these theories are not deterministic and thus do not allow for (even ideally) exact predictions concerning the outcomes of individual quantum experiments, they are only falsifiable by their probabilistically or statistically predictive capacity.[16] The falsifiability of QM and QFT, enabled by their mathematics, should, again, not be confused with the falsifiability of a conception of one form of the physical reality considered or a physical ontology, because a theory need not offer a physical ontology of the ultimate constitution of nature, and QM and QFT do not in the RWR view. Nor, in the RWR view, would they offer a purely mathematical ontology of this constitution, apart from any physical concepts, which type of ontology has, as noted, been sometimes claimed, for example, by Heisenberg in his later writing or (ontic) structural realism. The falsifiability of such purely mathematical ontology is, however, difficult to claim. Putting aside that the very assumption of the independent existence of material reality is not falsifiable, it remains a question whether either a falsifiable physical ontology of the ultimate constitution of nature or, again, a purely mathematical ontology of this constitution is possible. Thus, the ontologies provided by the many-worlds interpretation and most ontological interpretations of QM or Bohmian mechanics are

[15] I am grateful to D'Ariano for bringing to my attention the falsifiability angle on quantum theory and the difficulty of maintaining a falsifiable ontology of the ultimate constitution of reality there, the problematic addressed in his recent work (e.g., D'Ariano 2018, 2020).

[16] One might question whether Popper's concept of falsifiability is applicable in this way, because this concept appears to imply the possibility of a definitive falsification of a theory's claim by a single case, like in its proverbial case of the claim that there are no black swans, contradicted by the discovery of at least one black swan. I shall, however, assume that the concept can apply statistically, at least within certain limits, because it is difficult, although not impossible (e.g., D'Ariano 2020), to speak of the falsifiability of QM or QFT otherwise.

unfalsifiable (D'Ariano 2020).[17] On the other hand, while the RWR view is not falsifiable either, it offers no ontology of the ultimate nature of reality responsible for quantum phenomena, which is a crucial difference.

I am not claiming that falsifiability is a requirement for a given physical theory or an interpretation, but only argue that it is a useful practical criterion in assessing a physical theory, and that quantum theory, including in the RWR view, is no exception. This disclaimer also allows me to bypass the controversies surrounding the concept of falsifiability since it was introduced. The RWR view tells us that, while *unfalsifiable assumptions* may be unavoidable in quantum theory (and the assumption of a reality of RWR-type is, again, unfalsifiable), *unfalsifiable ontologies* of the ultimate constitution of matter are avoidable. Popper, a committed realist, who did not like Bohr's views (which he also profoundly misunderstood), would not have liked the RWR view or this implication of his concept either. His concept is, nevertheless, useful to the RWR view, an irony that is not uncommon.

2.5 Indeterminacy, Randomness, and Probability

This section offers a discussion of the concepts of indeterminacy, randomness, and probability in QM, or QFT. My discussion cannot do justice to the subject, extensively discussed in the literature (e.g., [Khrennikov 2009; Hájek 2014]), or to the history of these concepts. It is designed to address the points most relevant to my argument in this study. I will, however, continue to discuss these concepts throughout the remainder of this study. I define, first, the concepts of indeterminacy, randomness, chance, and probability, as they will be understood here, because, as other key terms used in this study, they can be understood otherwise.

In the present definition, indeterminacy is a more general category, while randomness or chance will refer to a most radical form of indeterminacy, when a probability cannot be assigned to a possible event, which may also occur unexpectedly. Randomness and chance may also be understood as different from each other. These differences are, however, not germane in the context of this study, and I shall only speak of randomness. Both indeterminacy and randomness only refer to possible future events and define our expectations concerning them. Once an event has occurred, it is determined. An indeterminate nature of events may either allow for assuming an underlying classically causal architecture (which is usually temporal) of the physical reality responsible for this nature, whether this process is accessible to us, or not, or disallow for making such an assumption. The first case defines indeterminacy in classical physics, in particular classical statistical physics or chaos theory, and the second in QM, in RWR interpretations. It is, it might be added, impossible to

[17] D'Ariano appears to be more open to the possibility of a purely mathematical ontology of that type, although the falsifiability of such an ontology remains an open question (D'Ariano 2017; D'Ariano and Faggin 2021). The second article, however, refers to a mathematical ontology (grounded in a mathematical model based on the formalism of QM) of thought and specifically consciousness, which is different from an ontology of matter in physics.

ascertain that an apparently random sequence of events, events that occurred apparently randomly, was in fact random, rather than connected by some rule, such as that defined by classical causality, and there is no mathematical proof that any sequence is actually random (e.g., Aaronson 2013, pp. 71–92). The sequences of indeterminate events that allow for probabilistic predictions concerning them is a different matter, although there is still no guarantee that such sequences are not ultimately underlain by classically causal connections in the case of quantum phenomena. This would imply that an RWR-type interpretation, which precludes such connections, does not correspond to the ultimate nature of reality. Experimentally, again, quantum phenomena only preclude determinism, because identically prepared quantum experiments, as concerns the state of measuring instruments, in general lead to different outcomes. Only the statistics of multiple (identically prepared) experiments are repeatable and predictable by QM or QFT. This is fortunate because, otherwise, it would be impossible to treat the corresponding data scientifically, as QM or QFT do.

A Bayesian view, based on dealing with probabilities of individual events, would require one to qualify this assessment, while still entailing the fundamental difference in question between quantum and classical physics. It is worth commenting briefly on the difference between probability and statistics. "Probabilistic" commonly refers to our estimates of the probabilities of either individual or collective events, such as that of a coin toss or of finding a quantum object in a given region of space. "Statistical" refers to our estimates concerning the outcomes of identical or similar experiments, such as that of multiple coin tosses or repeated identically prepared experiments with quantum objects, or to the average behavior of certain objects or systems. There are many versions of the Bayesian view (e.g., [De Finetti 2008; Jaynes 2003]). More generally, however, it defines probability as a degree of belief concerning a possible occurrence of an individual event on the basis of the relevant information we possess. This makes probabilistic estimates, generally, subjective, although there may be agreement (possibly among a large number of individuals) concerning them (and some Bayesian approaches are more objectively oriented, as is, for example, that of Edwin Jaynes, in contrast to that of Bruno de Finetti). The frequentist understanding, also referred to as "frequentist *statistics*," defines probability in terms of sample data by an emphasis on the frequency or proportion of these data, which is considered more objective. In quantum physics, exact predictions are, again, impossible even in dealing with elemental individual processes and events. This fact could, however, be interpreted either on Bayesian lines, under the assumption that a probability could be assigned to individual quantum events, or on frequentist lines, under the assumption that each individual effect is strictly random and hence cannot be assigned a probability at all.[18] An example of a Bayesian approach (following de Finetti), which

[18] The standard use of the term "quantum statistics" refers to the behavior of large multiplicities of identical quantum objects, such as electrons and photons, which behave differently, in accordance with the Fermi–Dirac and the Bose–Einstein statistics, for identical particles with, respectively, half-integer and integer spin. A brief qualification might be in order concerning two different uses of statistical, referring, respectively, to multiple repeated experiments and the average behavior of large systems. Just as one can make probabilistic or statistical estimates of finding a quantum object in a given region of space, one can find an electron gas occupying less than a given volume.

is of an RWR-type *in the present definition*, is quantum Bayesianism, or QBism (Fuchs et al. 2014; Fuchs 2016). I qualify because, as noted earlier, QBists themselves sometimes speak of realism by virtue of assuming the existence of exterior physical reality. Although most of my argument in this study would apply if one adopts a Bayesian, RWR-type view, I adopt the frequentist, RWR-type view, considered in detail in (Plotnitsky 2016, pp. 173–186; Plotnitsky and Khrennikov 2015).[19] Bohr appears to have been inclined to a statistical view as well, speaking of QM as "a proper, essentially statistical atomic mechanics" (Bohr 1987, v. 2, p. 18; Bohr 1987, v. 1, 110, v. 2, p. v. 3, p. 5, 12).[20] This study does, however, assume, in affinity with QBism, that our assignments of probabilities are subjective, although open to an agreement between different individuals. It might be preferable to see this assignment is *human*, which makes all probabilistic relationships between a given theory and observed phenomena *human* as well, in classical, relativistic, and quantum physics alike. In classical physics or relativity, however, these relationships may be assumed to be *physically* grounded in the behavior of the system considered, because these theories may be assumed to represent this behavior, which is no longer possible to assume in quantum theory, in RWR-type interpretations.

Finally, probability introduces an element of order into situations defined by the role of indeterminacy in them and enables us to handle such situations better. Probability or statistics is about the interplay of indeterminacy and order. This interplay takes on a unique significance in quantum physics, because of the existence of quantum correlations, such as the EPR or (in the case of discrete variables) EPR–Bell correlations. These correlations are properly predicted by QM, which is, thus, as much about an order as about indeterminacy, and about their unique combination

---

Accordingly, predicting the second type of phenomena can also be done along either Bayesian or statistical lines, in the latter case, under the assumption that each event is strictly random. One would need to repeat the experiment many times to establish their statistics, while no probability is assigned to each event as such.

[19] I have adopted a Bayesian view of QM in an earlier study (Plotnitsky 2009). In contrast, in (Plotnitsky 2016), I adopted a statistical view, as I do, at least as preferable view, in this study, although most of my arguments would apply to Bayesian approaches, such QBism—most, but not all, in part depending on which Bayesian approach one adopts.

[20] On one occasion, referring to the old atomic theory, he does speak of "estimating probabilities for the occurrence of the individual radiation processes" (Bohr 1987, v. 2, p. 34). Still, it does not appear that he ever expressed anything akin to the Bayesian view. See (Plotnitsky 2016, pp. 180–184) for a further discussion of Bohr's view on the subject, again revising the reading of Bohr, along more Bayesian lines (without strictly attributing the Bayesian view to Bohr), offered in (Plotnitsky 2009). See note 19 above. There have been statistical interpretations of QM, commonly on realist lines. Two instructive examples are those of Andrei Khrennikov (Khrennikov 2012; Plotnitsky and Khrennikov 2015) and Armen E. Allahverdyan, Roger Balian, and Theo Nieuwenhuizen (Allahverdyan et al. 2013). While Khrennikov's interpretation is expressly realist, that of Allahverdyan, Balian, Nieuwenhuizen may be seen as consistent with or allowing for the RWR view, even though they do not appear to adopte it. This is because they argue that one should only interpret outcomes of pointer indications, and leave the richer quantum structure, which has many ways of expressing the same identities, without interpretation. In RWR-type interpretations, this structure would be seen as enabling statistical predictions, without representing the ultimate constitution of the reality responsible for the outcomes of quantum experiments and thus pointer indications.

in quantum physics. The correlations themselves are collective, statistical, and thus do not depend on either the Bayesian or frequentist view of the individual events involved.

The circumstances just sketched and discussed in further detail later in this study imply a different reason for the recourse to probability in quantum physics, in RWR-type interpretations. According to Bohr, the idea of indeterminacy apart from a (classically) causal order has "hardly been seriously questioned until Planck's discovery of the quantum of action" (Bohr 1938, p. 94). As he said on a later occasion (in 1949): "[E]ven in the great epoch of critical [i.e., post-Kantian] philosophy in the former century, there was only a question to what extent a priori arguments could be given for the adequacy of space–time coordination and causal connection of experience, but never a question of rational generalizations or inherent limitations of such categories of human thinking" (Bohr 1987, v. 2, p. 65). That this statement occurs in "Discussion with Einstein" is not coincidental: it places his debate with Einstein in relation to the history of philosophy as well as physics. Even more radical philosophical questionings of causality, such as those by David Hume, are those of our epistemological capacity to perceive the underlying classically causal world, which would be presupposed at the ultimate level as inaccessible to us.[21] It is only with Charles Darwin's evolutionary theory and Friedrich Nietzsche's philosophy that this (classical) view of causality and randomness or indeterminacy *begins* to be questioned. Darwin's and Nietzsche's thoughts are contemporaneous with the emergence of the kinetic theory of gases and thermodynamics—both statistical theories, which, however, assume an ultimate underlying classical causality. How far either Darwin or Nietzsche were willing to go on this road is a complex question, which cannot be addressed here. In any event, in physics, this more radical questioning of causality does not appear before quantum physics. By extending the questioning of causality to, at least for now, its limit, quantum theory also revealed the deeper nature of the problem of causality, or again, reality. According to Bohr:

---

[21] There is some debate concerning Hume's and Kant's views on this issue, including in the context of the relationships between their view and the nature of quantum phenomena and quantum theory, and also Bohr's view (e.g., Palmquist 2013, and references there). It is possible, assuming Kant's principle of causality as an a priori form of cognition, to read, as Stephen Palmquist does, Kant as arguing "not that everything in nature must have some definite, objective cause, but that our expectation of everything having such a cause is a necessary component of our 'empirical knowledge' of phenomena" (Palmquist 2013, p. 96). Given that this book is not a study of Kant, I prefer to remain open concerning possible interpretations of his views. Even assuming Palmquist's reading, however, this expectation would still be defeated in the case of quantum phenomena, at least in the RWR view, which is, even in its weak form, more radical than that of Kant, as regards any claims concerning the ultimate nature of the reality responsible for quantum phenomena. So, Bohr's assessment still stands. Certainly, there is no generalization, such as complementarity, of the concept of causality in either Kant or Hume, who was more open to probability as a theoretical necessity rather than only a practical expedient (recognized by Kant). In any event, Palmquist's reading of Bohr, primarily guided by the Como lecture (without considering changes in Bohr's views), bypassed the more radical aspects of Bohr's thinking, which make it easier to see it as being closer to that of Kant.

> [I]t is most important to realize that the recourse to probability laws under such circumstances is essentially different in aim from the familiar application of statistical considerations as practical means of accounting for the properties of mechanical systems of great structural complexity [in classical physics]. In fact, in quantum physics we are presented not with intricacies of this kind, but with the inability of the classical frame of concepts to comprise the peculiar feature of indivisibility, or "individuality," characterizing the elementary processes. (Bohr 1987, v. 2, p. 34)

This statement is, again, seen here as expressing the strong RWR-type *interpretation* adopted by Bohr at this point, in 1949 (although, historically, the statement refers to the rise of quantum theory), even if Bohr was making a stronger than merely interpretive claim, by assuming this situation to be a definitive state of affairs in nature. For one thing, as noted, some interpretations of QM, such as those by Dirac (1958) and von Neumann (1932), or alternative theories, such as Bohmian mechanics, assume classically causal views of the behavior of quantum objects, with probability or statistics brought in by measurement. Both individuality and especially "indivisibility" appear to reflect the features of Bohr's concept of phenomena.

"The classical frame of concepts" may appear to refer to the concepts of classical physics, and it does include these concepts. By this time (in 1949), however, Bohr adopts the strong RWR view, at least the Q-RWR view and arguably the U-RWR view, which places the ultimate nature of reality responsible for quantum phenomena and possibly all physical phenomena beyond conception. This gives the phrase "the classical frame of concepts" a broader meaning: all representational concepts that we can form are classical or proto-classical. They may be seen as proto-classical insofar as the physical concepts of classical physics and, as noted, with significant limitations, relativity may be considered as refinements of our phenomenal intuition, a product of our evolutionary neurological machinery, intuition *embodied* in ideas like bodies and motion.[22] This refinement is no longer available for representing the ultimate nature of reality responsible for quantum phenomena, or possibly the ultimate constitution of nature, at least as things stand now. Classical physical concepts are still used in quantum physics in RWR-type interpretations, in particular that of Bohr or the present one, in dealing with the behavior of the observable parts of measuring instruments and, thus, data or information are found in these parts. But, in these interpretations, these concepts do not apply to the ultimate character of physical reality responsible for quantum phenomena. That need not mean that a realist interpretation of quantum phenomena or QM, or an alternative theory of quantum phenomena, using "the classical frame of concepts," is impossible. It is still possible, because, as stated in the Preface, the RWR view, possibly applicable (as the U-RWR view) to the ultimate constitution of nature, may become obsolete in quantum theory in its present form or whatever may replace it even for those who hold it and be replaced by a more realist view, based "on the classical frame of concepts." In other words, even if the ultimate constitution of nature will still be assumed to be beyond representation or

[22] See (Berthoz 2000, 2003), for a neurological analysis of, in the first book's title, "the brain's sense of movement," as classical, and for the brain's decision-making processes, also considered, from the Bayesian perspective, in (Doya et al. 2007).

conception because of our own evolutionary neurological constitution, there will be no physics that needs to take this limitation into account.

As discussed in Sect. 2.2, in RWR-type interpretations, two types of concepts are available for dealing with the ultimate nature of reality responsible for quantum phenomena, while the classical frame of concepts is used for describing the observable parts of measuring instruments and quantum phenomena, or data or information registered in them. The first type is purely mathematical concepts, which, as noted, eventually led Heisenberg to a form of mathematical realism, with a Platonist flavor, while assuming that QM or QFT does not represent quantum objects and behavior by physical concepts, at least as we conventionally understand them, for example, in classical physics or relativity. In contrast, in his ultimate (strong RWR-type) interpretation, Bohr rejected the possibility of a mathematical representation, along with a physical one, of the ultimate nature of reality responsible for quantum phenomena, at least as things stand now. The present interpretation is in accord with Bohr on this point.

The second type of concepts is physical concepts, such as Bohr's concept of phenomena and complementarity, when complementarity is that of phenomena. These concepts have both (classical) representational components and RWR components, and thus reflect that the ultimate nature of reality responsible for quantum phenomena is beyond representation or conception, which defines their RWR components. This structure is in accord with and is defined by the two-component structure of measuring instruments, consisting in their classical describable observable part and their quantum strata through which they interact with the ultimate, RWR-type physical reality responsible for quantum phenomena. Complementarity adds a new dimension to this situation. While each of the two mutually exclusive complementary phenomena involved in a given complementarity, say, that of the position and the momentum measurement, associated with a quantum objects, may be established alternatively at any given point in time, they cannot be combined so as to exist together at the same moment in time and in the same location in space, in the way they can in classical mechanics. Neither concept, however, phenomenon or complementarity, represents the ultimate nature of the reality responsible for quantum phenomena. They reflect the impossibility of representing it.

2.6 Measurement, Idealization, and Quantum Indefinitiveness

This section explains in detail and derives additional implications from the tripartite structure of the idealization of physical reality assumed by the present view of quantum phenomena as represented by the diagrams introduced at the outset of this study:

(Q-RWR): the ultimate constitution of the reality responsible for quantum objects and quantum phenomena → quantum objects (defined, as RWR-type concepts or idealizations, at the time of measurement) → measuring instruments (observed as classical physical objects) → quantum phenomena (defined by what is observed in measuring instruments).

As throughout this chapter, I shall mostly remain within the limits of the Q-RWR view. This tripartite structure is, again, not necessarily adopted by all RWR-type interpretations of quantum phenomena and QM or QFT, including that of Bohr. Even if, given some of his statements, it might be seen as a consequence of Bohr's interpretation in its ultimate form or even some previous versions after the Como lecture, Bohr never expressly stated this consequence.[23]

I begin by restating the defining aspect of the concept of quantum measurement adopted in this study, following Bohr (eventually leading him to his concept of "phenomenon"). Admittedly, measurement in quantum theory is a vast and multi-faceted subject, considered in numerous books and articles (quite a few cited in this study). Several important topics will, unfortunately, be bypassed here, including those concerning such mathematical features of quantum measurements as pure vs. mixed states, or positive operator-valued measures (POVMs). A further discussion of quantum measurement is offered in later chapters. Quantum measurement may be said, à la Wheeler, to be a measurement without measurement. As explained at the outset of this study, the term "measurement" is a remnant of classical physics, inherited by it from a still earlier history, which can be traced in particular to the rise of geometry in ancient Greece, without claiming it to be the origin, if there is such a single origin, of the idea of measurement. As understood here, a quantum measurement *does not measure* some preexisting property of reality. An act of quantum

[23] Bohr says, for example, that "the concept of stationary states may indeed be said to possess, within its field of application, just as much, or, if one prefers, as little 'reality' as the elementary particles themselves. In each case are concerned with expedients which enable us to express in a consistent manner essential aspects of the phenomena" (Bohr 1987, v. 1, p. 12). That may be close to the present view. The statement still appears, however, more likely to imply that either concept is an idealization but not that of the concept of the elementary particle or quantum object, as an idealization, applied only at the time of measurement. In commenting on the complexities of Bohr's argument in his reply to EPR (Bohr 1935), Arthur Fine asks: "But should we say that an electron is nowhere at all until we are set up to measure its position, or would it be inappropriate (meaningless?) even to ask?" (Fine 2020). In the present view, an electron, or what is idealized, is simply not assumed to have existed (that is, the corresponding is idealization is not assumed to apply) before the interaction between the ultimate constitution of the reality responsible for quantum phenomena and the measuring instrument, and this idealization is always introduced by us after this interaction. A similar view appears to be associated with Bohr in (D'Ariano et al 2017): "Bohr would have said that there is no particle before we reveal it [by an experiment]: only the outcome of an experiment is real" (p. 155). Bohr could have said it, but he did not, at least not expressly. On the other hand, I don't think that Bohr would subscribe to the second part of this statement. As indicated above and as will be discussed in detail in Chap. 6, he appears to have seen quantum objects as real, and only claimed that one cannot unambiguously refer to their properties but only to the effects of their interaction with measuring instruments. Quantum objects are entities of the RWR-type. Bohr would, of course, have agreed with the view expressed by (D'Ariano et al 2017) on the same occasion that "only the performed tests and their outcome have the objectivity status" (p. 155).

measurement *establishes* quantum phenomena by an *interaction* between the instrument and the quantum object, or, in the present view, the ultimate, RWR-type, physical reality, which, in this interaction, create a quantum phenomenon, as a classical entity, and a quantum object, as an RWR-type entity, or what is so idealized. My emphasis reflects the fact that, although our decision concerning which measurement to perform plays a key role in establishing what can be observed and then predicted, it cannot control the outcome of a measurement, such as the value of the observed variable, or the outcome of the predicted measurement when the latter is performed. Any such outcome is always a product of the *interaction* between the ultimate physical reality responsible for quantum phenomena, the reality that exists independently, and the measuring instruments. Then what is so observed, as the data or information, always after this interaction, can be measured classically, just as what is observed in classical physics, where, however, what is so measured could be associated with the object itself for all practical purposes. As concerns the data or information thus registered, always as classical, a quantum measurement is a number-generating process, creating new numbers, probabilistically predictable by means of quantum theory, such as QM or QFT, on the basis of numbers generated in the same way by a previous measurement. According to Bohr:

> This necessity of discriminating in each experimental arrangement between those parts of the physical system considered which are to be treated as measuring instruments and those which constitute the objects under investigation may indeed be said to form a *principal distinction between classical and quantum-mechanical description of physical phenomena.* It is true that the place within each measuring procedure where this discrimination is made is in both cases largely a matter of convenience. While, however, in classical physics the distinction between object and measuring agencies does not entail any difference in the character of the description of the phenomena concerned, its fundamental importance in quantum theory … has its root in the indispensable use of classical concepts in the interpretation of all proper measurements, even though the classical theories do not suffice in accounting for the new types of regularities with which we are concerned in atomic physics. In accordance with this situation there can be no question of any unambiguous interpretation of the symbols of quantum mechanics other than that embodied in the well-known rules which allow us to predict the results to be obtained by a given experimental arrangement described in a totally classical way. (Bohr 1935, p. 701)

Before I discuss the significance of this elaboration for understanding the idealization of physical reality in RWR-type interpretations, I would like to address two common misunderstandings to which this and related statements by Bohr have often led. First, Bohr's statement may suggest that, while observable parts of measuring instruments are described by means of classical physics, the independent behavior of quantum objects is described or represented by means of quantum–mechanical formalism. This type of view has been adopted by some, for example, as noted earlier, Dirac (1930, 1958) and von Neumann (1932), and in part under the impact of their books, especially that of von Neumann, is sometimes referred to as "the Copenhagen interpretation." It was not, however, Bohr's view, at least after he revised his Como argument, which, as noted, entertained (still ambivalently) this type of view and had influenced others, including Dirac and von Neumann, in this regard. Bohr's Como argument might, however, have been in turn influenced by Dirac's paper on transformation theory (Dirac 1927), which adopted this view (Plotnitsky 2009, pp. 191–218). Bohr *does say* above that the observable parts of measuring instruments are described by means of classical physics and that the classical theories cannot suffice to account for quantum phenomena. But he *does not say* that the independent behavior of quantum objects is described by the quantum–mechanical formalism. His statement only implies that quantum objects cannot be treated classically, for if they could be, classical theories would suffice in accounting for the new types of regularities in question. The "symbols" of quantum–mechanical formalism are assumed here, as elsewhere in Bohr, only to have a probabilistically or statistically predictive role.

The difference of this view of Bohr from that of Dirac or von Neumann, and those who have followed them is worth a brief further commentary. Thus, von Neumann, taking Schrödinger's equation as a postulate, defines, ontologically, the independent development of a quantum system, postulated to be classically causal, in terms of the so-called unitary evolution, followed by a von Neumann–Lüders measurement. Von Neumann even speaks of this evolution as "thermodynamically reversible," which implies a physical ontology (von Neumann 1932, p. 418). This is a (non-falsifiable) ontological postulate, which is not necessary, insofar as Schrödinger's equation could be seen, as it was by Bohr or is here, only as part of the probabilistically predictive machinery of QM. A "unitary evolution" defined by Schrödinger's equation, is, in the RWR or related views, not a physical process, which understanding would also allow one to avoid some of the difficulties von Neumann has under his assumptions, including in using his language. Thus, it is difficult to speak of operators as measuring quantities, as von Neumann does in considering the intervention of measurement into the unitary evolution defined by Schrödinger's equation (e.g., von Neumann 1932, p. 347). In the present view, or that of Bohr, the intervention, while its outcome can be probabilistically predicted mathematically by QM, cannot itself be represented mathematically by QM, and von Neumann's argument shows the difficulties of getting around the possibility that it might not. I shall discuss this aspect of Bohr's view in Chap. 6. For the moment, in Bohr's or the present view, Schrödinger's equation (cum Born's rule) establishes an "expectation-catalog" for the outcome of possible future measurements at any future point in time based on the outcome of an already performed measurement (Schrödinger 1935, p. 154). Once a new measurement is made, the equation, using the data obtained in this measurement, will define a new expectation-catalog, for which any data obtained prior to this measurement is no longer of any use.[24]

[24] See note 13 above. I have discussed the difficulties of von Neumann's argument in (Plotnitsky 2009, pp. 205–214). See (D'Ariano 2020), for a critique of von Neumann's unitarity ontology, especially on its non-falsifiable nature. The article put into question, "*the ontology of the unitary realization of quantum transformations*," is introduced by von Neumann but used more generally (D'Ariano 2020, p. 1925). D'Ariano's conclusion is as follows: "Finally, we have said that most interpretations of [quantum] theory (many-world, relational, Darwinism, transactional, von Neumann–Wigner, time-symmetric,...) are indeed interpretations of the unitarity-purity dogma, and not genuine interpretations of the theory strictly speaking. Such interpretations ... still play a role as models, helping our conceptual understanding and intuition. However, they should not be

Bohr's insistence on the indispensability of classical physical concepts in considering measuring instruments is often misunderstood as well, in particular by disregarding that measuring instruments contain both classical and quantum strata. Even though what is observed as phenomena in quantum experiments is beyond the capacity of classical physics to account for them, the classical description can and, in order for us to be able to give an account of what happens in quantum experiments, must apply to the observable parts of measuring instruments. The instruments, however, also have a quantum stratum, through which they interact with quantum objects or, in the present view, the ultimate constitution of the reality responsible for quantum phenomena, which interaction would not be possible without this quantum stratum. This interaction is quantum and thus cannot be observed as such or, in RWR-type interpretations, represented. It is "irreversibly amplified" to the classical level of observable effects, say, a spot left on a silver screen (e.g., [Bohr 1987, v. 2, p. 73]).[25]

---

taken too seriously. This is the main lesson of Copenhagen" (D'Ariano 2020, p. 1932). D'Ariano is right to trace these ontological interpretations to the unitarity-purity dogma, which he shows to be unnecessary, also by deriving quantum theory (admittedly, only in finite dimensions), by associating to it a Hilbert space (as von Neumann does), but by using different postulates, in his "minimal interpretation." (Elsewhere, in applying the formalism to the problem of consciousness, he also proposes a different, discrete or "atomic" non-unitary evolution of pure states [also D'Ariano and Faggin 2021].) D'Ariano only needs, in addition to a Hilbert space and a tensor product of Hilbert spaces, two additional postulates (vs. four of von Neumann), one of which is a form of Born's rule. Physics, he argues, does not require the unitarity-purity dogma, as the black-hole information paradox and other features of higher-lever quantum theory or potentially quantum gravity suggest as well (D'Ariano 2020, pp. 1925–1926). He is also right to see it as a lesson of Copenhagen. In Bohr's and the present view (which I am not attributing to D'Ariano), the main interpretative postulate, assuming the formalism of quantum theory, in either infinite or finite dimensions, or even QFT is as follows. No structure, for example, an axiomatic one, pertaining to the formalism is assumed or even is, in principle, allowed to have an ontological, realist, significance. Obviously, other postulates would be necessary for deriving quantum theory (in finite or infinite dimensions). What is put into question here is the ontological nature of *any* structure within the formalism or any postulate concerning the relationships between such structures, apart from a classical ontology of measuring instruments or phenomena to which such structures might relate, via Born's or analogous rule. I am, again, not attributing this (strong RWR-type) position to D'Ariano. In contrast, however, to von Neumann's axioms, the axioms proposed by D'Ariano are not ontological, and hence consistent with the RWR view. They are, he argues, also falsifiable, as opposed to "the unitarity-purity dogma," which is not. In any event, in Bohr or the present view, as a strong RWR view, any such structure could only be assumed to have a probabilistically predictive significance and is thus part of the symbolic and, hence, nonontological probabilistic machinery of quantum theory. No ontological interpretation of the ultimate constitution of the reality responsible for quantum phenomena is accordingly possible under this assumption, which is, admittedly, not falsifiable either, but is practically justifiable. This is the main lesson of "*Bohr's* Copenhagen," defined by the spirit of reality without realism.

[25] The physical nature of this "amplification" is a separate matter and is part of the problem, commonly, including by this author, seen as unsolved (although there are claims to the contrary, for example, on lines of the consistent histories approach), of the transition from the quantum to the classical, which and related subjects, such as "decoherence," are beyond my scope here. One could attempt to formalize this situation, as, for example, in (Ozawa 1997). One considers a compound quantum system, QO + QI, consisting of the quantum object under investigation, QO and the

The situation under discussion is sometimes referred to as the arbitrariness of the "cut" or, because the term cut [*Schnitt*] was favored by Heisenberg and von Neumann, the "Heisenberg-von Neumann cut." As Bohr noted, however, while "it is true that the place within each measuring procedure where this discrimination [between the object and the measuring instrument] is made is … largely a matter of convenience," it is true only largely, but not completely. This is because, as he said on the same occasion, "in each experimental arrangement and measuring procedure we have only a free choice of this place within a region where the quantum–mechanical description of the process concerned is effectively equivalent with the classical description" (Bohr 1935, p. 701).[26] In other words, the ultimate constitution of the physical reality responsible for quantum phenomena, including quantum objects (in the present view defined as an idealization applicable only at the time measurement) observed in measuring instruments is never on the measurement side of the cut. Neither are quantum strata of the instruments through which the latter interact with this reality. At one end, then, the effects observed in quantum experiments, as quantum phenomena, can be represented by means of classical physical concepts. Indeed, classical concepts are necessary already in order to establish the peculiar features of these effects and to ascertain their quantum emergence, which is beyond the grasp of these concepts. (This is, as noted, why classical physics cannot predict them either, for if it could, as a realist theory, represent this emergence as well.) This necessity of classical concepts and yet their inability to grasp the ultimate nature of reality responsible for quantum processes or allow room for the laws of quantum theory has been and remain a difficult point to confront. This difficulty was emphasized by Heisenberg in his unpublished response to EPR's paper, based on the concept of the shifting cut (Heisenberg 1935) and elsewhere (e.g., Heisenberg 1962, p. 56). Heisenberg's response was written before Bohr's reply to EPR, and influenced it, arguably especially Bohr's passage under discussion, as their correspondence indicates, although Heisenberg's views were in turn influenced by Bohr's earlier writings (e.g., Plotnitsky 2009, pp. 309–310). Heisenberg's view expressed there and his understanding of the cut appear to be somewhat different from that of Bohr (Plotnitsky 2009, pp. 309–310). Heisenberg's response was, however, never finished, perhaps because of Bohr's parallel work on his reply and its quick publication, within only a few weeks after EPR's paper. This makes it difficult to assess Heisenberg's ultimate view there, as opposed to that of

quantum part, QI, of the instrument I, QO + QI, which is isolated during the (short) time interval when the quantum interaction in question takes place. The rest of the instrument, I, performs the measurement, a pointer measurement, on QI, after the interaction has taken place. In realist schemes, such as that of Ozawa, the evolution of t QO + QI, the unitary evolution operator, $U(t) = e^{-i\Delta t H}$, where $H = H_{QO} + H_{QI} + H_{QOQI}$ is the Hamiltonian representing the internal behavior of the subsystems involved and H_{QOQI} the interaction between them. In Bohr's and the present view, as RWR views, no element of the formalism represents the ultimate nature of reality responsible for quantum phenomena, including its stratum involved in the interaction between QO and QI, responsible for the effects observed. Any such element only serves as part of the mathematics of QM that, with the help of Born's or analogous rule, predicts such effects.

[26] This could be seen as a manifestation of Bohr's correspondence principle, which says that in the classical limit the predictions of quantum and classical theory should coincide, even though the phenomena considered are still quantum and can exhibit quantum effects.

Bohr in his replay. Both, however, are manifestly based on the stratified structure of reality defining quantum phenomena, which is the main point of my discussion at the moment.[27]

It might be opportune to briefly comment on Schrödinger's cat-paradox from the perspective here outlined. His thought experiment is as follows:

> One can even set up quite ridiculous cases. A cat is penned up in a steel chamber, along with the following device (which must be secured against direct interference by the cat): in a Geiger counter there is a tiny bit of radioactive substance, so small, that perhaps in the course of the hour one of the atoms decays, but also, with equal probability, perhaps none; if it happens, the counter tube discharges and through a relay releases a hammer which shatters a small flask of hydrocyanic acid. If one has left this entire system to itself for an hour, one would say that the cat still lives if meanwhile no atom has decayed. The ψ-function of the entire system would express this by having in it the living and dead cat (pardon the expression) mixed or smeared out in equal parts. (Schrödinger 1935, p. 157)

In RWR-type interpretations, there is no superposition of the states of the cat, but only a superposition of state vectors, which allows us to predict the 50% probability of the quantum radioactive decay while the cat is in the box and hence for either outcome once we open the box. In fact, Schrödinger pretty much says as much, in part given that (which is often forgotten) his thought experiment is introduced in explaining the difficulty or even impossibility of assuming that variables considered are "blurred" (Schrödinger 1935, pp. 156–157). QM, in these interpretations, tells us nothing about the cat inside the box, which (unlike the radioactive decay) may be considered a classical macro object, whose behavior, as a cat, is not described by QM. Indeed, in these interpretations, QM does not describe the (RWR-type) reality responsible for the decay either but only predicts the probability of its occurrence. So, one could consider the cat as a (quantum) object under investigation in Bohr's scheme here outlined, in which case the cat could not be represented or even conceived of as such since we can only conceive of and experience cats classically. In RWR-type interpretations, QM does not represent cats as quantum objects either, any more than electrons: it only predicts the effects that quantum objects can have on the classical world as experience it, effects defining quantum phenomena, such as the one created when we open the box. Once we do so, there is a classical state of the reality of the observable part of the instrument, which includes the cat, as either dead or alive.

Now, in the present view, if not that of Bohr (or that of Heisenberg or von Neumann), a quantum object, while still an RWR-type idealization, is different from that of the ultimate, RWR-type, reality that is responsible for both quantum objects and quantum phenomena, and is assumed to have an independent existence. While a measuring instrument, which is, in its observable part, a classical object, or, at the other pole, the ultimate constitution of the reality considered, are assumed to be independent, a quantum object can, in view of these considerations, only be rigorously defined by a measurement and its setup, including the cut, and thus by our observation of the outcome of this measurement. By the same token, the corresponding

[27] An argument concerning the stratified character of the reality defining quantum phenomena, without adopting the RWR view was proposed in (Rovelli 1996).

idealization, say, an electron, only applies after the interaction between this reality and the measuring instrument has taken place, and the exact time of this interaction or its duration cannot be exactly known. Accordingly, there is no independent behavior of quantum objects either: there is only the interaction between the ultimate (RWR-type) nature of reality and measuring instruments, the interaction of which allows one to define quantum objects. As discussed in Chap. 6, this interaction actually takes place before the measurement qua measurement itself, which pertain to the state of the quantum stratum of the instrument after this interaction, and no longer to the state of the quantum object, which, or what is so idealized, thus no longer exists either in the present view (Bohr 1987, v. 2, p. 57). It is this state, rather than the state of quantum object or the independent reality with which the instrument had interacted but no longer does that is then "irreversibly amplified" to the macroscopic, classical level of observable effects, such as a spot left on a silver screen. If one assumes the independent existence of quantum objects, then one can say, as Bohr does, that "the particle [the quantum object] is already on its way from one instrument to another," the statement that, again, appears to suggest that Bohr assumes the independent existence of quantum objects (Bohr 1987, v. 2, p. 57). What is considered as a quantum object in a given experiment can be different in each case, including possibly something that, if considered by itself, could be viewed as classical, as in the case of Carbon 60 fullerene molecules, which were observed as both classical and quantum objects (Arndt et al. 1999). The quantum nature of any quantum object is still defined by its microscopic constitution, with which elementary particles are associated. The ultimate, RWR-type reality itself responsible for this situation is, again, assumed here to have an independent existence and is always, in any possible experiment, on the other side of the cut.[28]

The following question might, then, be asked. If a quantum object is an idealization that only refers to what exists is as defined by an experiment or measurement, rather than to something that exists independently, could one still speak of the same quantum object, say, the same electron, in two or more successive measurements? Consider (speaking first in more conventional terms) two position measurements, the first defined by a slit in a diaphragm through which an electron, emitted from some source, may be assumed to be registered to pass by some counter, and the second defined by a collision between it and a silver bromide screen at some distance from the diaphragm. Each of these two measurements defines an electron, with the same mass and charge, in two different positions at two different moments in time. As explained later, the case can be given a strictly RWR interpretation, insofar as all these properties (mass, charge, and position) are, physically, those of measuring devices, assumed to be impacted by quantum objects, rather than of these objects themselves, placed beyond representation or conception. For the moment, the question is: Do these two measurements register the same electron, or can the corresponding idealization refer to the same quantum object? Rigorously, if the idealization of quantum objects is

[28] The concept "quantum object" could be defined otherwise, as they would be in a different interpretation of QM or an alternative theory of quantum phenomena, as for example, on more realist lines, in (Jaeger 2014), via Abner Shimony's concept of "quantum indefiniteness" (Shimony 1988).

only applicable, as it is here, at the time of measurement, then a prediction based on a given measurement and the new measurement based on this prediction could only concern a new quantum object, arising in the interaction between the ultimate constitution of the reality responsible for quantum phenomena and measuring instruments, and not an object that we measured earlier in making a prediction. Accordingly, in the present view, rigorously, one deals with two different quantum objects, two different electrons, for example. To consider them as the same electron is, however, a permissible idealization in low-energy (QM) regimes, an idealization ultimately statistical in nature, because a collision with the screen, is not guaranteed, although the probability that it will not is low. Nevertheless, one could, within these limits, speak of the transition between (physical) states of the same quantum object, the states defined in terms of the effects observed in measuring instruments.

On the other hand, as discussed in Chap. 8, speaking of the same electron in any two successive measurements in high-energy (QFT) regimes is meaningless, which further justifies the present concept of a quantum object and the tripartite idealization scheme, as outlined here. The concept of quantum field, considered in Chap. 8, responds to this situation, also allowing one, applying the concept of quantum field (which is assumed to exist independently and thus is not a quantum object in the present definition), to see measurement in low-energy (QM) regimes as a limit case. The interaction between measuring instruments and quantum fields are, then, manifested in registering two quantum objects of the same type (or a quantum object and nothing) in two successive measurements, and speaking of the same quantum objects in two successive measurements is a statistically permissible idealization.

As noted from the outset of this study, the epistemological cost of the RWR view is not easily absorbed by most physicists and philosophers, and to some, beginning, famously, with Einstein, is unacceptable. This is not surprising because the features of quantum phenomena that are manifested in many famous experiments and that led to RWR views defy many assumptions concerning nature commonly considered as basic. These assumptions, arising, again, due to the neurological constitution of our brain, have served us for as long as human life, and within certain limits, are unavoidable, although, while fully respected by classical physics, their scope, as noted, was already challenged by relativity. QM has made this challenge much greater. The same neurological constitution may also prevent us from conceiving of the ultimate (RWR-type) nature of physical reality responsible for quantum phenomena. Thus, it is humanly natural and even unavoidable to assume that *something happens* between observations. The sense that something happened is one of the most essential elements of human thought, and in a certain sense defines it, especially as concerns temporality. However, in the RWR view, the expression "something happened" is ultimately inapplicable to the ultimate constitution of the reality responsible for quantum phenomena. According to Heisenberg:

> There is no description of what *happens* to the system between the initial observation and the next measurement. …The demand to "describe what happens" in the quantum-theoretical process between two successive observations is a contradiction in adjecto, since the word "describe" refers to the use of classical concepts, while these concepts cannot be applied in

> the space between the observations; they can only be applied at the points of observation. (Heisenberg 1962, pp. 57, 145)

The same would apply to the word "happen" or "system," or any word we use, whatever concept it may designate, including reality, although when "reality" refers to that of the RWR-type, it is a word without a concept attached to it. As Heisenberg says: "But the problems of language are really serious. We wish to speak in some way about the structure of the atoms and not only about 'facts'—the latter being, for instance, the black spots on a photographic plate or the water droplets in a cloud chamber. However, we cannot speak about the atoms in ordinary language" (Heisenberg 1962, pp. 178–179). Nor is it possible in terms of ordinary concepts, from which ordinary language is indissociable, or, in the RWR view, even in terms of physical concepts, assuming the latter can be entirely dissociated from ordinary concepts.[29]

This is a formidable problem even if one adopts the strong RWR view. The term "reality" in the phrase "reality without realism" does not pose a difficulty here, because this term has no concept associated with it, making it akin to a mathematical symbol. Of greater difficulty are expressions like "quantum objects interact with each other," used for example, in considering the EPR experiment and entanglement, or "the interaction between the independent RWR-type reality and a measuring instrument," which refer to something between or before observations. One can handle this difficulty as follows in the RWR view. Although one can provisionally speak of a "relation" between two or more quantum objects, including the quantum strata of measuring instruments, there is no term or concept, such as "interaction" or "relation," or "taking place," applicable to what "takes place." Any rigorous statements can only concern observable events, with which, moreover, and only with which the concept of a quantum object is associated. Accordingly, in this view, one cannot rigorously speak of an interaction between quantum objects and between experiments, with the concept quantum objects applicable, again, only at the time of measurement in the present interpretation. One can only speak of two quantum objects associated with two measurements performed initially and then two quantum objects associated with two measurements performed subsequently. These measurements may be related in one way or another, for example, in terms of entanglement, and predicted accordingly, in the case of entanglement by using the concept of an entangled state in the formalism. I shall consider this aspect of the situation in detail in Chaps. 6 and 7, in the context of the EPR-type experiments, entanglement, and quantum correlations, and in Chap. 8 in the context of the concept of virtual particles. Mathematical concepts are, in Heisenberg's view, a possible exception, a possibility that I shall address presently.

Before I do so, I would like to restate the quantum indefinitiveness postulate, defined in the Preface, which is a consequence of the RWR view and which especially reflects the situation just considered. It dictates *the impossibility of making definitive statements of any kind, including mathematical ones, concerning the relationship between any two individual quantum phenomena or events, indeed to definitively*

[29] Heisenberg's view of language invites a more extended treatment, which is beyond my scope. For a helpful analysis, see (Camilleri 2011, pp. 152–171).

ascertain the existence of any such relationship. It does allow for making definitive statements concerning individual phenomena or events, defined by measurements, and statements concerning the relationships between multiple events, statements statistical in nature. It only concerns events that have already happened, rather than possible future events, in which case one can make probabilistic statements, on Bayesian lines.

Precluding the possibility of any mathematical connections between individual events makes the postulate stronger than Heisenberg's claim, just cited. While prohibiting common language and in effect a *physical description* of what "happens" between quantum experiments, this claim in principle allows for the *mathematical representation* of what the ultimate constitution of physical reality, including, in Heisenberg's view in his later works, in terms of elementary particles defined mathematically through symmetry groups. The words "happens" or even "physical" need, accordingly, no longer apply to this representation. As Heisenberg said on an earlier occasion, mathematics is "fortunately" free from the limitations of ordinary language and concepts:

> It is very difficult to modify our language so that it will be able to describe these atomic processes, for words can only describe things of which we can form mental pictures, and this ability, too, is a result of daily experience. Fortunately, mathematics is not subject to this limitation, and it has been possible to invent a mathematical scheme—the quantum theory [e.g., QM]—which seems entirely adequate for the treatment of atomic processes. (Heisenberg 1930, p. 11)

Admittedly, this freedom is gained at the cost of an enormous reduction of our representational phenomenal thinking from which mathematics was born, in part by way of this reduction, but only in part, because, as noted at the outset of this chapter, mathematics is also and even primarily defined by its concepts, which need not be reductive, and, more often than not, are not. In physics, however, or in mathematics itself, mathematics enables us to *relate* to things in nature and mind which are beyond the reach of phenomenal thinking or all other thinking, including mathematical. In quantum theory it does so by enabling us to estimate probabilities or statistics of quantum events, to which Heisenberg refers here by speaking of this scheme, QM, as "entirely adequate for the treatment of atomic processes." At the time, Heisenberg, adopting the RWR view, used this freedom to construct QM as a theory only designed to predict the probabilities or statistics of events observed in measuring instruments. It is equally fortunate that nature allows us, in our interaction with it, just to have such a scheme, for the fact of its freedom from the limitations of common language and concepts, or in its abstract nature, even physical concepts, does not guarantee that it will work in physics. But it does in QM or QFT, making them, as I argue, fundamentally mathematical in this respect. In contrast, in his later writings, in part in view of QFT, Heisenberg assumed the possibility of a mathematical *representation* of the ultimate constitution of reality, while excluding physical concepts (at least in their customary sense found in classical physics or relativity) as applicable to this constitution (e.g., Heisenberg 1962, pp. 145, 167–186). Heisenberg, as noted, speaks of this representation in terms of symmetry groups and defines elementary particles accordingly, without considering them as particles in a physical sense. The concept of an elementary particle can be given a mathematical sense insofar as the corresponding representation of the group is irreducible (Wigner 1939; Newton and Wigner 1949). Heisenberg even suggests that Kant's "thing-in-itself is for the atomic physicist, if he uses this concept at all, finally a mathematical structure; but this structure is—contrary to Kant—indirectly deduced from experience [rather than is given to our thought a priori, as in Kant]" (Heisenberg 1962, p. 83).

In any event, such was the case for Heisenberg, given that, while there were others at the time or since who adopted this view, it was not universal as a philosophical view, although, as stated from the outset of this study, the practice of theoretical physics, especially QFT and elementary particle, has been defined by working primarily with abstract mathematical structures, beginning with Heisenberg's discovery of quantum mechanics. Bohr, for one, rejected the possibility of a mathematical representation of quantum objects and behavior, or the reality they idealize, along with a physical one, at least in his ultimate, strong RWR-type interpretation. Bohr often speaks of this reality as being beyond our phenomenal intuition, also involving visualization, sometimes used, including by Bohr, to translate the German word for intuition, *Anschaulichkeit* (e.g., [Bohr 1987, v. 1 p. 51, 98–100, 108; v. 2, p. 59]). It is clear, however, that, apart from the Como lecture, Bohr saw the ultimate nature of this reality as being beyond any representation or even conception, including a mathematical one, at least as things stand now.

Indeed, notwithstanding its dominant role in modern physics, amplified and even made unique in quantum theory, it is, as noted in the Preface, not clear why mathematics, which is the product of the same human thinking as ordinary language or physical concepts are, should be able to represent how nature ultimately works at its very small (or very large) scales. It is not clear either that, in contrast to its capacity to do so at the scales handled by QFT, mathematics will enable us to predict phenomena shaped by the workings of nature far at the smaller scales, such as Planck's scale, although the current consensus appears to be that it should be able to do so. But a consensus is not always a guarantee. Bohr, in speaking of "the special role … played by mathematics in development of [all] logical thinking" and "invaluable help [offered] by its well-defined abstractions" in quantum theory, nevertheless, added: "Still, … we should not consider pure mathematics as a separate branch of knowledge, but rather as a refinement of general language, supplementing it with appropriate tools to represent relations for which ordinary verbal expression is imprecise and cumbersome" (Bohr 1987, v. 2, p. 68). This refinement can take us very far from ordinary language and concepts, a distance that is, again, manifested in the mathematics of QM and QFT, as Bohr was of course aware, but it would still reflect its human nature. Is there any mathematics other than human, given that the mathematics done by computers is still an extension of human mathematics? It is of course possible to assume, as it has been sometimes, that mathematics is something trans-human, for example, something that extraterrestrials would possess or could understand, as opposed to being seen, as here, as a refinement of human thinking and language. These views will, however, be put aside here. There may also be trajectories in human thinking beyond those related to "general language" or, as it might be more accurate to say, general human thinking and language that led to the emergence of mathematics, for example, trajectories extending from philosophical thinking, admittedly in turn the product of human thinking in general, but already refined logically and conceptually to be closer to mathematics.[30] In any event, the deeper meaning or implications of Bohr's statement is that mathematics is still human, which is also a philosophical position concerning mathematics adopted in this study. As such, even mathematics may not ultimately be suited, because nothing human can, to deal with the ultimate constitution of nature, either in terms of representing or conceiving of it, in accord with the U-RWR view, or even in predicting probabilistically, as QM or QFT does, the outcome of the events considered, without offering a representation or conception of how they come about.[31] I am not saying that we cannot go further with our fundamental physical theories and their mathematics; quite the contrary, especially as concerns the role of mathematics there, building on quantum theory, which is the affirmation of mathematics and mathematical thinking in physics. This thinking is all the more remarkable because, beginning with Heisenberg's discovery or even Bohr's 1913 atomic theory, it connects us, by means of mathematics, to that which may be beyond the reach of thought.

2.7 Conclusion

As, I hope, the discussion in this chapter makes clear and, as I also hope, the rest of this study will confirm, the RWR view does not preclude the advancement of thought and knowledge in mathematics, science, or philosophy, in all of which the RWR view may apply. What changes is the character of our thinking and knowledge. They now include the assumption that there is something that is beyond knowledge or even thought, not only now but possibly ever, while at the same time being responsible, as its effects, for what we can think, know, or represent at the level of observed phenomena, in quantum physics and, in the U-RWR view, in all physics. To handle these effects one needed new experiments and new theories, such as QM and QFT, which also required and created new mathematics, and one will continue to need them to advance physics. Physics, theoretical, and experimental require much more than RWR-type thinking, which would have never been sufficient to create QM or QFT,

---

[30] The origin of geometry as born, as a single event, from philosophy in ancient Greece, was considered by Edmund Husserl, in philosophical rather than strictly historical terms, in one of his last works, "The Origin of Geometry" (Husserl 1970). Husserl similarly contemplated the origin of arithmetic in his first major work, *Philosophy of Arithmetic* (Husserl 2003). The singular nature of the origin of geometry or its necessary emergence from philosophy was unlikely historically, and Husserl's argument could be challenged on Husserl's own terms (e.g., Derrida 1989). It is possible to argue (an angle not considered by Jacques Derrida) that the emergence of philosophy was preceded by some form of mathematical thinking, even if not technical mathematics. The subject would, however, require a separate discussion.

[31] This argument is different from Eugene Wigner's much discussed argument for "the unreasonable effectiveness of mathematics in the natural sciences" (Wigner 1960). This effectiveness is not unreasonable, especially in classical physics (e.g., Plotnitsky 2011a). By contrast in question in the present argument are the limits of this effectiveness.

defined most fundamentally by their extraordinary mathematics. RWR-type thinking may, however, be helpful and possibly even necessary for advancing physical theories or their interpretations. Thus, as discussed in Chap. 4, although RWR-type thinking could not do so on its own accord, this thinking compelled Heisenberg to search for and discover the mathematics of QM. As Bohr said:

> This [RWR-type] argumentation does of course not imply that, in atomic physics, we have no more to learn as regards experimental evidence and the mathematical tools appropriate to its comprehension. In fact, it seems likely that the introduction of still further abstractions into the formalism will be required to account for the novel features revealed by the exploration of atomic processes of very high energy. (Bohr 1987, v. 3, p. 6)

The history of high-energy physics and QFT has confirmed this assessment, made in 1958, and continues to do so, without, thus far, contradicting Bohr's argumentation referred to here or the RWR view, in which this argumentation is based. So has the history of QM during the same period. It is true that, unlike mathematical breakthroughs in QFT, such as those that led to the standard model, there have been no major changes in the mathematics of QM. Even so, the exploration of quantum correlations from the 1960s on and the rise of quantum information theory have been major developments, which opened new possibilities for the future of quantum theory. It is also possible that quantum information theory will lead to new mathematical innovations, conceivably in helping to bring quantum theory and gravity together. Of course, a relative newcomer to this project, quantum information theory has many competitors in pursuing this task. It is difficult to predict, on either experimental or theoretical grounds, which current theory or approach, if any, is a viable starting point on this road or what kind of trajectory will lead to such a theory. It is equally difficult to predict what kind of theory, if found, such a theory will be. It may reveal itself to be neither realist nor of the RWR type, weak or strong. Such a view is difficult to imagine, given that the strong RWR view assumes that the ultimate constitution of the reality responsible for quantum phenomena, in the Q-RWR view, or, in the U-RWR view, the ultimate constitution of nature in general is beyond thought. What, then, could such an alternative be apart from one or another form of realism, if defined, as here, by assuming that the reality considered is at least conceivable? One might, however, be reluctant to exclude this possibility. Nobody had expected or imagined anything like the physical reality that relativity and quantum physics made us think as possible or even assume to exist. And yet, here we are.

References

Aaronson, S.: Quantum Computing since Democritus. Cambridge University Press, Cambridge (2013)

Allahverdyan, A.E., Balian, R., Nieuwenhuizen, T.M.: A sub-ensemble theory of ideal quantum measurement processes. Ann. Phys. (2016). https://doi.org/10.1016/j.aop.2016.11.001

Arndt, M., Nairz, O., Voss, J., Keller, C., van der Zouw, G., Zeilinger, A.: Wave particle duality of C60. Nature **401**, 680–682 (1999)

Berthoz, A.: The Brain's Sense of Movement, Trans. Weiss, G. Harvard University Press, Cambridge (2000)
Berthoz, A.: La Décision. Odile Jacob, Paris (2003)
Bohr, N.: Can quantum-mechanical description of physical reality be considered complete? Phys. Rev. **48**, 696–702 (1935)
Bohr, N.: Causality and complementarity. In: Faye, J., Folse, H.J. (eds.) The Philosophical Writing of Niels Bohr, Volume 4: Causality and Complementarity, 1999, pp. 83–91. Ox Bow Press, Supplementary Papers, Woodbridge, CT, USA (1937)
Bohr, N.: The causality problem in atomic physics. In: Faye, J., Folse, H.J. (eds.) The Philosophical Writings of Niels Bohr, Volume 4: Causality and Complementarity, 1999, pp. 94–121. Ox Bow Press, Supplementary Papers, Woodbridge, CT, USA (1938)
Bohr, N.: Niels Bohr: Collected Works, 10 vols. Elsevier, Amsterdam (1972–1996)
Bohr, N.: The Philosophical Writings of Niels Bohr, 3 vols. Ox Bow Press, Woodbridge (1987)
Borel, E.: La logique et l'intuition en mathématique. Revue De Métaphysique Et De Morale **15**(3), 273–283 (1907)
Born, M., Jordan, P.: Zur Quantenmechanik. Z. Phys. **34**, 858–888 (1925)
Butterfield, J., Isham, C.J.: Spacetime and the philosophical challenge of quantum gravity. In: Callender, C., Huggett, N. (eds.) Physics Meets Philosophy at the Planck Scale: Contemporary Theories of Quantum Gravity, 2021, pp. 33–89. Cambridge University Press, Cambridge (2001)
Cabello, A.: Interpretations of quantum theory: a map of madness. In: Lombardi, O., Fortin, S., Holik, F., López, C., What is Quantum Information? pp. 138–144. Cambridge University Press, Cambridge (2017)
Camilleri, K.: Heisenberg and the Interpretation of Quantum Mechanics: The Physicist as Philosopher. Cambridge University Press, Cambridge (2011)
Cartwright, N.: How the Laws of Physics Lie. Oxford University Press, Oxford (1983)
Cartwright, N.: The Dappled World: A Study of the Boundaries of Science. Cambridge University Press, Cambridge (1999)
D'Ariano, G.M.: Physics without physics. Int. J. Theor. Phys. **56**(1), 97–138 (2017)
D'Ariano, G.M.: Causality re-established. Philos. Trans. R. Soc. A **376**, 20170313 (2018). https://doi.org/10.1098/rsta.2017.0313
D'Ariano, G.M.: No purification ontology, no quantum paradoxes. Found. Phys. **50**, 1921–1933 (2020). https://doi.org/10.1007/s10701-020-00398-6
D'Ariano, G.M., Chiribella, G., Perinotti, P.: Quantum Theory from First Principles: An Informational Approach. Cambridge University Press, Cambridge (2017)
D'Ariano, G. M., Faggin, F.: Hard problem and free will: an information-theoretical approach. arXiv:2012.06580v2 [quant-ph] (2021)
D'Ariano, G. M., Manessi, F., Perinotti, P.: Determinism without causality. Physica Scripta **T 63** 14013 (9pp) (2014). https://doi.org/10.1088/0031-8949/2014/T163/014013
Dauben, J.F.: Georg Cantor: His Mathematics and Philosophy of the Infinite. Princeton University Press, Princeton, NJ (1990)
De Finetti, B.: Philosophical Lectures on Probability, Trans. Hosny, H. Springer, Berlin (2008)
Deleuze, G., Guattari, F.: What is Philosophy? Trans. Tomlinson, H., Burchell, G. Columbia University Press, New York (1994)
Derrida, J.: Edmund Husserl's "The Origin of Geometry": An Introduction, Trans. Leavy, J. University of Nebraska Press, Lincoln, NA (1989)
Dirac, P.A.M.: The physical interpretation of the quantum dynamics. Proc. R. Soc. London A **113**, 621–641 (1927)
Dirac, P.A.M.: The Principles of Quantum Mechanics. Clarendon, Oxford (1930)
Dirac, P. A. M.: The Principles of Quantum Mechanics, 4th ed, rpt. 1995. Clarendon, Oxford (1958)
Doxiadis, A., Mazur, B.: Circles Disturbed: The Interplay of Mathematics and Narrative. Princeton University Press, Princeton, NJ (2012)
Doya, K., Ishii, S., Pouget, A., Rao, R.P.N. (eds.) Bayesian Brain: Probabilistic Approaches to Neural Coding. MIT Press, Cambridge, MA (2007)

Einstein, A.: Physics and reality. J. Franklin Inst. **221**, 349–382 (1936)
Einstein, A. Autobiographical Notes, Trans., Schilpp, P. A. Open Court , La Salle, IL (1949)
Einstein, A., Podolsky, B., Rosen, N.: Can Quantum-Mechanical Description of Physical Reality be Considered Complete? In: Wheeler, J.A., Zurek, W.H. (eds.) Quantum Theory and Measurement, 1983, pp. 138–141. Princeton, NJ, USA, Princeton University Press (1935)
Falcon, A.: Aristotle on causality. In Zalta, E.N. (ed.) The Stanford Encyclopedia of Philosophy (Spring 2015 ed.) (2015). http://plato.stanford.edu/archives/spr2015/entries/aristotle-causality/
Fine, A.: The Einstein-Podolsky-Rosen argument in quantum theory. In: Zalta, E.N. (ed.) Stanford Encyclopedia of Philosophy (Summer 2020 Edition) (2020). https://plato.stanford.edu/archives/sum2020/entries/qt-epr/.
Frigg, R.: Theories and Models. Acumen, Slough (2014)
Frigg, R., Hartmann, S.: Models in science. In: Zalta, E.N. (ed.) Stanford Encyclopedia of Philosophy (Fall 2012 Edition). http://plato.stanford.edu/archives/fall2012/entries/models-science/
Fuchs, C.A.: On participatory realism. arXiv:1601.04360v3 [quant-ph] 28 Jun 2016
Fuchs, C.A., Mermin, N.D., Schack, R.: An introduction to QBism with an application to the locality of quantum mechanics. Am. J. Phys. **82**, 749–754 (2014). https://doi.org/10.1119/1.4874855
Galison, P.: Image and Logic: A Material Culture of Microphysics. University of Chicago Press, Chicago, IL (1997)
Gray, J.: Plato's Ghost: The Modernist Transformation of Mathematics. Princeton University Press, Princeton, NY (2008)
Hacking, I.: Representing and Intervening, Introductory Topics in the Philosophy of Natural Science. Cambridge University Press, Cambridge (1983)
Hacking, I.: The Social Construction of What? Harvard University Press, Cambridge, MA (2000)
Hájek, A.: Interpretation of probability. In: Zalta, E.N. (ed.) Stanford Encyclopedia of Philosophy (Winter 2014 edition) (2014). http://plato.stanford.edu/archives/win2012/entries/probability-interpret/
Healey, R.: The Quantum Revolution in Philosophy. Oxford University Press, Oxford (2019)
Heisenberg, W.: Quantum-theoretical re-interpretation of kinematical and mechanical relations. In Van der Waerden, B.L. (ed.) Sources of Quantum Mechanics, rpt. 1968, 261–277. Dover, New York (1925)
Heisenberg, W.: The physical content of quantum kinematics and mechanics. In: Wheeler, J.A., Zurek, W.H. (eds.) Quantum Theory and Measurement, 1983, pp. 62–84. Princeton University Press, Princeton, NJ (1927)
Heisenberg, W.: The Physical Principles of the Quantum Theory, Trans. Eckhart, K., Hoyt, F.C, rpt. 1949. Dover, New York, NY (1930)
Heisenberg, W.: Ist eine deterministische Ergänzung der Quantenmechanik möglich? Archive for the history of quantum physics (microfilm 45, section 11). (E. Crull &G. Bacciagaluppi, English Trans.) (1935). http://philsci-archive.pitt.edu/8590/1/Heis1935_EPR_Final_translation.pdf
Heisenberg, W.: Physics and Philosophy: The Revolution in Modern Science. Harper & Row, New York, NY (1962)
Heisenberg, W.: Quantum theory and its interpretation. In: Rozental, S. (ed.) Niels Bohr: His Life and Work as Seen by his Friends and Colleagues, pp. 94–108. North-Holland, Amsterdam, Netherlands (1967)
Heisenberg, W.: Encounters with Einstein, and other Essays on People, Places, and Particles. Princeton University Press, Princeton, NJ (1989)
Husserl, E.: The origin of geometry, in Husserl, E., The Crisis of European Sciences and Transcendental Phenomenology: An Introduction to Phenomenological Philosophy, Trans. Carr, D.). Northwestern University Press, Evanston, 353–378 (1970)
Husserl, E.: Philosophy of Arithmetic: Psychological and logical investigations with supplementary texts from 1887–1901 Husserliana: Edmund Husserl – Collected works (Book 10), Trans. Willard, D. Springer, Berlin (2003)
Jaeger, G.: Quantum Objects: Non-local Correlations, Causality and Objective Indefiniteness in the Quantum World. Springer, New York, NY (2014)

Jaynes, E.T.: Probability Theory: The Logic of Science. Cambridge University Press, Cambridge (2003)

Kant, I.: Critique of Pure Reason, Tran. Guyer, P., Wood, A.W. Cambridge University Press, Cambridge (1997)

Khrennikov, A.: Interpretations of Probability. de Gruyter, Berlin (2009)

Khrennikov, A.: Quantum probabilities and violation of CHSH-inequality from classical random signals and threshold type detection scheme. Prog. Theor. Phys. **128**, 31–58 (2012). https://doi.org/10.1143/PTP.128.31

Ladyman, J.: Structural realism. In: Zalta, E.N. (ed.) Stanford Encyclopedia of Philosophy (2016). https://plato.stanford.edu/archives/win2016/entries/structural-realism/

Latour, B.: Pandora's Hope: Essays on the Reality of Science Studies. Harvard University Press, Cambridge, MA (1999)

Newton, T.D., Wigner, E.: Localized states for elementary systems. Rev. Mod. Phys. **21**, 400–406 (1949)

Ozawa, M.: An operational approach to quantum state reduction. Ann. Phys. **259**, 121–137 (1997). https://doi.org/10.1006/aphy.1997.5706

Palmquist, S.: Kantian causality and quantum quarks: the compatibility between quantum mechanics and Kant's phenomenal world. Theoria **77**, 283–302 (2013)

Pincock, C.: Mathematics and Scientific Representation. Oxford University Press, Oxford (2012)

Plato: The collected dialogues of Plato. In: Hamilton E., Cairns, H. (eds.) Princeton University Press, Princeton, NJ (2005)

Plotnitsky, A.: Epistemology and Probability: Bohr, Heisenberg, Schrödinger and the Nature of Quantum-Theoretical Thinking. Springer, New York, NY (2009)

Plotnitsky, A.: On the reasonable and unreasonable effectiveness of mathematics in classical and quantum physics. Found. Phys. **41**, 466–491 (2011a). https://doi.org/10.1007/s10701-010-9442-2

Plotnitsky, A.: 'Dark materials to create more worlds': on causality in classical physics, quantum physics, and nanophysics. J. Comput. Theor. Nanosci. **8**(6), 983–997 (2011b)

Plotnitsky, A.: Bohr and Complementarity: An Introduction. Springer, New York, NY (2012)

Plotnitsky, A.: The Principles of Quantum Theory, from Planck's Quanta to the Higgs Boson: The Nature of Quantum Reality and the Spirit of Copenhagen. Springer/Nature, New York, NY (2016)

Plotnitsky, A.: "Comprehending the connection of things": Bernhard Riemann and the Architecture of mathematical concepts. In: Ji, L., Yamada, S., Papadopoulos, A. (eds.) From Riemann to Differential Geometry and Relativity, pp. 329–363. Springer, Berlin (2017)

Plotnitsky, A.: On the concept of curve: geometry and algebra, from mathematical modernity to mathematical modernism. In Dani, S.G., Papadopoulos, A. (eds.), Geometry in History. Springer/Nature, Berlin, pp. 153–212 (2019)

Plotnitsky, A.: The ghost and the spirit of Pythagoras: the twentieth and twenty-first century mathematics between and beyond geometry and algebra. In: Sriraman, B. (ed.) Handbook in the History and Philosophy of Mathematics. Springer/Nature, Berlin (2020). https://doi.org/10.1007/978-3-030-19071-2_7-1

Plotnitsky, A., Khrennikov, A.: Reality without realism: on the ontological and epistemological architecture of quantum mechanics. Found. Phys. **25**(10), 1269–1300 (2015). https://doi.org/10.1007/s10701-015-9942-1

Poincaré, H.: The Foundations of Science: Science and Hypothesis. University Press of America, The Value of Science, Science and Method, Lanham, MD (1982)

Pólya, G.: Mathematics and Plausible Reasoning, Volume 1: Induction and Analogy in Mathematics, rpt. 1990. Princeton University Press, Princeton, NJ (1958)

Popper, K.: Realism and the Aim of Science: From the Postscript to The Logic of Scientific Discovery (1st ed.). Routledge, London (1983)

Redhead, M.: Incompleteness, Nonlocality, and Realism. Oxford University Press, Oxford (1989)

Riemann, B.: On the hypotheses that lie at the foundations of geometry. In: Pesic, P. (ed.) Beyond Geometry: Classic Papers from Riemann to Einstein, 2007, pp. 23–40. Dover, Mineola, NY (1854)

Riemann, B.: The mechanism of the ear (translation of *Mechanik des Ohres*), by Cherry, D., Gallagher, R., Sigerson, J. Fusion **6**(3), 1984, 31–38 (1866)

Ross, D., Spurrett, D.: Notions of cause: Russell's thesis revisited. Br. J. Philos. Sci. **58**, 45–76 (2007)

Rovelli, C.: Relational quantum mechanics. Int. J. Theor. Phys. **35**(8), 1637–1678 (1996)

Ruetsche, L.: Interpreting Quantum Theories. Oxford University Press, Oxford (2011)

Russell, B.: On the notion of cause. Proc. Aristotel. Soc. New Series **13**, 1–26 (1913)

Schrödinger, E.: The present situation in quantum mechanics. In: Wheeler, J.A., Zurek, W.H. (eds.) Quantum Theory and Measurement, 1983, pp. 152–167. Princeton University Press, Princeton (1935)

Schrödinger, E.: Interpretation of Quantum Mechanics: Dublin Seminars (1949–1955) and Other Unpublished Essays. Ox Bow Press, Woodbridge, CT (1995)

Shimony, A.: The reality of the quantum world. Sci. Am. **258**(1), 46–53 (1988)

Smolin, L.: Einstein's Unfinished Revolution: The Search for What Lies Beyond the Quantum. Penguin, New York, NY (2018)

Sorkin, R.D.: Spacetime and causal sets. In: D'Olivo, J.C. et al. (eds.) Relativity and Gravitation: Classical and Quantum, 1991, pp. 150–173. World Scientific, Singapore (1991)

Spekkens, R.W.: Evidence for the epistemic view of quantum states: a toy theory. Phys. Rev. A **75.3** 032110 (2007)

Spekkens, R.W.: Quasi-quantization: classical statistical theories with an epistemic restriction. In Chiribella, G., Spekkens, R.W. (eds.) Quantum Theory: Informational Foundations and Foils. Springer/Nature, New York, pp. 83–136 (2016)

Svozil, K.: Physical (A)Causality: Determinism, Randomness, and Uncaused Events. Springer/Nature, Berlin (2018)

Timpson, C.: Quantum Information Theory and the Foundations of Quantum Mechanics. Oxford University Press, Oxford (2016)

Van Fraassen, B.: Scientific Representation: Paradoxes of Perspective. Oxford University Press, Oxford (2008)

Von Neumann, J.: Mathematical Foundations of Quantum Mechanics (Trans. Beyer, R.T.), rpt. 1983. Princeton University Press, Princeton, NJ (1932)

Werner, R.F.: Comment on 'What Bell did'. J. Phys. A **47**, 424011 (2014)

Wheeler, J.A.: The Participatory Universe. Science **81** 2(5), 66–67 (1981)

Wheeler, J.A.: Law without law. In: Wheeler, J.A., Zurek, W.H. (eds.) Quantum Theory and Measurement, pp. 182–216. Princeton University Press, Princeton, NJ (1983)

Wheeler, J.A.: Information, physics, quantum: the search for links. In: Zurek, W.H. (ed.) Complexity, Entropy, and the Physics of Information, pp. 3–28. Addison-Wesley, Redwood City (1990)

Wigner, E.P.: On unitary representations of the inhomogeneous Lorentz group. Ann. Math. **40**, 149–204 (1939)

Wigner, E.P.: The unreasonable effectiveness of mathematics in the natural sciences. Commun. Pure Appl. Math. **13**, 1–14 (1960). https://doi.org/10.1002/cpa.3160130102

Wilczek, F.: In Search of Symmetry Lost. Nature **423**, 239–247 (2005)

Wilson, A.: The Nature of Contingency: Quantum Physics as Modal Realism. Oxford University Press, Oxford (2020)

Wittgenstein, L.: Tractatus Logico-Philosophicus (Trans. Ogden, C.K.), rpt. (1985). Routledge, London (1924)

Chapter 3
Bohr's Breakthrough: Quantum Jumps, Quantum States, and Transitions Without Connections

That this insecure and contradictory foundation was sufficient to enable a man of Bohr's unique instinct and sensitivity to discover the principal laws of the spectral lines and of the electron shells of the atoms, together with their significance for chemistry, appeared to me as a miracle—and appears to me a miracle even today. This is the highest musicality in the sphere of thought.
—Albert Einstein, *Autobiographical Notes* (Einstein 1949, pp. 42–43; translation modified).

Abstract The chapter considers Bohr's 1913 atomic theory and takes advantage of its key ideas to define the *physical* concept of quantum state (which is different from the *mathematical* concept of "quantum state," as a vector in a Hilbert space, in the formalism of QM) and the concept of transitions between quantum states in accordance with the RWR view. Instead of representing the *motion* of electrons and predicting, even if only probabilistically, their future states on the basis on this representation, as in the previous, classical, electron theory, Bohr's theory and then QM were concerned with *transitions between states of quantum objects*. This was a radical conceptual shift. These transitions, observed only in measuring instruments, were conceived by Bohr as strictly discontinuous without allowing one to represent or even conceive how they come about, which made them "transitions without connections." After a general introduction given in Section 3.1, Section 3.2 considers the structure of Bohr's concept of a "quantum jump," leading to the concept of transitions between quantum states as "transitions without connections." Sections 3.3 and 3.4 discuss the key physical and philosophical features and implications of Bohr's approach, as inaugurating the RWR view.

Keywords Bohr's 1913 theory · Quantum jumps · Quantum postulates · Quantum states · Transitions without connections

A. Plotnitsky, *Reality Without Realism*,
https://doi.org/10.1007/978-3-030-84578-0_3

3.1 Introduction

Bohr's (1913) atomic theory (hereafter Bohr's theory), initially dealing with the hydrogen atom, was introduced in Bohr's paper, "On the Constitution of Atoms and Molecules," the first part of his 1913 "trilogy," eventually published together in his 1924 book *The Theory of Spectra and Atomic Constitution* (Bohr 1924).[1] While Bohr's work built on previous, already revolutionary, discoveries of Planck, Einstein, and Ernest Rutherford, it made new radical assumptions, reflected in its key postulates concerning the behavior of both light and, especially, electrons, which made his paper an inaugural event in the history of the RWR view. In accord with my general agenda in this study, I will, in this chapter, be primarily concerned with the conceptual fundamentals of Bohr's thinking and its significance for quantum theory, QM and QFT, now, as shaped by the RWR view, which Bohr's (1913) theory inaugurated. My historical account of Bohr's theory and its development prior to QM will be governed by this aim.[2]

Bohr's theory aimed to remedy the difficulties of his mentor Rutherford's "planetary model" of the atom, with electrons orbiting atomic nuclei. Although a revolutionary and important conception, this model was inconsistent with classical electrodynamics, which would dictate that the electrons would nearly instantly spiral down into the nucleus, and hence that atoms would not be stable, while they are manifestly stable. Bohr's theory avoided these difficulties, by means of radical new postulates, based on Planck's and Einstein's *quantum* theories, based on the possibility of the discontinuous emission of light in the form of light energy quanta, $h\nu$, eventually understood as photons, the hypothesis proposed by Einstein in 1906. Curiously, Bohr did not use the photon hypothesis, which he was reluctant to accept until the early 1920s, when the particle behavior of photons was established by Compton's experiment, which findings, finally, convinced Bohr. This is, however, a secondary matter, given that Bohr's theory did not depend on whether one treated photons as particles or as quanta of energy, and all of his theory's key features would have equally applied if he had accepted the photon hypothesis at the time. Making his own audacious move, Bohr postulated both the so-called stationary states of electrons in the atom, at which they could remain in orbital motion, and discontinuous "quantum jumps" between stationary states, resulting in the emission or from the absorption of Planck's quanta of radiation, without electrons radiating continuously while remaining in orbit. Accordingly, Bohr's theory was no less in conflict with classical electrodynamics than Rutherford's model was. In addition, again, in contradiction to the laws of classical electrodynamics, Bohr postulated that there would exist a lowest energy level at which electrons would not radiate, but would only

[1] My designation "Bohr's theory" includes its subsequent development, including in dealing with more complex atoms (in considering which the theory encountered difficulties), by Bohr and others, Arnold Sommerfeld in particular, until Heisenberg's discovery of QM in 1925.

[2] Among helpful accounts are (Kragh 2012), which offers a comprehensive treatment of Bohr's atomic theory in its historical development, and (Folse 2014), a more philosophical treatment.

absorb, energy.[3] Bohr abandoned, as, in his view, hopeless at the time, an attempt to offer a mechanical explanation for such transitions, as opposed to the stationary states themselves, thus in accord with the weak RWR view (there is no evidence that he entertained the strong RWR view at this point). Stationary states, he said, "can be discussed by help of the ordinary mechanics, while the passing of the system between different stationary states cannot be treated on that basis" (Bohr 1913, p. 7).

Bohr's theory was radical and appeared to be pretty unnatural at the time. It was a *quantum jump* from the orbit of classical theory toward quantum theory and nonrealism, a metaphor of which Bohr was undoubtedly aware. It took another decade before Heisenberg abandoned a classical treatment of stationary states as orbits, the concept that he saw as having its limitations from the very beginning (Heisenberg 1962, p. 41). For one thing, these orbital motions were unobservable. All that one could observe in a laboratory, as Bohr likely did in Rutherford's laboratory, were discrete phenomena, such as those corresponding to an electron's energy or quanta of radiation, observed as spectra. (In certain circumstances, one can observe continuous spectra in quantum physics, which are, however, ultimately manifestations of the underlying quantum phenomena.) Orbits were remnants of classical thinking. One does not need them. But it took Heisenberg to abandon them and treat all quantum phenomena in a suitably quantum way, including by understanding a *quantum* measurement as the establishment of a new phenomenon observed in a measuring instrument rather than measuring, in the way one does in classical physics, a property of the quantum object considered. Heisenberg also developed an entirely new mathematics, never previously used in physics (Heisenberg 1925). Both Bohr's theory and QM predicted the probabilities or statistics of transitions between them, but unlike Bohr's theory, which treated stationary states classically, Heisenberg's matrix mechanics did not treat the behavior of electrons in stationary states. Dirac's *q*-number scheme and then Schrödinger's equation were able to do so, also in probabilistically or statistically predictive terms, although Schrödinger initially aimed at a representational and, moreover, classically causal treatment of the behavior of electrons in terms of waves.

The history leading from Bohr's theory to QM was, of course, not so simple: it was not a single "quantum jump" from one state of quantum theory to another. This history contained several developments during the intervening decade, some of which were radical and akin to quantum jumps of their own. Three contributions are especially worth noting, not only given their importance in their own right but also their significance in the history of the RWR view, because they reflected quantum phenomena that challenged realism.

The first was Louis de Broglie's 1923 theory of matter waves, which conjectured that the same type of wave–particle duality that Planck and Einstein assumed in the case of radiation would apply to other elementary constituents of nature, such

[3] Bohr's 1913 postulates should be distinguished from Bohr's related but more general concept of "the quantum postulate," essentially combining QI (quantum individuality) and QD (quantum discreteness) postulates, introduced, along with the concept of complementarity, in 1927, following the discovery of QM, although the quantum postulate concerned quantum phenomena themselves and did not depend on QM (Bohr 1987, v. 1, pp. 52–53).

as electrons. The conjecture was soon experimentally confirmed by the discovery of electrons' diffraction in crystals. The presence of these wave-like aspects of electrons' behavior did not imply that electrons could now be treated merely as waves, any more than photons could be treated merely as particles. Both aspects of quantum behavior were equally unavoidable, but they manifested themselves in different and indeed mutually exclusive circumstances. The situation was ultimately handled by means of complementarity, although, while commonly invoked as an example of complementarity, wave–particle complementarity is a more complex matter discussed in Chap. 6. Bohr generally avoids speaking of it. In any event, de Broglie's theory posed major difficulties for representing the behavior of quantum objects, given that we cannot conceive of entities that are both continuous and discrete.

The second discovery was the 1925 discovery of spin by Ralph Kronig, George Uhlenbeck, and Samuel Goudsmit. Spin has no classical counterpart and is, thus, especially fitting for the RWR view, which, however, took a while to adopt. Initially, the idea, manifestly classical in nature, that an electron is somehow "rotating" deterred both Heisenberg and Wolfgang Pauli, whose criticism stopped Kronig from publishing his discovery, first published by Uhlenbeck and Goudsmit. Spin is only manifested as a number, which is the same for each particle of a given type, in certain specific effects of the interaction between a quantum object and a measuring instrument, effects properly predicted by QM (by using a finite-dimensional Hilbert space over $\mathbb{C}$). It is impossible to conceive of a rotation that would manifest itself in these effects.

The third discovery, actually preceding that of spin, was that of the Bose–Einstein statistics, based on the fact that one cannot treat any two photons having equal energy as being two distinct identifiable objects. The radical nature of this discovery was initially underappreciated by Einstein, who helped Satyendra Nath Bose to publish his paper, although Einstein took advantage of Bose's work in his own work on the Bose–Einstein statistics and the Bose–Einstein condensate. In any event, this feature and the concept of the indistinguishability of particles radically changed our understanding of the nature of quantum objects versus classical ones, with major implications for QFT, which, as discussed in Chap. 8, brought even more radical changes to this concept. "Particles" did not behave like "particles," a situation that became already apparent earlier because of wave–particle duality, but that was made more pronounced by Bose's discovery. This case is different from that of Einstein's relativity, where, as discussed in Chap. 2, while the relativistic motion with velocities close to the speed of light (defined by Lorentz's velocity addition formula) was beyond our phenomenal intuition, it still allowed for a representation by means of mathematized physical concepts. By contrast, in Bose's theory, the behavior of photons in a photon gas was not given a representation, but only the statistical counting for predicting a future state of the gas. Bose's paper also rederived Planck's law by using this new (Bose–Einstein) statistics (Bose 1924). Bose's counting was incompatible with the assumption of distinguishability of individual objects, an assumption considered basic at the time. Pauli's exclusion principle, in which spin figures are significant (although Pauli missed on discovering spin), contributed to this new situation, although its main implications, such as the Fermi–Dirac statistics, became apparent after the discovery of QM.

Bose's paper, thus, manifested the situation that has been confronted and debated in quantum physics ever since, and that is central to my argument in this study. This situation is defined by two fundamental (and perhaps ultimately insurmountable) splits between the mathematics of quantum theory and physical reality: the first is that between this mathematics and the nature of the physical reality responsible for observed phenomena (which are representationally connected in classical physics and relativity), and the second between this mathematics and the observed phenomena themselves, described by classical physics, which could not, however, predict them. Nor, it follows, could classical physics account for how quantum phenomena come about, for if it could, it would then be able to predict them. The mathematics of quantum theory would only predict, probabilistically or statistically, the data thus observed without having any *physical* connection to these data. As was indicated earlier and as discussed in detail in Chap. 6, the probabilistic or statistical connections between QM, or QFT, and quantum phenomena are essentially *human* because they are defined *only* by assignments of probabilities, which assignments are always human, even in classical physics. There, however, they or the relationships between classical theories and classical phenomena may be assumed to be physically grounded because classical physics may be assumed to represent the behavior of the systems considered. In QM or QFT, at least in RWR-type interpretations, this assumption is not possible, which precludes one from knowing or even conceiving of *how* these relationships and, hence, our assignments of probabilities are grounded in our interactions with nature, as they must be as effects of these interactions. But we do know the mathematics of QM or QFT and that this mathematics allows us to correctly assign these probabilities or statistics.

As the intervening history, exemplified by these discoveries (and there were several other relevant developments, including Heisenberg's work in the old quantum theory), would demonstrate, Heisenberg's thinking leading to him to his discovery of QM was far from merely an extension of Bohr's thinking in his 1913 theory. Bohr's theory was, nevertheless, the most decisive event of the history leading to QM and beyond, certainly as concerns the RWR view. On the mathematical side, QM was the greatest event in this history, although it was a more collective effort, with Born, Jordan, Dirac, Schrödinger, and Heisenberg himself, adding to Heisenberg's initial work. With Dirac's introduction of QED and especially his relativistic equation for the electron, a new dimension, that of multiplicity, was added to the RWR view, bringing together the irreducible unthinkable and the irreducibly multiple. One might, however, still see Dirac's theory as an extension of Bohr's concept of quantum jumps, if one understands this concept, as I shall do here, as that of transitions without connections between quantum states or events.

The remainder of this chapter proceeds as follows. The next section, Sect. 3.2, considers the structure of Bohr's concept of a quantum jump, leading to the concept of transitions between quantum states as "transitions without connections." Sections 3.3 and 3.4 discuss the key features and implications of Bohr's approach, as inaugurating the RWR view.

3.2 What is a Quantum Jump?: Quantum States and Transitions Without Connections

The deeper meaning and significance of Bohr's concept become more apparent if one properly considers its structure as a concept, defined by its various components, a structure imperfectly conveyed by the term "quantum jump." First of all, stationary states are not really *stationary*. "Stationary" only meant that the electrons remained in their orbits with the same energy, and were in the same "energy state," while continuously changing their position or their "position state" along each orbit. On the other hand, the electrons would discontinuously, by quantum jumps, change their energy states, or their other states, by moving from one orbit to another. Bohr's theory abandoned the aim of physically representing such transitions, as neither the time nor direction of each jump could be explained, although it could be predicted probabilistically or statistically. This makes the term "jump" misleading as suggesting some representation of what happens. Electrons do not jump: quantum states (as physical states) discontinuously change, and no representation of how they do this is available. What was responsible for these changes was assumed to be real but its reality was assumed to be at least beyond representation. It was a reality without realism, in accord, at this stage, at least with the weak RWR view, although intimating the strong RWR view, insofar as no concept of how these transitions appeared to be possible to form either. The same situation was eventually found in the case of electrons in stationary states. Electrons were not moving in orbits around nuclei: quantum states associated with them were changing, with these changes ultimately only observable in discrete phenomena.

I here take advantage of an angle on the development of quantum theory suggested by Laurent Freidel (Freidel 2016), although not in connection with Bohr's theory, which, I would argue, inaugurates this change, especially in its RWR potential, not considered by Freidel. Freidel's main focus is Born and Jordan's paper, which gave Heisenberg's new mechanics its proper matrix form and which Freidel (rightly) sees as underappreciated (Born and Jordan 1925). Freidel notes that, instead of representing the motion of electrons in terms of oscillators and predicting, even if only probabilistically, their future states on the basis on this representation, as in the previous, classical, electron theory of Hendrik Lorentz and his followers, quantum theory was *most essentially* concerned just with *transitions between states*, which Freidel sees as a deep conceptual shift. These transitions, manifested, as became eventually apparent, only as observed in measuring instruments, as quantum phenomena, were conceived by Bohr as strictly discontinuous, "quantum," without allowing one to represent or even conceive of how they come about, which makes these transitions "transitions without connections." This, in effect RWR, view, ultimately the strong RWR view, of these transitions defines its significance for this study. Freidel, by contrast, still speaks (perhaps only loosely) of *processes* of such transitions and is describing these processes. Bohr's concept was the first instance of the use of the concept of RWR type in physics, a concept never entertained or accepted as permanent by either Planck or Einstein at the time, and ultimately grounding Einstein's discontent with QM.

The probabilistic dimensions of Bohr's theory have rarely been focused on, but they were important, just as they became in QM, which extended Bohr's concept of these transitions between stationary states to those between all quantum states. The mathematics of QM was far beyond that of Bohr's theory, but conceptually QM (at least in the RWR view, assumed by Heisenberg in his derivation of QM) was an extension of Bohr's theory of quantum jumps, defined strictly as transitions between states without representing them, specifically in terms of the motion of electrons. Hence, these transitions may be seen as "transitions without connections," keeping in mind the overall structure of this concept, especially insofar as these transitions can only be observed in measuring instruments, as effects of their interactions with quantum objects. These effects represent quantum *discreteness*, as that of quantum phenomena, defined by these effects. By contrast, quantum objects or their quantum states cannot be observed and, in the RWR view, represented or known or even conceived of; only their effects on measuring instruments can be observed, making the transitions in question transitions between such effects, from one measurement to another. This view emerged later, beginning with Heisenberg's understanding that his "scheme" of QM, "reduced *all interactions* between atoms and the external world … *to transition probabilities*," that is, between quantum objects and measuring instruments defining quantum phenomena (Heisenberg, Letter to Ralph Kronig, 5 June 1925; cited in Mehra and Rechenberg 2001, v. 2, p. 242; emphasis added). This implies the redefinition quantum measurement as establishing such phenomena, rather than measuring the properties of quantum objects, a view eventually crystalized in Bohr's ultimate, strong RWR-type interpretation.

The situation invites a new concept of physical state implicit in Bohr's work, in contradistinction to classical mechanics and relativity, where the concept of state is coextensive with the concept of motion, within the same mathematical representation, and where it can, by the same token, be defined independently of our interaction with the object considered. By contrast, quantum states cannot be unambiguously considered apart from the interaction between quantum objects and measuring instruments, and are defined strictly in terms of effects of these interactions on these instruments. As explained in Chap. 1, both concepts, quantum objects and quantum states, are idealizations that are due to the RWR-type reality (a concept that is also an idealization) ultimately responsible for these effects, by interacting with the quantum strata of measuring instruments. This interaction is quantum but it leads to effects observed in the classical strata of measuring instruments, effects that define quantum phenomena. In the present interpretation, this interaction gives rise to quantum objects, defined, as an idealization, on the basis of these effects at the time of measurement, for example, as electrons, photons, and so forth, although quantum objects need not be elementary. This means that a "quantum jump" or any transition between quantum states in effect refers to a measurement that, in general, involves a new quantum object, arising in the interaction between the ultimate constitution of the reality responsible for quantum phenomena and measuring instruments, and not an object that we measure before a quantum jump happened. As explained in Chap. 2, however, in low-energy (QM) quantum regimes, considered by Bohr or Heisenberg, viewing these two quantum objects as the same object is a statistically permissible idealization. The concept of transitions without connection between quantum states, as manifested in the corresponding effects on measuring instruments, is fully compatible with and even fitting to this view, because, technically, they only deal with measurements associated with two quantum *states*, rather than with two quantum *objects*. This situation is given a better grounding by the concept quantum field, introduced in Chap. 8.

Heisenberg, adopting the RWR view (in its weak version), abandoned the association of stationary states with orbits or, more generally, any quantum behavior represented by a *geometrical* mechanical picture, and gave a rigorous mathematics, in particular *algebra*, to this new physics, the mathematics of QM. As will be seen in Chap. 4, however, this algebra also leads to a new form of geometry. This move revealed that the most essential feature of Bohr's theory was its concept of the transitions between stationary states, rather than the orbital picture of stationary states. According to Heisenberg (in his later account of this history), the proper *quantum* features of Bohr's theory just outlined, "had to be interpreted as a limitation to the concept of the electronic orbit" (Heisenberg 1962, p. 41). In Heisenberg's scheme, there were only states and transitions between them, and probabilities of these transitions, which QM could predict, without representing either these states or these transitions. If one were to use the language of Bohr's theory, in Heisenberg's theory, there were only quantum states and quantum jumps between them. This was a decisive shift in our understanding of the nature of physical reality, which laid the foundation for Bohr's interpretation of quantum phenomena and QM, which reached its ultimate, strong RWR form by the late 1930s. This is where Bohr was ultimately led by the trajectory initiated by the concept of quantum jumps of his 1913 theory, a trajectory that was given a new impetus by Heisenberg's discovery of QM.

3.3 "This Insecure and Contradictory Foundation": Bohr's 1913 Theory as an RWR-Type Theory

Already Bohr's 1913 theory, however, abandoned classical causality and, as a consequence determinism, was dealing with the transitions between stationary states. While determinism was not possible on experimental grounds (which is still the case), these discrete transitions could in principle, as was hoped at the time, be assumed to be underlain by a classically causal process. The situation became more acute with QM, and it has not changed since. As noted, in Bohr's scheme neither the time of the emission of an energy quantum (when jumping to a lower energy state) nor direction of a quantum jump (either up or down) was determinable. There was no mechanical law for these transitions. This worried most at the time, including both Rutherford (on reading of Bohr's paper, which he published nevertheless) and Einstein (on reading its published version), but evidently not Bohr or, later, Heisenberg. At least both had reasons overriding such concerns, very good reasons, for they led to their remarkable inventions. They were greatly helped by the fact that, regardless of the absence of a mechanical law and, thus, from the classical viewpoint, mysteriously, one could still mathematically estimate probabilities of transitions between states, which Bohr's theory was able to do and Heisenberg's QM could do even better. Accordingly, one could focus not on discrete (or, as Schrödinger tried later, continuous, wave-like) quantum objects and their behavior, as one would in classical physics, but on discrete states of these objects and probabilities of predicting the transitions between these states. In yet another audacious move, Bohr *dissociated* the frequency of the light emitted by the atom from the frequency at which the electron orbited the atom. In Bohr's theory, the electron would absorb or emit energy only by changing its orbital state from energy E_1 to energy E_2. The frequency of the absorbed or emitted energy was defined in accordance with Planck's and Einstein's rule as $h\nu = E_1 - E_2$. In order, however, to get his theory to correspond with the experimental data (spectral lines) in question, Bohr combined this postulate with another quantization postulate, which allowed that energies for orbiting electrons were whole number multiples of h multiplied by half of the final orbital frequency, $E = \frac{1}{2}nh\nu$. It was thus half of the energy, $E = nh\nu$, that Planck, in deriving his black-body radiation law assumed for his oscillators. These two assumptions, combined with classical formulas that related the frequency of an orbit to its energy, gave Bohr the Rydberg frequency rules for hydrogen spectral lines, well established by then. In Bohr's scheme, only certain frequencies of light could be emitted or absorbed by a hydrogen atom, in correspondence with the Rydberg rules. Bohr's theory was a more radical departure from classical electrodynamics than Einstein's work ever dared to be. According to A. Douglas Stone:

> Bohr did something so radical that even Einstein, the Swabian rebel, had found it inconceivable: Bohr *dissociated* the frequency of the light emitted by the atom from the frequency at which the electron orbited the atom. In the Bohr formula, [$h\nu = E_1 – E_2$], there are two electron frequencies, that of the electron in its initial orbit and that of the electron in its final orbit; *neither* of these frequencies coincides with the frequency, ν, of the emitted radiation. This was a pretty crazy notion to a classical physicist, for whom light was *created* by the acceleration of charges and must necessarily mirror the frequency of the charge motion. Bohr admitted as much: "How much the above interpretation differs from an interpretation based on the ordinary electrodynamics is perhaps most clearly shown by the fact that we have been forced to assume that a system of electrons will absorb radiation of a frequency different from the frequency of vibration of electrons calculated in the ordinary way" (Bohr 1913, p. 149). However, he noted, using his new rule, "obviously, we get in this way the same expression for the kinetic energy of an electron ejected from an atom by photo-electric effect as that deduced by Einstein." (Bohr 1913, p. 150) (Stone 2015, pp. 177–178)

Bohr's theory proved to be correct. In essence, Bohr's postulates have remained part of quantum theory. They were given a more rigorous physical meaning by Bohr's interpretation of QM and quantum phenomena, beginning with Bohr's 1927 concept of the quantum postulate (Bohr 1987, v. 1, p. 53). Bohr eventually reconceived quantum discreteness strictly in terms of quantum phenomena, rather than the Democritean atomicity of quantum objects themselves. While the view of quantum theory as a probabilistic or statistical theory predicting the transitions between states was there to stay and while it still governs quantum theory, the idea of orbits for stationary states soon ran into major difficulties (such as some of these orbits falling into the nucleus). Accordingly, as I argue here, the most essential part of Bohr's theory was his concept of transitions, as transitions without connections, between stationary states, a concept that Heisenberg extended to that of transitions between quantum states in general, without giving a classical-like geometrical representation to any of them. This was yet to come. In the meantime, Einstein, convinced by major experimental confirmations (e.g., the Pickering–Fowler spectrum), accepted Bohr's theory and used it in his important 1916 papers, in which he rederived Planck's law yet again (Einstein 1916a, b). Conceptually, however, Bohr's theory could not satisfy Einstein any more than later on QM could. His predilection for a classical-like field theory might well have played a role. At the time he was also working and publishing on his general relativity (completed in 1915), a theory of continuous fields, an ideal never relinquished by Einstein (Einstein 1949, pp. 83–85). According to George Hevesy, Einstein himself had "similar ideas, but did not dare to publish them" (Stone 2015, p. 178). As Stone notes: "Bohr's atomic theory was hardly the new mechanics for which Einstein had been searching. There was still no underlying principle to replace classical mechanics, just another ad hoc restriction on classical orbits, a variant of Planck's desperate hypothesis" (Stone 2005, p. 178).

Bohr's theory was clearly much more, and more radical, than a variant of Planck's hypothesis, as Einstein realized. First of all, there was in fact an underlying principle—that of transitions without connections between quantum states—to replace classical mechanics as a theory of (trajectories of) motions of physical objects. Things were only to get worse, as far as Einstein was concerned, with only a few glimmers of hope with de Broglie's theory and then Schrödinger's wave mechanics, to quickly fizzle. Einstein, as noted in the Introduction, never abandoned his "search for a more complete conception," ideally a realist field theory, which would avoid probability at the ultimate level (Einstein 1936, p. 375, 1949, pp. 83–85). He never accepted that Bohr's type of thinking (either in his 1913 theory or in his interpretation of QM) was based on anything more than "insecure and contradictory foundations," even in praising Bohr's (1913) theory. According to Einstein's comment made 30 years later:

> That this insecure and contradictory foundation was sufficient to enable a man of Bohr's unique instinct and sensitivity to discover the principal laws of the spectral lines and of the electron shells of the atoms, together with their significance for chemistry, appeared to me as a miracle—and appears to me a miracle even today. This is the highest musicality in the sphere of thought. (Einstein 1949, pp. 42–43; translation modified)

Although beautiful and reflecting the magnitude of Bohr's achievement in a way that was undoubtedly gratifying to Bohr, the comment still reflects Einstein's unease concerning the "foundations" on which Bohr built his theory. These foundations never became secure for, or even accepted as foundations by, Einstein, even though he admitted that QM was a consistent and impressive theory that "seized hold of a beautiful element of truth about nature":

> There is no doubt that quantum mechanics has seized hold of a beautiful element of truth and that it will be a touchstone for a future theoretical basis in that it must be deducible as a limiting case from that basis, just as electrostatics is deducible from the Maxwell equations of the electromagnetic field or as thermodynamics is deducible from statistical mechanics. I do not believe that quantum mechanics will be the starting point in the search for this basis, just as one cannot arrive at the foundations of mechanics from thermodynamics or statistical mechanics. (Einstein 1936, p. 361)

For Bohr, by contrast, his 1913 theory proved only to be the first step on a road toward making these foundations logical and secure in his interpretation of quantum phenomena and QM. While they may not have been entirely secure initially, the foundations of Bohr's theory were not contradictory. They were only in conflict with classical physical theories and realist philosophical imperatives.

3.4 "Symbols Taken from the Mechanics," the Choice of the Observer, and the Being of the Photon

As Bohr observed a few months before Heisenberg's discovery, quantum processes "cannot be estimated within the ordinary space–time description" (Letter to Heisenberg, 18 April 1925, Bohr 1972–1996, vol. 5, pp. 79–80), a statement that Heisenberg must have taken to heart given that it literally defines his approach. In a letter to Born, Bohr goes further:

> [Quantum experiments] preclude the possibility of a simple description of the physical occurrences [at the quantum level] by means of visualizable pictures. . . . [S]uch pictures are of even more limited applicability than is ordinarily supposed. This is of course almost a purely negative assertion, but I feel that . . . one must have recourse to symbolic analogies to an even greater extent than hitherto. Just recently I have been racking my brain to dream up such analogies. (Letter to Born, 1 May 1925, Bohr 1972–1996, vol. 5, p. 311)

The ultimate *symbolic* analogy, indeed the symbolic theory, that Bohr wanted was provided by Heisenberg's new mechanics, about to be discovered then. Bohr's well-known, even best known, sentence in his first 1913 paper merits additional attention in this context, because of his appeal to "*symbols* taken from [classical] mechanics": "While, there obviously can be no question of a mechanical foundation of the calculation given in this paper, it is, however[,] possible to give a very simple interpretation of the result of the calculation on p. 5 [concerning stationary states] by help of *symbols* taken from the mechanics" (Bohr 1913, p. 15; emphasis added). The sentence is famous for its first part: "there obviously can be no question of a mechanical foundation of the calculation given in this paper." This statement radically challenges the applicability of classical causality and in the first place realism to quantum jumps. Heisenberg echoes this statement in his paper introducing QM: "a geometrical interpretation of such quantum-theoretical phase relations by analogy with those of classical theory seems at present scarcely possible" (Heisenberg 1925, p. 265). Heisenberg's paper is, however, also a response to Bohr's sentence as a whole, and Heisenberg's approach is a full-scale (rather than limited, as in Bohr's theory) enactment of the program implicit in this sentence, even if Bohr might not have fully realized these implications or their scale at the time. As, however, Bohr immediately grasped in assessing Heisenberg's discovery, Heisenberg's approach enabling it amounted to taking "*symbols...* from the ordinary mechanics," where they represent classical physical variables (such as position and momentum) and equations connecting these symbols, and giving a very different mathematical form to these variables and a new physical meaning to the resulting mathematical scheme. In Heisenberg's theory, these symbols became (unbounded) infinite matrices with complex elements, replacing real functions of coordinates and time, as in classical physics. It was a combination of these symbols, a combination defined by formally classical equations, that became the architecture of QM. Physically, these new variables were linked to the probabilities or statistics of the occurrences of certain observable phenomena, manifested in atomic spectra, instead of describing the motion of quantum objects on the model of classical mechanics. Heisenberg's "new kinematics," as he called it, was nothing else. Accordingly, Heisenberg's mechanics was *symbolic* mechanics, as Bohr had referred to it (as he later also did to Schrödinger's version), thus echoing his earlier thinking concerning his 1913 atomic theory, and extending it to his interpretation of QM, and to his philosophical thinking in general.

Bohr offers a helpful elaboration in his 1929 article, echoed in his 1937 remark, cited earlier, to the effect that "an artificial word like 'complementarity' which does not belong to our daily concepts serves only … to remind us of the epistemological situation …, which at least in physics is of an entirely novel character" (Bohr 1937, p. 87). He says:

> Moreover, the purpose of such a technical term [as complementarity] is to avoid, so far as possible, a repetition of the general argument as well as constantly to remind us of the difficulties which, as already mentioned, arise from the fact that all our ordinary verbal expressions bear the stamp of our customary forms of perception, from the point of view of which the existence of the quantum of action is an irrationality. Indeed, in consequence of this state of affairs even words like "to be" and "to know" lose their unambiguous meaning. In this connection, an interesting example of ambiguity in our language is provided by the phrase used to express the failure of the [classically] causal mode of description, namely, that one speaks of a free choice on the part of nature. Indeed, properly speaking, such a phrase requires the idea of an external chooser, the existence of which, however, is denied already by the use of the word nature. We here come upon a fundamental feature in the general problem of knowledge, and we must realize that, by the very nature of the matter, we shall always have last recourse to a word picture, in which the words themselves are not further analyzed. (Bohr 1987, v. 1. pp. 19–20)

Bohr's "example" is not accidental and has its history beginning with the question "How does an electron decide what frequency it is going to vibrate at when it passes from one stationary state to the other?" asked by Rutherford in his initial response to Bohr's paper (A Letter to Bohr, 20 March 1913, reproduced in "The Rutherford Memorial Lecture," Bohr 1987, v. 3, p. 41). Rutherford and others, in particular Dirac (who even spoke of an electron as having a "free will"), used such expressions without any further explanation, even though they might have been aware of the pitfalls of doing so. By contrast, Bohr's use of "a free choice on the part of nature" and similar locutions must, at least after his 1927 discussion with Einstein at the Solvay Conference in Brussels, be considered with this passage in mind. Heisenberg offered a penetrating response to Dirac's appeal to "choice on the part of nature," which Heisenberg questioned, on experimental grounds, in the course of a discussion that took place at the same conference. Dirac's comment implied the classically causal nature of independent quantum behavior undisturbed by observation in accord with his transformation-theory paper (Dirac 1927) and Bohr's Como argument, which appears to have been influenced by that paper. Dirac also helped Bohr to edit the Como lecture. Dirac, as noted in Chap. 2, adopted his view in his classic book, published soon after the Como lecture (Dirac 1930), as did von Neuman (1932), arguably influenced by both Bohr's Como argument. By contrast, Heisenberg's uncertainty-relations paper (Heisenberg 1927), which used the *mathematics* of Dirac's transformation theory, was ambivalent as concerned the assumption that the independent behavior of quantum objects was classically causal. On the occasion under discussion, Heisenberg, referring to this paper, expressed a view that was closer to Bohr's post-Como thinking, which no longer made this assumption. Heisenberg said (according to the available transcript):

> I do not agree with Dirac when he says that in the [scattering] experiment described nature makes a choice. Even if you place yourself very far from your scattering material and if you measure after a very long time, you can obtain interference by taking two mirrors. If nature had made a choice, it would be difficult to imagine how the interferences are produced. Obviously we say that nature's choice can never be known until the decisive experiment has been done; for this reason we cannot make any real objection to this choice because the expression "nature makes a choice" does not have any physical consequence. I would rather say, as I have done in my latest paper [on the uncertainty relations], that the *observer himself* makes the choice because it is not until the moment when the observation is made that the "choice" becomes a physical reality. (Bohr 1972–1999, v. 6, pp. 105–106)

The technical details of the experiment are not important at the moment, apart from noting the significance of scattering experiments in the development of QM. The experiment itself here mentioned is essentially equivalent to the double-slit experiment. The crucial point is that one in effect deals here with the complementary character of certain quantum experiments and with our choice of which of the two mutually exclusive or complementary experiments we want to perform, rather than with a choice of nature. Without realizing it, Heisenberg described the so-called delayed choice experiment of Wheeler (Wheeler 1983, pp. 190–192; Plotnitsky 2009, pp. 65–69). Heisenberg also suggested that in considering quantum phenomena only what has already occurred as the outcome of a measurement could be assigned the status of reality, which view eventually came to define Bohr's concept of phenomenon (e.g., Bohr 1987, v. 2, p. 64).

Bohr's comment cited above was written in 1929, following the Solvay Conference and further exchanges, and it refers specifically to his article, "The Quantum of Action and the Description of Nature" (Bohr 1987, v. 1, pp. 92–101). At stake is not merely stressing the metaphorical or "picturesque," rather than physical use of expressions like "a choice on the part of nature," but, as Heisenberg's remark makes clear, also a deeper philosophical point, which eventually made Bohr either avoid speaking in these terms or to qualify their metaphorical use. In the RWR view, toward which Bohr begins to move in this article, the ultimate constitution of the *reality* responsible for quantum phenomena is assumed to have an independent existence, without our being able to represent or even conceive of the ultimate character in this existence, including by means of such terms or concepts as to "be" or to "know," or "reality." These terms or concepts, invoked by Bohr in his elaboration, are not accidental. They are common in our everyday language but, as Bohr implies, they refer to philosophical problems that have been the subject of profound philosophical reflections from the pre-Socratics to Heidegger and beyond. These problems acquired radically new dimensions with quantum theory.

Bohr is reported to have replied, after the rise of quantum physics but before quantum mechanics, to H. Høffding's question "Where can the photon be said to be?" with "To be, to be, what does it mean to be?" (cited in Wheeler and Ford 1998, p. 131). Bohr might have been echoing the most famous sentence of Shakespeare's *Hamlet*, "To be, or not to be, that is the question," (Act 3, ll. 1749), realizing that in quantum physics one might want to or even must ask first "What does it mean to be?" Høffding's and Bohr's questions are still unanswered and, in Bohr's ultimate, RWR-type views are rigorously unanswerable, or even unaskable when it comes to quantum objects, such as photons. Quantum objects are idealizations (still beyond representation or conception), in the present interpretation, ultimately only applicable at the time of measurements, even if Bohr himself did not go that far. One not only cannot say or think anything about what they are or where they are independently of observations, but one cannot even apply such names as photons, electrons, and so forth apart from measurements and effects observed in them. According to Bohr, such questions as "Where can something be said to be?" or "When had something happened?" can only be rigorously asked about quantum phenomena, observed in measuring instruments. Nature has no photons or electrons, any more than being or reality, including that of the RWR type, all of which are human concepts, admittedly created by our biological and neurological *nature*, and thus by nature, which, however, is not the same as to say that nature uses these concepts through us. Nature, however, allows us to use our concepts, physical, philosophical, and mathematical, and sometimes our daily concepts, in considering or interactions with it by means of our technology, beginning with that of our bodies. It is also this interaction that enables us to idealize some part of the constitution of nature, even its ultimate constitution, as a reality without realism.

Under these assumptions, what is "sometimes picturesquely described as a 'choice of nature' [between different measurement outcomes]" will be given by Bohr a different meaning as well. As he says in 1954, "needless to say, such a phrase implies no allusion to a personification of nature, but simply points to the impossibility of ascertaining on accustomed lines directives for the course of a closed indivisible phenomenon" (Bohr 1987, v. 2, p. 73). In other words, it points to the impossibility of a classically causal or any representation, or even conception of the ultimate nature of reality, RWR, ultimately responsible for each quantum phenomenon, because this reality can never be extracted from this enclosure of phenomena observed in measuring instruments. In fact, while each phenomenon (by this point, in Bohr's sense) is closed and indivisible insofar as the quantum object involved cannot be considered apart from it, there is no "course" of a phenomenon, but only discrete transitions, "quantum jumps," from one phenomenon to another. Quantum objects could, however, be related or, in the present view, defined in terms of certain effects observed in measuring instruments, effects that do not always allow us to distinguish some of these objects, specifically elementary particles, from one another in their individual characteristics. Photons or electrons could be distinguished by such local effects like position, momentum, or energy, but not by their permanent characteristics, such as charge, mass, or spin, even though these characteristics, too, can, in the RWR view, only be manifested in effects observed in measuring instruments. Thus, rigorously, one only speak of *a charge measurement of an electron*, but *not a measurement of an electron's charge*.

This, more radical, understanding was yet a decade away in 1929, by which time, however, Bohr already managed to overcome the difficulties of using classical causality that plagued the Como argument of 1927. By 1928 or even already in 1927, following his exchanges with Einstein at the Solvay conference, Bohr returns to Heisenberg's approach that led to Heisenberg's discovery QM and his own initial understanding of the theory. Although the road ahead was to be long and difficult, the trajectory leading toward Bohr's ultimate, strong RWR interpretation of quantum phenomena and QM was now firmly established.

3.5 Conclusion

I close this chapter with Shakespeare's *Hamlet*, a connection brought in by Høffding's question or rather Bohr's counterquestion—"To be, to be, what does it mean to be?"—a very human question. But then, there are no questions other than human. Science, too, is human. It is a human project, even when it is concerned with nature apart from humanity. Even if one accepts that computers could do science on their own, it would still be a continuation of the previous human project of science, at least insofar as remains science. While, however, we cannot prevent science from being human, we should be concerned with making it "all too human" (Nietzsche 1996), by grounding it too much in extraneous philosophical principles or even, as Nietzsche would see it, prejudices, rather than in what physics must test, subjecting "everything to the inexorable test of experiment," to adopt Heisenberg's expression in his assessment of Bohr's philosophy (Heisenberg 1967, p. 95). Nietzsche even summons physics against our philosophical prejudices, realism and (classical) causality, among them, anywhere: "Therefore: *long live physics*! And even so what *compels* us to turn to physics—our honesty," which helps us to combat our prejudices (Nietzsche 1974, p. 266). In other words, Nietzsche urges us to subject all our assumptions, as hypothesis, in any domain, to "the inexorable test of experiment." When it comes to realism and classical causality, which Nietzsche questioned in the domain of human thinking and action, quantum physics would be the kind of physics Nietzsche wants to celebrate here. Coincidentally, the first ever university course on Nietzsche was taught in Copenhagen by Georg Brandes, a friend of Bohr's father. Bohr greatly admired Brandes.

A few decades before Nietzsche, Riemann, an important influence on Bohr, including, as discussed in Chap. 6, in Bohr's thinking leading him to complementarity, issued a similar warning in commenting on a possible role in physics of his rethinking of the foundations of geometry in his Habilitation lecture, as part of the passage cited in the Introduction: "Investigations like the one just made here [concerning geometry], which begin with general concepts, can … serve to insure that [developing new physics] is not hindered by unduly restricted concepts and that progress in comprehending the connection of things is not obstructed by traditional prejudices," such as those of Euclidean geometry or Newtonian physics (Riemann 1854, p. 33). This progress can, of course, also be obstructed by newly formed prejudices. Nevertheless, new experimental findings and new unhindered theoretical concepts are our best allies in our struggle against our prejudices in comprehending the connection of things, as both Bohr's discovery of his 1913 atomic theory and Heisenberg's discovery of QM teach us.

This, of course, also means that one can equally question the RWR thinking advocated by this study. But it does not mean that the only alternative is a return to realist thinking. As yet unknown and even unimagined alternatives may await, hard as they are to expect, if you believe Sydney Coleman: "if thousands of philosophers spent thousands of years searching for the strangest possible thing, they would never find anything as weird as quantum mechanics" (cited in Randall 2005, p. 117). Perhaps poets can do better. Shakespeare may be given the last word:

> Horatio: O day and night, but this is wondrous strange!
> Hamlet: And therefore as a stranger give it welcome.
> There are more things in heaven and earth, Horatio,
> Than are dreamt of in your philosophy.
> *The Tragedy of Hamlet, Prince of Denmark* (Act I.4, 165–166)

Some editions have "our philosophy." "Your philosophy" makes Hamlet more suspicious of philosophy. It makes him more akin to a physicist, perhaps, even a quantum physicist, who can only estimate the probabilities of future events defined by experiments he stages at the castle of Elsinore. Physics helps us to discover these things and helps philosophy to understand them, and keeps its honesty, as Nietzsche said. "And therefore, long live physics!" *Hamlet*, a play (it may be shown) with many transitions without connections, takes place in Denmark, too. "To be, to be, what does it mean to be?"—Bohr asked. Quantum physics changed not only possible answers to this question, but also what it means to ask them. This may also be a lesson for *our* philosophy when it deals with the human world. Bohr certainly thought so, as is testified to by his famous image: "one must never forget that in the drama of existence we are ourselves both actors and spectators" (Bohr 1987, v. 1, p. 119, v. 2, p. 81). We are, as Bohr clearly implies, equally both in staging and observing quantum experiments. Many of them, beginning with those of Rutherford, which led to Bohr's 1913 theory, and ending (for now) with those, such as the discovery of the Higgs boson, in LHC, have been nothing less than dramatic.

References

Bohr, N.: On the constitution of atoms and molecules (Part 1). Phil. Mag. **26**(151), 1–25 (1913)

Bohr, N.: The Theory of Spectra and Atomic Constitution. Cambridge University Press, Cambridge, UK (1924)

Bohr, N.: Causality and complementarity. In: Faye, J., Folse, H.J. (eds.) The Philosophical Writing of Niels Bohr, Vol 4, Causality and Complementarity, 1999, pp. 83–91. Ox Bow Press, Supplementary Papers, Woodbridge, CT, USA (1937)

Bohr, N.: *Niels Bohr: Collected Works*, 10 vols. Elsevier, Amsterdam, Netherland (1972–1996)

Bohr, N.: The Philosophical Writings of Niels Bohr, 3 vols. Ox Bow Press, Woodbridge, CT, USA (1987)

Born, M., Jordan, P.: Zur Quantenmechanik. Z. Phys. **34**, 858–888 (1925)

Bose, S.N.: Plancks Gesetz und Lichtquantenhypothese. Z. Phys. **26**, 178–181 (1924)

Dirac, P.A.M.: The physical interpretation of the quantum dynamics. Proc. Royal Soc. London A **113**, 621–641 (1927)

Dirac, P.A.M.: The Principles of Quantum Mechanics. Clarendon, Oxford, UK (1930)

Einstein, A.: Strahlungs-emission und -absorption nach der Quantentheorie, Physikalische Gesellschaft. Verhandlungen **18**, 318–323 (1916a)

Einstein, A.: Zur Quantentheorie der Strahlung. Physikalische Gesellschaft Zurich **18**, 173–177 (1916b)

Einstein, A.: Physics and reality. J. Franklin Inst. **221**, 349–382 (1936)

Einstein, A.: Autobiographical Notes (tr. Schilpp, P. A.). Open Court, La Salle, IL, USA (1949)

Folse, H.J.: The methodological lesson of complementarity: Bohr's naturalistic epistemology. Physica Scripta T163 (2014). http://m.iopscience.iop.org/1402-4896/2014/T163

Freidel, L.: On the discovery of quantum mechanics by Heisenberg, Born, and Jordan (Unpublished) (2016)

Heisenberg, W.: Quantum-theoretical re-interpretation of kinematical and mechanical relations. In Van der Waerden, B.L. (ed.) Sources of Quantum Mechanics, rpt. 1968, pp. 261–277. Dover, New York, NY, USA (1925)

Heisenberg, W.: The physical content of quantum kinematics and mechanics. In: Wheeler, J.A., Zurek, W.H. (eds.) Quantum Theory and Measurement, 1983, pp. 62–84. Princeton University Press, Princeton, NJ, USA (1927)

Heisenberg, W.: Physics and Philosophy: The Revolution in Modern Science. Harper & Row, New York, NY, USA (1962)

Heisenberg, W.: Quantum theory and its interpretation. In: Rozental, S. (ed.) Niels Bohr: His Life and Work as Seen by his Friends and Colleagues, pp. 94–108. North-Holland, Amsterdam, Netherlands (1967)

Kragh, H.: Niels Bohr and the Quantum Atom: The Bohr Model of Atomic Structure 1913–1925. Oxford University Press, Oxford, UK (2012)

Mehra, J., Rechenberg, H.: The Historical Development of Quantum Theory, 6 vols. Springer, Berlin, Germany (2001)

Nietzsche, F.: The Gay Science (tr. Kaufmann, W.), Vintage, New York, NY, USA (1974)

Nietzsche, F.: Human, All Too Human: A Book for Free Spirits (tr. M. Fabet, M., Lehman, S.), Lincoln, NE, USA: Bison Books (1996)

Plotnitsky, A.: Epistemology and Probability: Bohr, Heisenberg, Schrödinger and the Nature of Quantum-Theoretical Thinking. Springer, New York, NY, USA (2009)

Randall, L.: Warped Passages: Unraveling the Mysteries of the Universe's Hidden Dimensions. Harpers Collins, New York, NY, USA (2005)

Riemann, B.: On the hypotheses that lie at the foundations of geometry. In: Pesic, P. (ed.) Beyond Geometry: Classic Papers from Riemann to Einstein, 2007, pp. 23–40. Dover, Mineola, NY, USA (1854)

Stone, A.D.: Einstein and The Quantum: The Quest of The Valiant Schwabian. Princeton University Press, Princeton, NJ, USA (2015)

Von Neumann, J.: Mathematical Foundations of Quantum Mechanics (tr. R. T. Beyer). Princeton University Press, Princeton, NJ, USA , Princeton, NJ, rpt. 1983 (1932)
Wheeler, J.A.: Law without law. In: Wheeler, J.A., Zurek, W.H. (eds.) Quantum Theory and Measurement, pp. 182–216. Princeton University Press, Princeton, NJ, USA (1983)
Wheeler, J.A., Ford, K.: Geons, Black Holes, and Quantum Foam: A Life in Physics. W. W. Norton, New York, NY, USA (1998)

Chapter 4
"The Heisenberg Method": Algebra, Geometry, and Probability in Quantum Mechanics

Perhaps the success of the Heisenberg method points to a purely algebraic method of description of nature, that is, to the elimination of continuous functions from physics. Then, however, we must give up, in principle, the space–time continuum.
—Albert Einstein, "Physics and Reality" (1936)

Abstract This chapter reconsiders, from the RWR viewpoint, Heisenberg's discovery of QM and QM itself, in terms of two diagrams:

$$QUANTUMNESS \rightarrow PROBABILITY \rightarrow ALGEBRA$$

and

$$QUANTUMNESS \rightarrow PROBABILITY \rightarrow ALGEBRA \rightarrow GENERALIZED\ GEOMETRY.$$

The first arrow, *QUANTUMNESS→PROBABILITY* represents the fact that the physical character of quantum phenomena implies that our predictions concerning them are irreducibly probabilistic or statistical. This in turn implies, defining the mathematical character of Heisenberg's approach and QM, the second arrow, *PROBABILITY→ALGEBRA*, that QM is primarily algebraic, in contrast to more geometrical classical or relativistic theories, reflecting Einstein's assessment of "the Heisenberg method" as "algebraic." This assessment, however, requires qualifications, which shed new light on the relationships between algebra and geometry in physics, represented in the second diagram. This diagram entails a new concept of geometry, "generalized geometry," part of "modernist mathematics," explained in this chapter as well. Section 4.1 offers an introduction to the chapter. Section 4.2 discusses the formalism of QM and the role of fundamental physical principles in Heisenberg's derivation of this formalism. Section 4.3 revisits Heisenberg's discovery of QM. Section 4.4 considers the relationships between algebra and geometry in this discovery. Section 4.5 addresses the role of algebra and geometry in modernist mathematics and physics.

Keywords Algebra · Geometry · Heisenberg method · Hilbert space · Matrix mechanics · Modernist mathematics · Noncommutativity

A. Plotnitsky, *Reality Without Realism*,
https://doi.org/10.1007/978-3-030-84578-0_4

4.1 Introduction

This chapter reconsiders, from the RWR viewpoint, the key features of Heisenberg's discovery of QM, and of QM itself, in terms of two diagrams:

$$(QPA)\ QUANTUMNESS \rightarrow PROBABILITY \rightarrow ALGEBRA$$

and

$$(QPGG)\ QUANTUMNESS \rightarrow PROBABILITY \rightarrow ALGEBRA \rightarrow GENERALIZED\ GEOMETRY.$$

The first arrow, *QUANTUMNESS→PROBABILITY,* represented the fact that the physical character of quantum phenomena implies that our predictions concerning them are irreducibly probabilistic or statistical. This fact in turn implies, defining the mathematical character of Heisenberg's approach and QM, and the second arrow, *PROBABILITY→ALGEBRA*, that QM is primarily algebraic, in contrast to more geometrical classical or relativistic theories. This arrow reflects Einstein's assessment of "the Heisenberg method" as "algebraic." Einstein's assessment, however, requires qualifications, which also shed new light on the relationships between algebra and geometry in physics, represented in the second diagram, which, however, entails a new concept of geometry, "generalized geometry." This concept emerged in "modernist mathematics," which, as explained in this chapter, defined twentieth- and twenty-first-century mathematics and physics, including both relativity and quantum theory, in the latter case, beginning with Heisenberg's discovery of QM, in conjunction with the RWR view.

Section 4.2 discusses the mathematical formalism of QM and of the role of fundamental principles in Heisenberg's derivation of this formalism. Section 4.3 revisits Heisenberg's discovery of QM. Section 4.4 considers the relationships between algebra and geometry in this discovery, as complicating Einstein's assessment of Heisenberg's method as algebraic. Section 4.5 addresses the role of algebra and geometry in modernist mathematics and physics.

4.2 Quantum Mechanics as a Fundamental Theory: Principles, Postulates, and Formalism

For nearly a century now, since the publication of von Neumann's *The Mathematical Foundation of Quantum Mechanics* (von Neumann 1932), QM (and subsequently QED and QFT) most commonly use as their mathematical formalism the Hilbert-space formalism over complex numbers, $\mathbb{C}$. There are other versions, some more abstract ones, such as those of C*-algebras and, more recently, sheaf theory and category theory, all of which are, thus far, generally, albeit not always entirely, equivalent mathematically. Hilbert-space formalism remains dominant, however. It is worth summarizing the key features of this formalism and how it is used:

(1) The Hilbert space (over $\mathbb{C}$), used in QM, is an abstract vector space of any dimension, finite or infinite (in QM countably infinite), which possesses the structure of an inner product that allows lengths and angles to be measured, analogously to a n-dimensional Euclidean space (which is a Hilbert space over real numbers, $\mathbb{R}$);
(2) The feature of that formalism, arguably most crucial for QM and never used in physics previously, is the *noncommutativity* of the Hilbert-space operators, also referred to as "observables," which are *mathematical* entities associated, in terms of probabilistic or statistical predictions, with *physically* observable quantities, by means of 3);
(3) The presence of Born's rule or an analogous rule (such as von Neumann's projection postulate or Lüder's postulate), which is added to the formalism and establishes the relation between "quantum amplitudes," associated with complex Hilbert-space vectors as complex entities and probabilities as real numbers, by using square moduli or, equivalently, the multiplication of these quantities and their complex conjugates (technically, these amplitudes are first linked to probability densities);
(4) The probabilities involved are nonadditive: the joint probability of two or more mutually exclusive alternatives in which an event might occur is, in general, not equal to the sum of the probabilities for each alternative, as in classical probability theory; instead, it obeys the law of the addition of amplitudes for these alternatives, to the sum of which Born's rule is then applied.

While keeping in mind alternative postulates just mentioned, I shall primarily refer to Born's rule from now on. In technical terms, Born's rule is defined as follows. First, in the case of a self-adjoint Hilbert-space operator A ($\langle Av, w\rangle = \langle v, Aw\rangle$, for all vectors in this space) with a discrete spectrum, Born's rule states that, if a physical observable associated with A is predicted by means of a (normalized) wave function $|\psi\rangle$, then the outcome, when measured, will be one of the eigenvalues λ of A; the probability that it will correspond to a given eigenvalue λ_i is equal to $\langle\psi|\mathrm{P_i}|\psi\rangle$, where $\mathrm{P_i}$ is the projection onto the eigenspace of A corresponding to λ_i. When this eigenspace is one-dimensional and spanned by the normalized eigenvector $|\lambda_i\rangle$, $\mathrm{P_i} = |\lambda_i\rangle\langle\lambda_i|$, so the probability $\langle\psi|\mathrm{P_i}|\psi\rangle = \langle\psi|\lambda_i\rangle\langle\lambda_i|\psi\rangle$. The probability amplitude is $\langle\lambda_i|\psi\rangle$, and Born's rule means that the corresponding probability is the square of the amplitude or is the amplitude multiplied by its own complex conjugate, or $\mathrm{P_i} = \langle\lambda_i|\psi\rangle^2$. If the spectrum has a continuous part, then the spectral theorem (which deals with cases when a linear operator or matrix can be diagonalized) tells us that there exists a projection-valued measure Q, called the spectral measure of A. In this case, the probability that the measurement outcome lies in a measurable set M is $\langle\psi|Q(M)|\psi\rangle$. In the simplest case, when ψ is a wave function for a point particle in position space, the probability density function $p\ (x, y, z)$ for predicting a measurement of the position at time t_1 equals to $|\psi(x, y, z, t_1|^2$. Born's rule could be generalized for POWM (positive operator-valued measures), the values of which are positive semidefinite operators in a Hilbert space.

Although Born's or analogous rules are, as defined by the complex conjugation, connected rather naturally mathematically to the formalism of QM, they are, nevertheless, added to this formalism rather than derived from it. We do not know why these postulates work, which makes it tempting to argue that why they work is the greatest mystery of QM. But then we do not know why the formalism works either. There is neither one without the other.[1]

The features just outlined were, however, not initially assumed, but, beginning with Heisenberg, were inferred from certain physical features of quantum phenomena and principles arising from these features. While Heisenberg's mathematical scheme, matrix mechanics, was not formally defined by him as a Hilbert-space formalism, which was introduced by von Neumann soon thereafter, it essentially amounted to a Hilbert-space formalism, especially in Born and Jordan's full-fledged formulation of matrix mechanics (Born and Jordan 1925). (Heisenberg initially did not use the term matrix mechanics.) As a physical theory, QM, was developed by Heisenberg as a principle rather than constructive theory, in the sense of Einstein, explained below.

It might be noted first that, similarly to the case of the term of concept, as discussed in Chap. 2, although one could sometimes surmise the meaning of the term principle from its use, it is rarely defined or explained in physical or philosophical literature. Thus, his title notwithstanding, Heisenberg did not do so in his first book, *The Physical Principles of the Quantum Theory* (Heisenberg 1930). Nor did Dirac, in his *Principles of Quantum Mechanics*, first published in the same year (Dirac 1930). Terms like "principle," "postulate," and "axiom," are often used in physics somewhat indiscriminately, and it is difficult to entirely avoid overlapping between the concepts designated by these terms, or those designated as "laws," especially because physical principles often derive from (or give rise to) postulates or laws. Thus, conservation laws are sometimes seen as conservation principles. These terms are usually better clarified in mathematics. Thus, Euclid distinguished between "axioms" and "postulates." Axioms were thought to be something manifestly self-evident, such as the first axiom of Euclid ("things equal to the same thing are also equal to each other"). A postulate, by contrast, is *postulated*, in the sense of "let us assume that A holds" and see what follows from it according to established logical rules. Euclid's postulates may be thought of assumptions necessary and sufficient to derive the truths of geometry, of which one might already be intuitively persuaded. The famous fifth

[1] There have been attempts to provide reasons for Born's rule (or equivalent rules) by using, presumably, more basic principles or postulates, rather than seeing it as a primary postulate, beginning with Andrew Gleason's celebrated theorem, a remarkable mathematical result (Gleason 1957). I shall put these attempts aside, because, while some of them are conceptually important, none of them (to my knowledge) *derives* Born's rule from the formalism of QM. In one form or another, Bohr's rule, or some equivalent rule, is still an additional independent postulate. Its origin and necessity is, again, in the circumstance that, on the one hand, QM is an irreducibly probabilistic theory for all quantum systems, no matter how elementary, and probabilities are real numbers, while, on the other hand, all current versions of the formalism of QM (for both continuous and discrete variables) are over $\mathbb{C}$.

postulate is a case in point. It defines Euclidean geometry alone, which in part explains millennia of attempts to derive it as a theorem.

Given that my subject is physics, I shall primarily refer to postulates, assumed on the basis of experimental evidence (as it stands now and hence as potentially refutable) and often grounding principles, as "empirically discovered ... general characteristics of natural processes ... that give rise to mathematically formulated criteria which the separate processes or the theoretical representations of them have to satisfy" (Einstein 1919, p. 228). It is not easy to speak of axioms in the sense just defined in physics, although the term is used by physicists, as it was by Hardy (2001) or D'Ariano (2020), in the latter case following von Neumann's use of axioms in von Neumann (1932). In von Neumann's case, the use of axioms is motivated by his aim to establish, in the words of his title, "the mathematical foundations of quantum mechanics." Little has the self-evidence of axioms even in classical mechanics, and most of the uses of the term "axiom" in physics are closer to that of "postulate," as just defined. For example, von Neumann's assumption of Hilbert spaces over $\mathbb{C}$ as defining the formalism of QM is a postulate in this sense, grounded in the already established mathematics of QM, rather than, as in Heisenberg, established on the basis of certain grounding physical principles, "empirically discovered ... general characteristics of natural processes." Von Neumann also postulated that Schrödinger's equation represented the independent behavior (a unitary evolution) of a quantum system. This, as discussed in Chap. 2, is a nonfalsifiable ontological postulate, which has been often adopted but the efficacy of which can be questioned (D'Ariano 2020).

Bohr, more careful and etymologically attuned than most in using his terms, preferred both postulates, such as the quantum postulate of the Como lecture (Bohr 1987, v. 1, pp. 52–53), and principles, such as the correspondence principle. As will be seen, in Heisenberg's hands, the correspondence principle gave rise to a mathematically expressed postulate. Complementarity, I would argue, functions more as a concept than a principle in Bohr, although it is sometimes referred to (although not so much by Bohr himself) as a principle. "First principles" commonly refer, from Plato and Aristotle on (e.g., Aristotle, *Metaphysics* 1013a14-15, Aristotle 1984, v. 2, p. 1601), as they will do here, to foundational assumptions that are not deduced from any other assumptions, in this case, akin to axioms in mathematics. Einstein's concept of a "principle theory," corresponds to the use of principles by Bohr and Heisenberg, and some quantum-information theorists, as in D'Ariano et al. (2017), discussed in Chap. 9.

According to Einstein: "constructive theories [aim at] build[ing] up a picture of the more complex phenomena out of the materials of a relatively simple formal scheme from which they start out" (Einstein 1919, p. 228). "A relatively simple formal scheme" represents, in a mathematically idealized way, a more, or even the most, elementary underlying reality responsible for these phenomena. Einstein's example of a constructive theory in classical physics was the kinetic theory of gases, which "seeks to reduce mechanical, thermal, and diffusional processes to movements of molecules—i.e., to build them up out of the hypothesis of molecular motion," described by the laws of classical mechanics (Einstein 1919, p. 228). The assumption that this motion obeys the laws of classical mechanics was *in effect* abandoned by Planck's black-body radiation theory, which inaugurated quantum physics. It was, however, Einstein who was the first to realize the incompatibility between Planck's quantum hypothesis and this assumption, still made by Planck himself (Einstein 1906). As noted in Chap. 3, a derivation of Planck's law by using a proper statistical counting (the Bose–Einstein statistics) was given by Bose (1924), confirming the difficulty of a physical representation of the behavior of photons compatible with our phenomenal intuition (Bose 1924). In any event, one could build quantum theory independently of constructing any scheme representing an underlying physical reality. Instead, as Heisenberg did in the case of QM, or Bohr in the case of his 1913 atomic theory, on which Heisenberg built, one can develop quantum theory as a principle theory. The quantum behavior of photons requires QED, which was born, with Dirac, as a principle theory as well (Plotnitsky 2016, pp. 207–225). Neither theory required a realist and classically causal representation of the behavior of individual quantum objects, including the ultimate individual constituents of matter, elementary particles. On the other hand, none of them excluded a realist or classically causal interpretation of this behavior either. Only determinism was excluded on experimental grounds.

In contrast to constructive theories, principle theories, according to Einstein, revealing the Kantian genealogy of his distinction, "employ the analytic, not the synthetic, method. The elements which form their basis and starting point are not hypothetically constructed but empirically discovered ones, general characteristics of natural processes, *principles* that give rise to mathematically formulated criteria which the separate processes or the theoretical representations of them have to satisfy" (Einstein 1919, p. 228). Thermodynamics, Einstein's example of a classical principle theory (parallel to the kinetic theory of gases, as a constructive theory), is a principle theory because it "seeks by analytical means to deduce necessary conditions, which separate events have to satisfy, from the universally experienced fact that perpetual motion is impossible" (Einstein 1919, p. 228). I would add the following qualification, which is likely to have been accepted by Einstein: Principles are not empirically discovered but are formulated on the basis of empirically established evidence.The formulations defining principle theories are synthetic, constructed already by virtue of the role mathematics plays in them. This construction is, of course, different from the construction of "the materials of a relatively simple formal scheme" defining a constructive theory because it only has to satisfy the mathematically formulated criteria or postulates established by principles, but not to represent a more fundamental reality defining the phenomena considered. "The impossibility of perpetual motion" could hardly be seen as empirically given; it was formulated, as a principle, on the basis of empirically established evidence. Principles, thus, need not have the self-evidence of axioms or, at least initially, the assumptive character of postulates, although, once introduced, they may function as or lead to postulates from which a given theory is built by means of logical deductions. One might, as I would like to do here, amplify this understanding of principles by seeing them as a foundation and guidance for inventing and building new theories.

Einstein's language of "theoretical representations of natural processes" is shaped by his realist thinking, grounding his view of both constructive and principle theories.

One can, however, think, along RWR lines, of principle theories that satisfy the mathematically formulated criteria supplied by principles without assuming a theoretical representation of the ultimate nature of the reality considered, although one would need such a representation at other levels of this reality. This is the case I argue for QM, which explains my appeal to Einstein's distinction between constructive and principle theories as my starting point. Einstein's 1919 paper, introducing this distinction, was written before QM. One could, however, see Bohr's (1913) atomic theory as a principle theory, which adopted the RWR view in the case of quantum jumps.

An appeal to principles need not imply that there is some permanent, Platonist, metaphysical essence to them, although there are those who take this view of principles. In the present view, one is not dealing here with a Platonist mimesis of some primordial or eternal original forms, but rather with thinking of principles, no matter how fundamental, in the way William James thinks of the truth of an idea, and a principle is always based in an idea or concept (also in the sense defined in Chap. 2): "Truth HAPPENS to an idea. It BECOMES true, is MADE true by events" (James 1978, p. 73). Principles change as our experimental findings and theories change (without asymptotically converging to some unique set of principles, thus, again, assuming their metaphysical permanence or essence), and we cannot always anticipate or control these changes. A theory could be deduced, in one way or another, from a set of first principles (and other assumptions), but it cannot be confirmed by them; it can only be confirmed or refuted by experiments the outcomes of which the theory predicts, or falsified by experimental findings. The principles of QM replaced, within a new scope, some among the main principles of classical physics, which continue to remain operative within the proper scope of classical physics and some of them extend to QM and QFT. There could also be such changes within the same physical scope, as in the case of general relativity theory vs. Newton's theory of gravity. Some principles of quantum theory have been abandoned or modified and new principles have been added, both in view of extending the scope of the theory to QFT and within the scope of QM itself. On the other hand, certain among these principles, such as the quantum probability or statistics (QP/QS) principle, have remained in place throughout the history of quantum theory. This is also true as concerns the quantum discreteness (QD) and quantum individuality (QI) principles and the postulates they lead to in RWR-type and certain other interpretations, but, unlike the QP/QS principle, these principles have been seen as not fundamental but as possibly circumventable when assuming QM or QFT to be a correct theory of quantum phenomena. In the RWR view, they are fundamental. But that does not mean that they, or the QP/QS principle, may not be abandoned at some point, as many, beginning, with Einstein, have hoped and still do.

While, however, I shall discuss the principles that grounded Heisenberg's invention of QM, I shall not be concerned with the question of a derivation of QM from first principles, apart from emphasizing the role of conceptual invention, such as that of a new type of physical variables, matrix variables, by Heisenberg. I have considered this subject previously, from Heisenberg's invention of QM to reconstruction projects in quantum information theory (Plotnitsky 2016, pp. 68–98, 238–248). I shall discuss the latter projects in Chap. 9. Instead, as stated at the outset, I would like to reconsider, on RWR lines, Heisenberg's approach in terms of two diagrams

$$QUANTUMNESS \rightarrow PROBABILITY \rightarrow ALGEBRA$$

and

$$QUANTUMNESS \rightarrow PROBABILITY \rightarrow ALGEBRA \rightarrow GENERALIZED\ GEOMETRY$$

The first arrow, *QUANTUMNESS→PROBABILITY*, captures the epistemological-experimental structure of Heisenberg's approach. It represents the fact that the *quantumness* of physical phenomena, that is, the character of physical phenomena known as quantum, implies, in accord with the QP/QS principle, that our predictions concerning them are irreducibly probabilistic or statistical. This is the case even in dealing with quantum phenomena resulting from what is assumed to be elementary individual quantum behavior, such as that of elementary particles. As indicated in Chap. 2, this fact is not in itself sufficient to define quantum phenomena, which are assumed in this study to be characterized by the following set of interrelated features: (1) the role of h, (2) the irreducible role of measuring instruments in defining quantum phenomena, (3) discreteness, (4) complementarity, (5) entanglement, (6) quantum nonlocality, and (7) the irreducibly probabilistic or statistical nature of quantum predictions. "Discreteness" plays a special role in the present context, coupled to the individuality or even uniqueness of each quantum phenomenon, in accord with the QD and QI postulates (Bohr 1987, v. 1, p. 53). Quantum phenomena are individual and discrete in relation to each other, which is, again, not the same as the atomic, Democritean, discreteness of elementary quantum objects, such as elementary particles (e.g., Bohr 1987, v. 2, p. 33). As explained in Chap. 8, their character as elementary could be ascertained on the basis of effects such objects have on measuring instruments, keeping in mind that some particles originally considered elementary can reveal themselves to be composite, as it happened in the case of hadrons that were found to be composed of quarks and gluons. The QPA structure would hold for most of realist and classically causal interpretations of quantum phenomena and QM, possibly, in contrast to the RWR view, under the assumption of a continuously connected underlying reality. Probability always concerns discrete events, which may or may not be connected by a classically causal process. In the RWR view, they are not and one only deals with probabilities of transitions, transitions without connections, between quantum events observed in measuring instruments. As stated from the outset, that we can only make probabilistic or statistical predictions concerning all phenomena thus far known as quantum is an experimentally established, "objective," fact because the repetition of identically prepared quantum experiments in general leads to different outcomes. This difference cannot be diminished (beyond the limits defined by Planck's constant, h) by improving the precision of our measuring instruments, as is manifested in the uncertainty relations, which would remain valid even if we had perfect instruments.

The situation just outlined in turn implies, defining the theoretical character of the second arrow of QPA diagram, *PROBABILITY→ALGEBRA*, that our theories concerning these phenomena, quantum theories, are fundamentally algebraic, in contrast to more geometrical classical or relativistic theories. This implication is reflected in Einstein's assessment of "the Heisenberg method" as essentially algebraic (Einstein 1936, p. 375). This is a more complex claim than it might appear, beginning with the fact that this distinction is not unconditional, because quantum theories do have geometrical aspects. Conversely, geometrical physical theories, such as classical physics and relativity, have algebraic aspects. Accordingly, Einstein's assessment requires qualifications, considered in detail below. Nevertheless, this assessment helps, in part against Einstein's own grain, to understand the relationships between algebra and geometry in physics, or in mathematics. In particular, the following diagram obtains, keeping in mind that a new concept of geometry is involved, defined by what I shall call "spatial algebra:"

$$\textit{QUANTUMNESS} \rightarrow \textit{PROBABILITY} \rightarrow \textit{ALGEBRA} \rightarrow \textit{GENERALIZED GEOMETRY}\,(\textit{QPAGG}).$$

More precisely the chain is: *QUANTUMNESS→PROBABILITY→ALGEBRA→ SPATIAL ALGEBRA→GENERALIZED GEOMETRY*, but it is a "return of geometry" via algebra that is most crucial. Accordingly, I shall speak of *QPAGG*. This is mostly a mathematical matter, and probabilistic predictions themselves are, while involving this generalized geometry, still essentially algebraic. Accordingly, the QPA diagram is part of the QPAGG diagram.

Probability theory has been primarily algebraic, too. Its origin, in the work of Gerolamo Cardano, Blaise Pascal, and Pierre Fermat coincides with the emergence of algebra. As Ian Hacking argued in explaining why the theory emerged in the seventeenth century rather than earlier, some form of algebra was necessary for it (Hacking 2006). Analytic geometry and calculus were introduced around the same time, the first by Fermat and Descartes, and the second by Newton and Gottfried Leibniz (although Fermat also made important contributions to the development of calculus). They, too, were the product of the algebraization of mathematics, a defining feature of the mathematics and physics of modernity (roughly from the sixteenth century on), even though geometry had continued to dominate both until the nineteenth century.

The situation is, however, more complex. There have been "returns of geometry," even parallel ones, in both theories, probability theory and then QM, due to the fact that analysis came to play a major role in both from their inception. In QM, where analysis, including complex analysis, was never any less important that in classical physics and relativity, functional analysis came to play a key role, as it did in probability theory. Hilbert spaces of QM are those of square-integrable complex functions (over $\mathbb{C}$), and in effect the concept itself of Hilbert space (although the term was not used by them) was introduced by Hilbert and Erik I. Fredholm in the theory of integral equations. Eventually, this concept become part of functional analysis, which was, as noted in Chap. 2, the main field of Bohr's brother, Harald, who was at Göttingen during that time. So, (Niels) Bohr could easily consult Harald on all mathematical matters related to quantum mechanics, although plenty of mathematical expertise was available in Copenhagen, given the presence there of Hendrik Kramers, Oscar Klein, or Léon Rosenfield (all Bohr's assistants at various points), or of course Heisenberg. The concept of Hilbert space was then used by von Neumann to recast the formalism of QM, although in effect it was in place already in Heisenberg's and Schrödinger's work. QM was an eigenvalue theory of Hermitian operators.

In a parallel development, probability theory, too, came to use spatialized mathematical concepts, such as "probability space," introduced by Andrei N. Kolmogorov as part of his axiomatization of probability theory (Kolmogorov 1956). This concept follows the concepts of "space" developed in functional analysis and measure theory (which Kolmogorov used to axiomatize probability theory). It should be kept in mind that quantum-mechanical probability is not Kolmogorovian (at least in most views). The Hilbert spaces of QM are not the spaces of probabilities, which would have to be over $\mathbb{R}$, but spaces, over $\mathbb{C}$, which allow one to establish probability (density) amplitudes. One then needs Born's or analogous rule to get to probabilities.

The algebra of probability in classical statistical physics, the primary domain of using probability in classical physics, was commonly, as in the kinetic theory of gases, underlain by a geometrical picture of the behavior of the individual constituents of the systems considered, assumed to follow the laws of classical mechanics. By contrast, as became apparent earlier on, in quantum physics, even individual objects, no matter how elementary, and their behavior and the events they give rise to had to be treated probabilistically. One needed, accordingly, to find a new theory, representational or not, to make correct predictions concerning them, a task that quantum theory pursued from its inception. As discussed in Chap. 3, Bohr's (1913) theory offered a new way of thinking about quantum phenomena in the case of transitions, quantum jumps, between stationary states (still represented geometrically, as orbits). The theory was defined by a shift from considering, even probabilistically, the motion of electrons, to predicting the probabilities of transitions between quantum states, without providing a physical mechanism for these transitions, thus making them transitions without connections. Heisenberg extended this thinking by making QM fully a theory of such transitions and, mathematically, transition probabilities between all quantum states, without offering any geometrical representation of the behavior of electrons in stationary states, which became merely energy states.

Heisenberg's use of his matrix variables, in retrospect as operators in linear vector spaces (infinite-dimensional Hilbert spaces over $\mathbb{C}$), as the main part of the mathematical machinery of predicting such transition probabilities, defined the algebraic nature of "the Heisenberg method." This was in contrast to Schrödinger's method in his wave mechanics, accompanied by a geometrical conception of the ultimate constitution of the reality responsible for quantum phenomena in terms of a continuous vibrational process, which, initially, did not involve probabilities, at least, not fundamentally. This conception was never worked out by Schrödinger to accord with the experimentally established discrete features of quantum phenomena. On the other hand, these were physical and mathematical demands of accounting for these features, against the grain of his agenda, that led Schrödinger to a mathematically equivalent scheme. Probability was made fundamental by several subsequent investigations, most especially Born's probabilistic interpretation of the wave function. I shall discuss Schrödinger's work in Chap. 6. My main point at the moment is that Schrödinger's equations themselves (especially, his time-dependent equation) may be seen, as they came to be by Bohr and Heisenberg, in RWR-terms, as a probabilistically predictive mathematical technology, which did not represent the ultimate constitution of the reality responsible for quantum phenomena. In any event, Schrödinger's "geometrical" method proved to be unable to provide a true alternative to the Heisenberg "algebraic" method, as Einstein, who initially welcomed Schrödinger's wave mechanics as such an alternative, came to realize long before his assessment of the Heisenberg method in 1936.

4.3 From Geometry to Algebra, and from Algebra to Geometry

Heisenberg's discovery and then Bohr's interpretation of QM were grounded in three main principles leading to the corresponding postulates, building on those of Bohr's (1913) theory, with Bohr's principle of complementarity added in 1927 (although this study views complementarity primarily as a concept rather than a principle).[2] These postulates were as follows:

(1) The postulate of quantum discreteness, the QD postulate, according to which all observable quantum phenomena are individual and discrete in relation to each other (which is different from the discreteness of quantum objects);
(2) The postulate of the probabilistic or statistical nature of quantum predictions, the QP/QS postulate, maintained, in contrast to classical physics, even in considering individual quantum objects, and accompanied by the nonadditive character of quantum probability and rules, such as Born's rule (a version of which was used in Heisenberg's derivation), for predicting them; and
(3) The correspondence *postulate*, based in Bohr's correspondence *principle*, which, as initially understood by Bohr, required that the predictions of quantum theory must coincide with those of classical mechanics in the classical limit, but was given by Heisenberg a mathematical form, *postulating* that the equations and variables of QM convert into those of classical mechanics in the classical limit.

Implicit in Heisenberg's approach, especially given the QD postulate, was the quantum individuality (QI) postulate. The postulate first appeared explicitly, even if

[2] A view of Bohr's interpretation of QM as a principle theory was suggested in Bub (2000), which, however, does not consider Bohr's interpretation as an RWR-type interpretation, and does not distinguish between Bohr's different interpretations. As indicated above, a principle theory or interpretation may be realist.

not under this name, in Bohr's "quantum postulate," introduced in his Como lecture of 1927 and grounding his first interpretation of quantum phenomena and QM. The earlier form of this postulate was implicit in Bohr's (1913) quantum postulates, used by Heisenberg in his derivation of QM. Applying Bohr's concept of transitions without connections to all quantum states (rather than only stationary states), Heisenberg was not concerned with representing or even predicting, either (ideally) exactly or probabilistically, the motion of electrons, in effect abandoning the very concept of motion as applicable to electrons. He was only concerned with predicting the probabilities of discrete transitions between the quantum states of electrons, transitions without connections.

He added a new twist: "What I really like in this scheme is that one can really reduce *all interactions* between atoms and the external world ... to transition probabilities" (Heisenberg, Letter to Kronig, 5 June 1925; cited in Mehra and Rechenberg 2001, v. 2, p. 242). By speaking of the "interactions between atoms and the external world," this statement suggests that QM was only predicting the effects of these interactions observed in measuring instruments. This procedure, thus, replaced measurement in the classical sense (of measuring some pre-existing properties of quantum objects) with establishing, by using measuring instruments, quantum phenomena, which can be treated classically without classically measuring the properties of quantum objects. This view, not found in Bohr's (1913) theory, was adopted by Bohr, to the point of becoming the single defining feature of his interpretation in all of its versions. As discussed in Chap. 2, in Bohr's interpretation, the classical treatment of the observed parts of measuring instruments meant that the data registered, as part of quantum phenomena, in these instruments, could be measured as classical properties just as one measures such properties in classical physics. Measuring instruments were also assumed to contain a quantum stratum through which they interacted, quantumly, with quantum objects, which stratum or this interaction were places beyond representation and, in Bohr's ultimate interpretation, conception. My point at the moment is that the key ingredients of this view of quantum measurement are found in Heisenberg's thinking leading him to his discovery of QM. None of these ingredients were, by contrast, considered by Schrödinger, who was not concerned with measurement at all, in his approach to his wave mechanics, in contrast to his later papers (Schrödinger 1935a, b, 1936), which offered a subtle analysis of quantum measurement.

One can also view Heisenberg's approach in quantum-informational terms, discussed in Chap. 9. While one could not say that this approach was, technically, quantum-informational, it could be viewed as quantum-informational in spirit, and conversely, quantum information theory as Heisenbergian in spirit (Plotnitsky 2002, 2016, pp. 72–73). The reason for this view is that the quantum-mechanical situation, as Heisenberg conceived of it, was defined by:

(A) certain already obtained information, derived from spectral lines (due to the emission of radiation by the electron), observed in measuring instruments; and

(B) certain possible future information, to be obtainable from spectral lines to be observed in measuring instruments and, hopefully, predictable in probabilistic or statistical terms by the mathematical formalism of a quantum theory.

Heisenberg's aim was to develop such a formalism, without assuming that it needed to represent a spatiotemporal process connecting these two sets of information or how each comes about, a representation that Heisenberg thought of, echoing Bohr's assessment of quantum jumps (Bohr 1913, p. 15), as "scarcely possible" at the time (Heisenberg 1925, p. 265). This information is, in each case, determined by what type of experiment one decides to perform, rather than by arbitrarily selecting one or another pre-existing properties of physical reality. As became quickly apparent, the formalism entailed this aspect of the situation in view of the noncommutativity of the operators associated with the variables defining such alternative decisions, such the position Q and momentum P operators, and the corresponding equation, $PQ - QP \neq 0$, through which QM connects to the uncertainty relations.

Heisenberg's theory was, thus, dealing with quantum information defined by a particular structure of bits of classical information obtainable in measuring instruments, the structure physically described by classical physics, but not predictable by it. Because this structure was defined by the effects of the interactions between quantum objects and measuring instruments, the theory was still concerned with quantum objects or the reality thus idealized, although this reality was not represented by the theory and could be beyond representation or even conception. Heisenberg did not make definitive claims in this regard, and, as indicated in Chap. 2, he eventually adopted the view that this reality could be represented mathematically, without using physical concepts, at least as such concepts are understood in classical physics or relativity.

The correspondence principle, made into the mathematical correspondence postulate, motivated Heisenberg's decision to retain the equations of classical mechanics, while introducing different variables to enable correct predictions for all energy levels of electrons. The correspondence with classical theory could be maintained because new variables could be replaced by conventional classical variables in the classical limit, as in the case of large quantum numbers, when the electrons were far away from the nuclei and when classical concepts, including orbits, could apply. The electrons' behavior itself is still quantum and could have quantum effects. The old quantum theory was defined by the strategy of retaining, on realist lines, the variables of classical mechanics while adjusting the equations to achieve better predictions. Heisenberg's reversal of this strategy was unexpected, as was a radical change in the role of these equations: they no longer represented the motion of electrons, but served as mathematical means for probabilistic or statistical predictions concerning effects of the interaction between quantum objects and measuring instruments. Heisenberg was initially concerned with spectra, so the objects interacting with measuring instruments were photons assumed to be emitted by electrons.

Heisenberg's new variables were, unlike his equations, completely independent of any classical theory. They were infinite unbounded matrices with complex elements, although it is not clear that he initially thought of them as matrices, rather than, as he called them, (double indexed) "ensembles of quantities" (Heisenberg 1925, p. 263). Their multiplication, which Heisenberg, who was famously unaware of the existence of matrix algebra and reinvented it, had to define to use them in his equations, is in general not commutative. Such mathematical objects had never been used in physics previously. Technically, tensors of the second rank, used in relativity, are matrices, and, accordingly, in general they do not commute. This noncommutativity, however, does not play the same role in relativity, and it did not attract attention. It is only with QM that noncommutative mathematics becomes used physics in an essential way. In fact, while matrix algebra, in finite and infinite dimensions, was developed in mathematics by then, *unbounded* infinite matrices, used by Heisenberg and then by Born and Jordan, had not been previously studied. As became apparent later, such matrices are necessary to derive the uncertainty relations for continuous variables. The Hilbert-space formalism of QM was, as noted, introduced by von Neumann shortly thereafter. There are further details: for example, as unbounded self-adjoint operators, these matrices do not form an algebra with respect to the composition as a noncommutative product, although some of them satisfy the canonical commutation relation. These details are, however, secondary. Most crucial is that the concept was used in a radically new, essentially (weak) RWR-type, way. Heisenberg's variables were algebraic entities enabling probabilistic or statistical predictions concerning *quantum phenomena*, observed in measuring instruments, without providing a mathematically idealized representation of the spacetime behavior of *quantum objects* responsible for these phenomena. It was a new way of using and, given its purely abstract nature (rather than a representational relation to physical objects and their behavior), a new way of *inventing* mathematics in physics.

In his 1925 paper, Heisenberg began his derivation of QM with an observation that reflected a radical departure from the classical ideal of continuous mathematical representation of individual physical processes. He said: "in quantum theory it has not been possible to associate the electron with a point in space, *considered as a function of time*, by means of observable quantities. However, even in quantum theory it is possible to ascribe to an electron the emission of radiation" [the effect of which is observed in a measuring instrument] (Heisenberg 1925, p. 263; emphasis added). As explained in Chap. 2, referring to what happened between experiments, via a classical concept of emission, this statement would pose difficulty for the RWR view (Heisenberg 1962, pp. 178–179). These considerations were, however, to come later. Also, a measurement could associate an electron with a point in space, with QM capable of predicting the probability for finding its position in given area. But it is not possible to do so by linking this association to a function of time (as a real variable) representing the continuous motion of this electron, as in classical mechanics, which would then allow one to predict this position ideally exactly. Matrix mechanics did not offer a treatment of electrons in stationary states, only in which one could speak of the position of an electron in an atom. An instantly repeated measurement could give the value, *the same value*, of its position, which instant repetition is, however, an idealization (e.g., Schrödinger 1935a, pp. 158–159). Schrödinger's time-dependent equation made it possible to predict a (physical) quantum state, for any variable, such as position, momentum, or energy, including for electrons in stationary states. I here understand the situation in accordance with the RWR-type concept of a quantum state defined in Chap. 3 as manifested only in its effects observed in measuring instruments, rather than in the way Schrödinger initially conceived of this situation, as against his 1935 paper just cited.

In his paper, Heisenberg states next: "In order to characterize this radiation we first need the frequencies which appear as functions of two variables. In quantum theory these functions are

$$\nu(n, n-\alpha) = 1/h\{W(n) - W(n-\alpha)\}$$

and in classical theory in the form.

$\nu(n, \alpha) = a\nu(n) = a/h(dW/dn)$" (Heisenberg 1925, p. 263).

This difference, central to Bohr's (1913) theory, leads to the difference between classical and quantum theories as regards the combination relations for frequencies, which, in the quantum case, correspond to the Rydberg–Ritz combination rules, reflecting "the discrepancy between the calculated orbital frequency of the electrons and the frequency of the emitted radiation," which, as discussed in Chap. 3, was a key new feature of Bohr's theory, incompatible with classical electrodynamics (Heisenberg 1925, p. 263). However, "in order to complete the description of radiation [in conformity, by the correspondence principle, with the classical Fourier representation of motion] it is necessary to have not only frequencies but also the amplitudes" (Heisenberg 1925, p. 263). On the one hand, then, the equations of QM must formally contain amplitudes as well as frequencies. On the other hand, these amplitudes could no longer serve their classical physical function (as part of a continuous representation of motion) and were instead related to discrete transitions between stationary states. In Heisenberg's theory and in QM since then, these "amplitudes" are no longer amplitudes of physical motions, but are instead formal mathematical entities, "probability (density) amplitudes," linked, via Born's rule, to the probabilities of transitions between stationary states, manifested in the spectral data observed in quantum experiments. As explained above, in the language of Hilbert-space formalism, the probability amplitude is just $\langle \lambda_i | \psi \rangle$ (λ_i is an eigen value and ψ is the wave function) and Born's rule says that the corresponding probability is the square of the amplitude or is the amplitude multiplied by its own complex conjugate, or $\mathrm{P_i} = \langle \lambda_i | \psi \rangle^2$. This makes the term "amplitude" symbolic, as these amplitudes are not anything physical, as amplitudes would be in classical physics, say, in a Fourier representation, formally or, again, symbolically used by Heisenberg (Bohr 1987, v. 1, p. 18). In commenting on linear superposition in quantum mechanics in his classic book, Dirac emphasized this difference: "*the superposition that occurs in quantum mechanics is of an essentially different nature from any occurring in the classical theory*" (Dirac 1958, p. 14). In RWR-type interpretations, this superposition is not physical: it is only mathematical. In classical physics the mathematics of (wave) superpositions represent physical processes; in QM, at least in RWR-type interpretations, it does not. The corresponding probabilities were derived by Heisenberg by a form of Born's rule, which he postulates for this limited case.[3] Heisenberg then argues as follows:

> The amplitudes may be treated as complex vectors, each determined by six independent components, and they determine both the polarization and the phase. As the amplitudes are also functions of the two variables n and α, the corresponding part of the radiation is given by the following expressions:
>
> Quantum-theoretical:
>
> $$\mathrm{Re}\left\{A(n, n-\alpha)e^{i\omega(n,n-\alpha)t}\right\}$$
>
> Classical:
>
> $$\mathrm{Re}\left\{A_\alpha(n)e^{i\omega(n)\alpha t}\right\}$$
>
> (Heisenberg 1925, p. 263).

The problem—a difficult and, "at first sight," even insurmountable problem—is that "the phase contained in A would seem to be devoid of physical significance in quantum theory, since frequencies are in general not commensurable with their harmonics" and, as a result, "a geometrical interpretation of such quantum-theoretical phase relations in analogy with those of classical theory seems at present scarcely possible" (Heisenberg 1925, pp. 263–265). If one wants to have a geometry here, a new, quantum-theoretical, form of geometry, the geometry of Hilbert spaces over $\mathbb{C}$ becomes necessary. This geometry is, however, no longer analogous to the geometry used in classical theory (or relativity) insofar as its aim now is to predict the probabilities of the outcomes of quantum experiments and not to represent how these outcomes come about. As discussed in Chap. 3, this incommensurability, which is in irreconcilable conflict with classical electrodynamics, was one of the most radical features of Bohr's (1913) atomic theory, on which Heisenberg builds. His strategy, too, is based, just as Bohr's was, on the shift from calculating the probability of finding a moving electron in a given state to calculating the probability of an electron's transition from one state to another, without describing the physical mechanism responsible for this transition. Heisenberg's theory is more in harmony with this approach because there are no longer orbits, where the classical approach would still apply.

Heisenberg says next: "However, we shall see presently that also in quantum theory the phase has a definitive significance which is *analogous* to its significance in classical theory" (Heisenberg 1925, p. 264; emphasis added). "Analogous" could only mean here that, rather than being analogous physically, the way the phase enters

---

[3] Born begins his paper with a reference to Heisenberg, but notes that a version of his rule "has so far been applied exclusively to the calculation of stationary states and vibration amplitudes associated with transitions [between such states]" (Born 1926a, p. 863). Born extends the idea to all QM—the discovery that justly, although belatedly (in 1954), brought him a Nobel Prize. The right formula itself (Born's first paper on the subject (Born 1926a) did not use square moduli) famously occurred in a footnote (Born 1925, p. 865, note). This extension was crucial. It gave the probabilistic character of the formalism a much greater generality and transformed, to Schrödinger's chagrin, our understanding to the wave function.

mathematically is analogous to the way the classical phase enters mathematically in classical theory, in accordance with the *mathematical* form of the correspondence principle, now the correspondence postulate, insofar as quantum-mechanical equations are formally the same as those of classical physics. Heisenberg only considered a toy model of an anharmonic quantum oscillator, and thus needed only a Newtonian equation for it, rather than the Hamiltonian equations required for a full-fledged theory, developed by Born and Jordan (Born and Jordan 1925; Born et al. 1926). As Heisenberg explains, if one considers a given quantity *x(t)* [a co-ordinate as a function of time] in classical theory, this can be regarded as represented by a set of quantities of the form

$$A_\alpha(n)e^{i\omega(n)\alpha t},$$

which, depending on whether the motion is periodic or not, can be combined into a sum or integral which represents *x(t)*:

$$x(n,t) = \sum_{-\infty}^{+\infty} A_\alpha(n)e^{i\omega(n)\alpha t}$$

or.

$$x(n,t) = \int_{-\infty}^{+\infty} A_\alpha(n)e^{i\omega(n)\alpha t}d\alpha$$

(Heisenberg 1925, p. 264).

Heisenberg next makes his most decisive and most extraordinary move. He notes that "a similar combination of the corresponding quantum-theoretical quantities seems to be impossible in a unique manner and therefore not meaningful, in view of the equal weight of the variables n and $n - \alpha$." However, he says, "one might readily regard the ensemble of quantities $A(n, n - \alpha)e^{i\omega(n,\, n-\alpha)t}$ [an infinite square matrix] as a representation of the quantity *x(t)*" (Heisenberg 1925, p. 264). The arrangement of the data into these ensembles, in effect square tables (although it is not clear that Heisenberg thought about them in this way, unlike Born, who on reading Heisenberg's paper, realized that they were matrices), was a remarkable way to handle the transitions between stationary states. In retrospect, once one deals with the transitions between stationary states, matrices appear naturally, with rows and columns linked to these states, respectively. This naturalness, however, became apparent, or one might say, became natural, only in retrospect.

At the time, it was an extraordinary guess, and a great conceptual invention, not unmotivated, but little prepared by the preceding history of physics. However, it does not by itself establish an *algebra* of these arrangements, for which one needs to find rigorous rules for adding and multiplying these elements. Otherwise, Heisenberg cannot use these variables in the equations of his new mechanics. For a quantum-theoretical version of the classical equation of motion considered, which would apply (no longer as an equation of motion) to these variables, Heisenberg needed to construct the powers of such quantities, beginning with $x(t)^2$, which was actually all he needed. The answer in classical theory is obvious and, for the reasons just explained, obviously unworkable in quantum theory, where, Heisenberg proposed, "it seems that the simplest and most natural assumption would be to replace classical [Fourier] equations … by

$$B(n, n-\beta)e^{i\xi(n,n-\beta)t} = \sum_{-\infty}^{+\infty} \alpha A(n, n-\alpha)A(n-\alpha, n-\beta)e^{i\omega(n,n-\beta)t}$$

or

$$= \int_{-\infty}^{+\infty} A(n, n-\alpha)A(n-\alpha, n-\beta)e^{i\omega(n,n-\beta)}d\alpha\text{''}$$ (Heisenberg 1925, p. 265).

This is the main mathematical postulate, the (matrix) multiplication postulate, of Heisenberg's theory, "an almost necessary consequence of the frequency combination rules" (Heisenberg 1925, p. 265). Although it is commutative in the case of x^2, this multiplication is in general noncommutative, expressly for position and momentum variables, and Heisenberg, without quite realizing it, used this noncommutativity in solving his equation, as Dirac was the first to notice (Mehra and Rechenberg 2001, v. 4, p. 129). Heisenberg spoke of his new algebra of matrices as the "new kinematics." This was not the best choice of term because his new variables were no related to motion as the term kinematic would suggest, one of many, historically understandable, but potentially confusing terms. Planck's constant, *h*, which is a dimensional, dynamic entity, has played no role thus far. Technically, as Einstein was to lament later, the theory wasn't even a mechanics: it did not offer a representation of individual quantum processes, but only predicted, probabilistically or statistically, what is observed in measuring instruments. This assessment may, however, depend on how one understands mechanics as a mathematical-experimental theory. Bohr, accordingly, spoke of Heisenberg's discovery as inaugurating "a new epoch of mutual stimulation of mathematics and mechanics" (Bohr 1987, v. 1, p. 51).

That in general his new variables did not commute, $PQ–QP \neq 0$, was, again, an especially novel feature of Heisenberg's theory, defining its algebraic structure. This feature proved to be momentous physically. Most famously, it came to represent the uncertainty relations constraining certain simultaneous measurements, such as those of the momentum (*P*) and the co-ordinate (*Q*), associated with a given quantum object in the mathematical formalism of QM and (correlatively) the complementary character of such measurements. Their noncommutative nature was, initially, off-putting for some, including Heisenberg himself and Pauli, although not for Dirac, who immediately realized its centrality, nor for Born and Jordan (Plotnitsky 2009, pp. 90, 111, 116). In retrospect, given the nature of the situation to which Heisenberg's new mechanics responded, this noncommutativity is not surprising. While one should, again, be cautious about a retrospective sense of things, it is worth considering the physical reasons for this noncommutativity from a later vantage, that of Schwinger's comments in his unpublished lecture, cited at length in Schweber (1994). Schwinger notes a commonly stated physical feature corresponding to this noncommutativity: if one measures two physical properties in one order, and then in the other, the outcome would in general be different. But his aim is to explain the reasons for why this is the case, which is why his comments are of interest here. He says:

> Once we recognize that the act of measurement introduces in the [quantum] object of measurement changes which are not arbitrarily small, and which cannot be precisely controlled ... then every time we make a measurement, we introduce a new physical situation and we can no longer be sure that the new physical situation corresponds to the same physical properties which we had obtained by an earlier measurement. In other words, if you measure two physical properties in one order, and then the other, which classically would absolutely make no difference, these in the microscopic realm are simply two different experiments ...
>
> So, therefore, the mathematical scheme can certainly not be the assignment, the association, or the representation of physical properties by numbers because numbers do not have this property of depending upon the order in which the measurements are carried out. ... We must instead look for a new mathematical scheme in which the order of performance of physical operations is represented by an order of performance of mathematical operations. (Cited in Schweber 1994, p. 361)

The last sentence is not entirely precise: *mathematical* operations, including multiplication (commutative or not), upon any quantities or symbols, are not *physical* measurements, which are, on the other hand, not mathematical, at least not inherently. This is an important point, of which Schwinger was undoubtedly aware. I shall return to it presently. The passage contains echoes of Bohr's writings, especially in invoking "changes that cannot be controlled." As Bohr said in the Como lecture: "It must not be forgotten... that in the classical theories any succeeding observation permits a prediction of future events with ever increasing accuracy, because it improves our knowledge of the initial state of the system. According to the quantum theory, just the impossibility of neglecting the interaction with the agency of measurement means that every observation introduces a new uncontrollable element" (Bohr 1987, v. 1, p. 68). Classically, one can continue to perform measurements of both the position and the momentum of an object at any point along its continuous and (classically) causally determined trajectory. This is not possible in quantum measurements, even if one assumes that such a trajectory is possible for a quantum object (as for example, one does in Bohmian mechanics). Heisenberg made the same point in his uncertainty relations paper and elsewhere (Heisenberg 1927, p. 36; 1930, pp. 66, 72–77). So did Schrödinger in his cat paradox paper (Schrödinger 1935a, pp. 152, 54, 57–158). The role of "the finite [and hence never arbitrarily small] and uncontrollable interactions between the object and the measuring instruments" was emphasized by Bohr in his reply to EPR (Bohr 1935, pp. 697, 700).

Schwinger's reasoning was not the same as that of Heisenberg in his discovery of QM. The considerations invoked by Schwinger emerged later. Indeed, the type of thinking described by Schwinger is more in accord with quantum-informational approaches to deriving quantum theory, in finite dimensions, from the (formalized) structure of quantum measurements, as discussed in Chap. 9. The noncommutativity of his matrix variables was not Heisenberg's starting point but a consequence of the multiplication rule for his matrices, which he needed to establish to be able to use them in his equations, formally borrowed from classical physics, by the correspondence principle or, again, in this case the correspondence postulate. Thus, the structure of his mathematics scheme was partly borrowed from classical physics and partly invented by finding the variables needed and constructing their algebra. That Heisenberg did not know about the existence of matrix algebra and reinvented it is a testimony to his mathematical creativity. However, even if he knew about it, he would still have had to invent how to relate this algebra to the probabilities or statistics of quantum predictions, in the absence, assumed by him, of a representation of the behavior of quantum objects responsible for quantum phenomena.

The main difficulty, shown by Schwinger's comments, is that, given this situation, any mathematical scheme defining the formalism of quantum theory requires mathematics that does not appear to be representationally connected, in the way it would be in classical physics or relativity, to the "measurement algebra," as Schwinger came to call it (Schwinger 2001; Jaeger 2016). Consider Schwinger's statement: "The order of performance of physical operations is *represented* by an order of performance of mathematical operations." But, if one follows Bohr, as Schwinger appears to do, these physical operations are measurements, with their outcomes, manifested in measuring instruments, measurements *predicted* by means of these mathematical operations, rather than represented by then in the way they would be in classical physics. Schwinger's argument, which, again, follows that of Bohr, is in accord with the view of quantum measurement adopted here: a quantum measurement is not a measurement of some pre-existing quantity of the quantum object considered but an establishment of a new quantum phenomenon, an entirely "new physical situation," as Schwinger says. In the present view, moreover, even the quantum object considered is defined by this measurement. Accordingly, the term "represented" (perhaps only loosely used by Schwinger) is not strictly accurate insofar as there is no homomorphic, let alone isomorphic, mapping from the algebra of QM, "the new mathematical scheme," to this measurement algebra. One needs additional elements of structure to arrive at this scheme and to relate it to measurements and their structure, "algebra," to the degree that they form an algebra, a point I shall address presently.

The assessment just given reflects a very different understanding of the structure of quantum measurements from that of assuming that the noncommutative nature of the multiplication of the operator variables involved, $PQ - QP \neq 0$, *represents* the order of performance of physical operations. Schwinger might have agreed with this understanding (and as I said, he might have used the term "representation" loosely in this lecture, never published by him), given his affinities with Bohr, who would virtually certainly have agreed with this view. This understanding arises from the key premise of Schwinger's or, again, Bohr's argument: "Once we recognize that the act of measurement introduces in the [quantum] object of measurement changes which are not arbitrarily small, and which cannot be precisely controlled ... then every time we make a measurement, we introduced a new physical situation and we can no longer be sure that the new physical situation corresponds to the same physical properties which we had obtained by an earlier measurement." By the QI postulate, each quantum measurement is a unique and unrepeatable event defining a new physical situation. As such, each new measurement, M_2 at a later moment in time t_2, even of the same variable, unavoidably, in Schrödinger's terms, requires a new expectation-catalog, enabled by QM, cum Born's rule, and the data obtained in this measurement. This new measurement, as a new unique event, even if one measures the same variable, say, the position, makes the previous expectation-catalog, defined by previous measurement, M_1, at time t_1 (which could have been used for predicting possible outcomes of M_2) meaningless as concerns possible predictions after M_2 is made.

If one considers the case of the measurement of two complementarity variables, say, the position and the momentum, associated with a quantum object, the case, in which the noncommutative algebra of the quantum-mechanical formalism comes into play, the situation is as follows. These measurements, observed in the measuring instruments used, are mutually exclusive as concerns the possibility of performing both simultaneously, by virtue of the uncertainty relations, which are experimentally established laws independent of any theory, reflected in the corresponding complementary quantum phenomena. If the first measurement, M_1, at time t_1, is that of a co-ordinate (observed in a measuring instruments), one can make predictions, by using QM cum Born's rule, concerning the probability or statistics of a future position measurement at any future moment in time, t_n, but can make no predictions whatsoever concerning future momentum measurements at any future moment in time. But if we then make the momentum measurement (in the same direction), say, M_2, at time t_2, our earlier predictions concerning the position measurement become meaningless as no longer verifiable, while this new measurement precludes making any predictions concerning future position measurements. One can only make predictions concerning future momentum measurements, concerning the value of the momentum with a given probability within a certain range.

The noncommutativity of the corresponding operator variables, Q and P, is just part of the mathematics that enables either prediction, rather than *representing* this situation of measurement, or this "algebra of measurement," to the degree that one could speak of algebra here, rather than of the "structure" of measurement, defined by this situation or by quantum measurement in general. It is true that, if in the experiment with the initial preparation of measuring instruments at time t_{01}, one makes first the position measurement, M_{1Q}, at time t_{11} and then the momentum measurement, M_{2P} at time t_{21} and then, with the same initial preparation of measuring instruments at time t_{02} (which preparation is possible because we can control the instruments classically) reverse the order of the quantities we measure, by first measuring the momentum, M_{1P}, at time t_{12} and then the position at time t_{22}, M_{2Q}, the outcome will be different. The double indexing of time is necessary because each set of measurements happens at a different set of time intervals and in fact requires a different quantum object, a point that as will be seen in Chap. 7, becomes important in considering the EPR experiment and Bohr's reply to EPR. This situation is the physical meaning of Schwinger's appeal to two different situations of measurements, the meaning defined by the fact that *each situation consists of two separate measurements*. One is dealing with two pairs of different physical entities, (M_{1Q}, M_{2P}) and (M_{1Q}, M_{2P}) observed in measuring instruments, and not with performing a reverse ("algebraic") operation with different outcomes on the same two entities, M_1 and M_2, as one would mathematically in noncommutative algebra, with $M_1 M_2 \neq M_2 M_1$. Of course, $PQ - QP \neq 0$, for all complementary variables considered, but this mathematical fact only pertains to the formalism of QM, enabling probabilities and statistics of predictions, correctly estimated by QM, by using these *Ps* and *Qs* and not to measurements. These variables are, at least in the RWR view, independent of the physical quantities measured, while the latter are, conversely, independent of any specific theory, as are, again, the uncertainty relations.[4] This independence is key to Bohr's interpretation, especially in its ultimate version as a strong RWR-type interpretation, and as will be seen in Chap. 6, missing this independence can lead to misunderstanding Bohr's argumentation. As, however, I have argued from the outset of this study, this independence was part of the new way in which mathematics was used by Heisenberg, thus transforming the nature of theoretical physics.

4.4 How Algebraic is the Heisenberg Algebraic Method?

The preceding discussion suggests that, whether one accepts or not Einstein's skeptical attitude toward "the Heisenberg method" and, by the time of Einstein's (1936) assessment, to QM, exploring his characterization of both as algebraic helps one better to understand Heisenberg's thinking, leading him to his discovery of QM and QM itself. According to Einstein:

> [P]erhaps the success of the Heisenberg method points to a purely algebraic method of description of nature, that is, to the elimination of continuous functions from physics. Then, however, we must give up, in principle, the space–time continuum [at the ultimate level of reality]. It is not unimaginable that human ingenuity will some day find methods which will make it possible to proceed along such a path. At present however, such a program looks like an attempt to breathe in empty space. (Einstein 1936, p. 378)

4 Because the interference of measurement cannot, in principle, be neglected in considering quantum phenomena, some of these considerations would still apply even if one assumes, as some interpretations of QM or alternative theories, such as Bohmian mechanics, do, that quantum objects do possess such properties as position and momentum, independently. As stated from the outset of this study, in classical physics the interference of measurement may be assumed to be "arbitrarily small," both the position and the momentum can always be measured and defined simultaneously at any given point and predicted ideally exactly at any future point. This is not so in Bohmian mechanics, even though it is realist and classically causal in its representation of quantum objects and behavior, because it is grounded in the assumption that a measurement actually disturbs the objects and, as a result, changes the pre-existing position or the momentum of this object, which is why the uncertainty relations still apply and the predictions of Bohmian mechanics (an expressly Einstein-nonlocal theory) are the same as those of QM. Moreover, in classical mechanics it is, in principle, possible (for individual or small system) to repeat a given physical situation exactly with the same outcome of any measurement or sequence of measurements in any given order, which is, in general, not possible in quantum physics.

For Heisenberg, or for Bohr, this method, once introduced by Heisenberg, was more like a breath of fresh mountain air. Einstein, as noted, admitted, on the same occasion "that there is no doubt that quantum mechanics has seized hold of a beautiful element of truth and that it will be a touchstone for a future theoretical basis in that it must be deducible as a limiting case from that basis." But he also reiterated, in accord with the assessment of the Heisenberg method just cited, that he did "not believe that quantum mechanics will be the starting point in the search for this basis, just as one cannot arrive at the foundations of mechanics from thermodynamics or statistical mechanics" (Einstein 1936, p. 361). Earlier, he referred to QM as a magical trick, "Jacob's pillow" of Göttingen and Copenhagen: it was not "the real thing" and did "not really bring us any closer to the secret of the 'old one,'" who, Einstein, equally unhappy with the recourse to probability in dealing even with the simplest possible systems, added in his famous pronouncement, "at any rate is... not playing at dice" (Born 2005, p. 88). Einstein meant nature, but using God helped to immortalize his statement, one especially famous but still one of those of rhetorical gestures common to Einstein, like that describing QM above, as "an attempt to breathe in empty space." Einstein was, again, most concerned with the absence of realism in considering the ultimate constitution of nature, while realizing that the irreducible recourse to probability would be automatic as a result. By 1936, Einstein's hopes for Schrödinger's wave mechanics as a more geometrical alternative have been long abandoned. Accordingly, the proper basis for his "search for a more complete conception," which he could not "forego," had to be elsewhere (Einstein 1936, pp. 361, 375).

He might have been right as concerns the type of theory he wanted, a more geometrical theory on the model of Maxwell's electrodynamics, as a classical field theory, and then his own general relativity, grounded in Riemannian geometry, which gave a realist mathematical idealization of gravity. Such a theory is unlikely to emerge from QM or QFT. Neither of these two theories is, however, "a *description* of nature," because, at least in the RWR view, neither offers a representation of the ultimate constitution of nature, which was Einstein's imperative for a fundamental theory. Accordingly, Einstein would not have viewed, and had not viewed, QM or QFT as a proper "method" of the *description* of nature, and one can understand why he did not believe that either theory offered a "useful point of departure for future development," which he associated with such a description (Einstein 1949, p. 83).

It is not coincidental either that Einstein's (1936) paper was published in the immediate wake of the EPR paper, which argued for the incompleteness, or else nonlocality of QM, and Bohr's reply, which, as discussed in Chap. 7, Einstein misread as allowing for nonlocality (as an action at distance). EPR's argument and related arguments by Einstein gave new impetus to his vision. This vision and the hope for its eventual fulfillment, the Einsteinian hope, have inspired and guided many physicists and philosophers and still do even in the face of the fact that QM and QFT have remained our standard theories of quantum phenomena in the corresponding energy regimes. One the other hand, we still don't have a quantum theory of gravity, which is one of the reasons (albeit far from the only one) that keeps alive *the Einsteinian hope*, against *the Heisenberg method*, with general relativity as the main theory supporting this hope.

It is worth reiterating that, Einstein, who was, as discussed in Chap. 2, no naïve realist, was aware that, just as general relativity was, a theory of the kind he wanted would be a (suitably mathematized) *conceptual* idealization. In his view, however, a fundamental theory would not be possible apart from such concepts, or in his language "a free conceptual construction," which defined his realism as conceptual (Einstein 1949, p. 47). The main problem of the Heisenberg method for Einstein was that it only provided the *algebra* of probabilities for the outcomes of quantum experiments, without a proper realist idealization, essentially geometrical in character (it contains algebra, such as that of tensor calculus, as well), provided by general relativity. Or so it appeared before singularities entered the theory within merely a year from the discovery of the theory with Karl Schwarzschild's 1916 work, although it took much longer to accept them as a permanent feature of the theory. Einstein, ever a thinker of continuity, never quite reconciled to the idea of singularities, and thought they should ultimately be avoided. Would he have been convinced by the current evidence and Roger Penrose's (1965) proof (a decade after Einstein's death in 1955), for which Penrose was awarded the 2020 Nobel prize in physics, that singularities of black holes were a consequence of general relativity, without any esoteric assumptions made in previous demonstrations (Penrose 1965)? It took nearly another half a century to show their actual existence in nature. Penrose shared the prize with Andrea Ghez and Reinhard Genzel, who experimentally confirmed the existence of a massive black hole (or what is generally assumed to be one) at the center of the Milky Way. It is, again, difficult to know, assuming that such a theory is possible at all, how far we are from quantum gravity, which, as a quantum theory, could be the ultimate triumph of the Heisenberg method, or whether a theory bringing gravity into harmony with other fundamental forces will be conversely more akin to general relativity, and thus a fulfilment of the Einsteinian hope, or be something else altogether.

All our physical theories, however, from Kepler and Galileo on, have been defined by the interplay of algebra and geometry, equally found in general relativity and QM or QFT. Accordingly, while Einstein's assessment of the Heisenberg method and QM is not out of place, it requires a more careful examination, giving it a greater complexity than Einstein's brief statement conveys, a complexity that also brings geometry back into the Heisenberg method and QM or QFT. The Heisenberg method might have been fundamentally algebraic, and, as an RWR-type method, it excluded realism, even though it could not preclude realism as in principle possible by means of other theories or by means of alternative interpretations of QM. This method, however, did not exclude geometry. Instead, it brought with it a new way of geometrical thinking and new form of geometry, that of Hilbert spaces, of finite and infinite dimensions (over $\mathbb{C}$), to fundamental physics. By doing so, it led to a new synergy of algebra and geometry in physics, a synergy by then already in place in mathematics itself.

Some qualifications of Einstein's assessment are immediately necessary in view of the basic *mathematical* structure of QM or QFT. Thus, in saying that "we must give up, in principle, the space–time continuum," Einstein must have had in mind the spacetime continuum in *representing* the ultimate character of physical reality by a geometrical theory, such as Maxwell's electrodynamics, as a classical field theory, and then general relativity, grounded in Riemannian geometry of differential manifolds. Einstein was of course aware that the idea that this character may be discrete had been around for long time by then. As discussed in the Introduction, it was, in particular, proposed as a possibility by Riemann as early as 1854 in his Habilitation lecture, which Einstein new well. Riemann brought it up, on Kantian lines, in connection with the difference between our continuous phenomenal representation of physical space and "the reality underlying space," which may be discrete. A subtler part of Riemann's argument, which still has significance for fundamental physics, such as quantum gravity, concerned the difference between the discrete and continuous nature of the reality underlying space in the infinitely small. If it is discrete then the ground of the metric relation of the corresponding manifold is given by the *mathematical* concept of this manifold itself. On the other hand, if it is continuous, this ground is given by the concept of phenomena justified by experience and defined by the physical principles arising from these phenomena, principles that may be modified or changed in view of new evidence. Einstein's general relativity is a manifestation of the second case, both as an example of this grounding in general and as an example of a change in these physical principles vis-à-vis those of Newton's theory of gravity.

In the RWR view, because the ultimate constitution of physical reality responsible for quantum phenomena is beyond representation and even conception, this constitution may not be seen as either continuous or discrete, any more than either spatial or temporal. Discreteness only pertains to quantum phenomena, observed in measuring instruments, insofar as each quantum phenomena is assumed to be singular (by the QI principle) and discrete in relation to any other quantum phenomena (by the QD principle). On the other hand, each phenomenon is perceived and described classically in continuous space and time as they are considered in classical physics or relativity. Accordingly, the spacetime continuum is retained at this level. Continuity is also a mathematical feature of the formalism of QM, which relates to *discrete phenomena* by predicting the probabilities or statistics of their occurrence. RWR-type interpretations, again, strictly maintain the difference between the discreteness of quantum phenomena, defined by what is observed in measuring instruments, and the discreteness of quantum objects, given that, in these interpretations, quantum objects cannot be described or even thought of as discrete, particle-like, entities, any more than continuous, wave-like, entities.

This situation gives a greater complexity and, with it, geometry to "the algebraic method" of QM or QFT. It is true that the continuous functions (over $\mathbb{R}$) used in classical physics or relativity (for variables such as position or momentum) are replaced by operators and their algebra in Hilbert spaces over $\mathbb{C}$. As explained in more detail below, however, Hilbert spaces are geometrical concepts, rigorously defined by what I call spatial algebra, and the distance (or norm), which this algebra establishes, is essential to the use of Hilbert spaces in QM. Hilbert spaces give one the possibility to *think geometrically*, by using both the rigor of spatial algebra and our phenomenal geometrical intuition as heuristic help and guidance, as discussed in detail below.

QFT may deal with Hilbert spaces whose continuity is denser than that of regular continua such as the (real number) spacetime continuum of classical physics or relativity. Besides, continuous functions are retained, because these Hilbert spaces are those of continuous functions, considered as infinite-dimensional vectors in dealing with continuous variables such as position and momentum, although these variables themselves are represented by operators. The nature of these functions (as those of complex variables) and their role in QM is of course different from those of classical physics and relativity. Born and Jordan developed a differential calculus, "symbolic differentiation," as they called it, for matrices used in quantum mechanics (Bohr and Jordan 1925, p. 862; Mehra and Rechenberg 2001, v. 3, p. 69). So did Dirac in his first paper on quantum mechanics (Dirac 1925). In the style of Leibniz, this differentiation was defined algebraically by using the noncommutation rules. It enables one to retain the differential equations of classical mechanics and their accompanying machinery, such as and in particular the Poisson bracket, while using new quantum variables, as Hilbert space operators. The quantum-mechanical analogue of the Poisson bracket is the expression.

$$\frac{2\pi i}{h}(pq - pq),$$

as Dirac was first to realize. Dirac's starting point, again, in the style of Leibniz, was the quantum-mechanical analogue of the rule for the differential of the product of two functions, which may be seen as a linear operator and which may be suitably quantized (Dirac 1925).

These functions, vectors, are those of complex (rather than, as in classical physics, real) variables and the vector spaces that they comprise or associated objects, such as operator algebras, have special properties, such as noncommutativity. The most famous example is Schrödinger's wave function, ψ, which proved to correspond to no physical waves, as Schrödinger initially hoped it would (hence, its name wave function), but only to a distribution of probability densities, which may be seen as wave-like in its discrete pattern. Born, in his initial probabilistic interpretation of the wave function states the case as follows: "[T]he motion of particles follows the probability law but the probability itself propagates [in a wave-like manner] according to the law of [classical] causality" (Born 1926b, p. 804; 1949, p. 103). First, if one adopts Heisenberg's or already Bohr's (1913) view, the probability law in question concerns not the motion of particles but the discrete transitions between quantum states associated with particles. Born adopted this view in his earlier papers on matrix mechanics (Born and Jordan 1925; Born et al. 1926), and in his paper on the probabilistic interpretation of the wave function. So, the appeal to the motion of particles can be misleading. The second part of Born's formulation was important, even though it was somewhat vague as well, forgivably, as we are in 1926. One cannot speak of a propagation of probability otherwise than metaphorically. All one can claim is a wave-like, but discrete pattern, to the distribution of probabilities. Probabilities do not propagate: they are assigned by us, as human agents, to the outcomes of certain possible future experiments on the basis of certain other previously performed experiments and the use of QM, and they discontinuously change with each new measurement. In Bohr's or the present view what Born refers to here as the law of causality is not the law of classical causality but the law according to which these probabilities, as discrete "expectation-catalogs," are given by the wave function cum Born's rule. From this viewpoint, one can discard the idea of waves in quantum theory altogether. There are no waves, but only, in certain specified cases, correlated patterns predictable by QM.

4.5 Geometry and Algebra in Modernist Mathematics and Physics

Rather than being merely algebraic, then, as Einstein saw it, "the Heisenberg method" exhibits complex relationships between algebra and geometry, which is not surprising given the mathematics he used, some of which he, again, reinvented, unaware of its existence. The complexity and sometimes tensions accompanying these relationships can, however, be traced all the ways to the ancient Greeks, beginning at least with the Pythagoreans' discovery of incommensurable entities, such as the diagonal and the side of the square, a discovery that occurred at the intersection of arithmetic and geometry. It would not be possible to trace this longer history here, in particular, in the history of classical physics, contemporaneous with the rise of algebra. Thus, Descartes's analytic geometry was one of the defining events in the history of these relationships in mathematics and physics, and interplay of algebra and geometry in Kepler's, Galileo's, and Newton's theories was equally exemplary of this interplay.

I shall only consider, as especially pertinent to this chapter, a more recent history of these relationships in *modernist mathematics*, which characterizes most of the mathematics of the twentieth and twenty-first century. The development of this mathematics was in part and even most fundamentally defined by its divorce from physics, specifically classical physics (as there was no other at the onset of this development), and its move toward its independent and, in this sense, abstract nature. This mathematics was, nevertheless, adopted, including in its newly found abstractness, by relativity and, in an especially new way (allowing for the RWR view), quantum theory, and was then reciprocally shaped by this new physics in geometry, functional analysis, group theory, and other fields.

"Modernist" is not a common term in referring to mathematics or science. It has been standard in designating an aesthetic category, describing certain artistic developments of the first half of the twentieth century, represented by such figures as James Joyce, Franz Kafka, and Virginia Woolf, in literature; Pablo Picasso, Wassily Kandinsky, and Paul Klee, in art; and Arnold Schoenberg, Anton Webern, and Igor Stravinsky in music. On occasion, although not commonly, it has been applied to the philosophy of the same period, such as that of Friedrich Nietzsche, Edmund Husserl, and Martin Heidegger. Contemporary mathematics and physics, specifically relativity and quantum theory, or earlier theories, such as thermodynamics, had a significant impact on modernist literature and art, and (although not quite as extensively) vice versa. The subject would require a separate treatment. In considering mathematics and science, "modern," has been used more commonly, but usually with a different periodization. In mathematics, "modern" tends to refer to the mathematics that had emerged in the nineteenth century, with such figures as Karl Friedrich Gauss, Niels Henrik Abel, Augustine-Louis Cauchy, and Évariste Galois, while in physics it refers to all mathematical-experimental physics, from Galileo and Descartes on.

This characterization is in accord with an understanding of modernity as a broad cultural category, although still linked to the history of mathematics and science. Modernity refers to the period of Western culture extending from about the sixteenth century to our own time: we are still modern, although during the last fifty years or so, modernity entered a new stage, postmodernity, shaped by the rise of digital information technology.[5] Modernity is defined by several interrelated transformations, sometimes known as revolutions, although each took a while. Among them are scientific (defined by the new cosmological thinking, beginning with the Copernican heliocentric view of the Solar system, and the introduction of physics as a mathematical-experimental science by Kepler, Descartes, Galileo, and Newton); industrial or, more broadly, technological (defined by the transition to the primary role of machines in industrial production and beyond); philosophical-psychological (defined by the rise of the concept of the individual human self, beginning with Descartes's concept of the Cogito); economic (defined by the rise of capitalism); and political (defined by the rise of Western democracies). One might add to this standard list the mathematical revolution, rarely discussed as such, although sometimes considered as part of the scientific revolution.

Although algebra was crucial to modern mathematics and physics, geometry remained dominant for a long time and has never lost its significance. Thus, while the laws of classical mechanics, embodied in its equations, are algebraic, they are grounded in a geometrical picture of the world, including the motion of bodies, such as, paradigmatically, planets moving around the Sun. Modern geometrical thinking had continued to define physics, including relativity (although it has modernist aspects), until QM, in which algebra came to play a dominant role. After the discovery of relativity and quantum theory, the term "classical physics" was adopted for the preceding physics, still considered modern by virtue of its mathematical-experimental character. I also adopt the designation "modern" for the mathematics emerging at the same time. If modernity is scientific, it is because it is also mathematical.

Modernist mathematics and physics are, by contrast, essentially twentieth-century developments, extending modern mathematics and science in new directions. Arguably the most prominent recent example of using the term modernism in referring to mathematics is Jeremy Gray's *Plato's Ghost: The Modernist Transformation*

---

[5] Some postmodernist philosophical thinking was, however, shaped by modernist developments in mathematics and science and their epistemological underpinnings (e.g., Lyotard 1984).

of Mathematics (Gray 2008). The book covers developments in mathematics (such as topology, set theory, abstract algebra, and rethinking the foundations of geometry) that had reached their *modernist* stage around 1900. Gray characterizes modernist mathematics most essentially by movement toward the independence of mathematics from physics and, thus, from relation to nature, *physis*, a feature difficult to deny in mathematical modernism. By contrast, new (modernist) physics, relativity and quantum theory, that was developed during the same period not only never divorced itself from mathematics, but also, ironically, depended on mathematics, such as that of tensor calculus in relativity and Hilbert spaces in QM, that emerged as a result of this divorce. Gray focuses primarily on calculus and geometry, with Hilbert's *Foundations of Geometry* (Hilbert 1999) as, arguably, the conceptual center of his argument, and on post-Cantorian foundations of mathematics, which had a major impact on the mathematics of continuity and analysis, areas where Hilbert, again, was a key figure. Gray gives far less attention to algebra, including its role in modernist geometry and topology, and, as a result, what I see as the Pythagorean aspects of modernism (Plotnitsky 2020). He bypasses Hilbert's invention of a Hilbert space (even though, as noted, Hilbert did not use the term, introduced by von Neumann), a return of geometry, in its new incarnation, to analysis and then QM and QFT. Hilbert, who also made major mathematical contributions to general relativity, becomes a uniquely important figure in this history. His student Weyl is the only one who comes close, given his major contributions to all fields just mentioned. It is hardly coincidental that Hilbert develops the concept of Hilbert space, especially in its geometrical aspects, on the heels of his work on the foundations of geometry, which helps one to properly think about this concept or its predecessors, such as vector spaces, and conversely, adds new dimensions to our understanding of geometry. This reciprocity was not lost on von Neumann, also in connection with QM, which he formalized in terms of Hilbert space in the late 1920s, thus thinking about it geometrically.

I shall understand modernist mathematics by the new complexity of relationships between geometry and algebra in it, along with the trend toward independence of mathematics from its connections to the representation of natural objects and thus physics. On the other hand, modernist physics, such as relativity and QM, can adopt this mathematics and thus make it part of physics. This view does not exhaust the development of mathematics during that period, not all of which is modernist by any definition I am familiar with.

While realizing that mathematics itself or its basic fields cannot be given a single definition, I shall merely state an understanding of these fields, which is sufficient for my purposes and is, I would contend, in accord with how they are generally viewed. I understand algebra as the mathematical formalization of the relationships between symbols, arithmetic as dealing specifically with numbers, geometry as the mathematical formalization of spatiality, especially in terms of measurement, and topology as the mathematization of the structure of spatial or spatial-like objects apart from measurements, through their continuity or discontinuity. The corresponding mathematical fields are algebra, number theory, geometry and topology. Analysis deals with questions of limit, and related concepts, such as continuity, change, differentiation, integration, and so forth. There are multiple intersections between these fields, and fields that branch off these basic fields.

Defining algebra as the mathematical formalization of the relationships between symbols makes it part of all modern mathematics. Ancient Greek geometry was more self-contained, although one finds elements of proto-algebraic symbolism there and, of course, in ancient Greek mathematics in general, especially at later stages of its development, as in Diophantus (third century AD), who shares the honor of "the father of algebra" with Muhammad ibn Mūsā al-Khwārizmī (eighth to ninth century). Geometrical and topological mathematical objects always have algebraic components, while algebraic objects may, but need not, have geometrical or topological components. The term algebra carries other meaning, such as that (standing at the origins of algebra as a mathematical discipline) of algebra as the study of algebraic equations, or referring to algebraic structures such as groups. All these forms of algebra are important to mathematical modernism, even in fields like geometry and topology. It is not only a matter of having an algebraic component as part of the mathematical structure of geometrical or topological objects but also of defining these objects algebraically.

An emblematic case of the role of algebra in modernist mathematics is algebraic topology, which also led to category theory, used in quantum information theory, as discussed in Chap. 9. As part of topology, algebraic topology does have an earlier history, extending from Leibniz, with Leonhard Euler as the main precursor before the rise of the discipline as such with Riemann, Poincaré, and others. General or point-set topology, a major part of mathematical modernism as well, has a much longer history, arguably, extending, at least proto-mathematically, to Plato and Aristotle, and (e.g., Thom 1988; Papadopoulos 2019). What makes algebraic topology a mathematical discipline is the fact that one can associate algebraic structures (initially numbers, eventually groups and other abstract algebraic structures) to the architecture of spatial objects that are invariant under continuous transformations, independently of their geometrical properties, such as those associated with measurements. This makes topology topo-*logy* versus geo-*metry*. While keeping in mind this difference between topology and geometry, and referring to each separately whenever necessary, I shall group them together here under the rubric of geometry as the mathematics of spatiality.

Is, then, modernist geometry, and indeed modernist mathematics in general, divorced from its connections to nature and physics, a form of algebra, on the model of its famous Cartesian predecessor, analytic geometry, in some of its areas accompanied by the rise of the RWR view of the ultimate nature of reality, both in mathematics itself and in the use of this new algebra in physics? It is tempting to argue such a case, and the present author has done so previously, even if in a qualified way, in part in conjunction with the Heisenberg method in QM (Plotnitsky 2019). Here, however, I take a different view, which is more nuanced and more faithful to the richness and complexity of modernist mathematics, or mathematics in general, as well as its use in physics (Plotnitsky 2020). This view is based on the interplay of geometry and algebra in mathematics and physics. I want to place "the Heisenberg method" in this more complex historical and conceptual landscape, making as much a new geometrical as new algebraic, as well as a new epistemological, RWR-type, method in physics.

Even at the most immediate level, geometry has retained its significance in modernist mathematics on several counts, two of them in particular. First, the algebra defining modernist geometrical and topological objects has a special form that may be called "spatial algebra," which, from the other side of the same coin, may be seen as a new form of geometry. Indeed, its genealogy includes such developments as projective geometries, geometries in dimensions higher than three, and discrete and finite geometries. Spatial algebra arises from algebraic structures that *mathematically* define geometrical or topological objects and reflect their proximity to $\mathbb{R}^3$ and mathematical objects there, which are more (but still not entirely) accessible to our phenomenal intuition. This proximity may be left behind in rigorous mathematical definitions and treatments of such objects, beginning with $\mathbb{R}^3$ itself. The same type of algebra may also be used to define mathematical objects that are no longer available to our phenomenal intuition. Among examples of such objects are, again, a projective space (a set of lines through the origin of a vector space, such as $\mathbb{R}^2$ in the case of the projective line, with projective curves defined algebraically, as algebraic varieties) and an infinite-dimensional Hilbert space, the points of which are typically square-integrable functions or infinite series, although a Euclidean space of any dimension is, technically, a Hilbert space, too. In sum, spatial algebra is an algebraization of spatiality that makes it rigorously mathematical, topologically or geometrically (as opposed to something that is phenomenally intuitive as spatial) even in the case of more conventional spatial objects in $\mathbb{R}^3$.

In addition, which is the second count on which spatial algebra retains it connection to geometrical or topological thinking (including when it is more qualitative in character), analogies with $\mathbb{R}^3$ and corresponding visualizations continue to remain useful and even indispensable, and perhaps unavoidable. Thus, the analogues of the Pythagorean theorem or parallelogram law in Euclidean geometry, which holds in infinite-dimensional Hilbert spaces, are important, including in applications to physics, especially quantum theory, the mathematical formalism of which is, again, based in Hilbert spaces over $\mathbb{C}$. Indeed, our *thinking* concerning geometrical and topological objects is not entirely translatable into algebra. This was well understood by Hilbert in his axiomatization of Euclidean geometry, although this axiomatization had and was defined by a spatial-algebraic character (Hilbert 1999). One can gain further insight into this situation by considering the principle, due to Joseph Silverman and John Tate, "Think Geometrically, Prove Algebraically," introduced in their book on "the rational points of elliptic curves" (a manifestly modernist subject, part of arithmetic algebraic geometry):

> It is also possible to look at polynomial equations and their solutions in rings and fields other than $\mathbb{Z}$ or $\mathbb{Q}$ or $\mathbb{R}$ or $\mathbb{C}$. For example, one might look at polynomials with coefficients in the finite field $\mathbb{F}_p$ with p elements and ask for solutions whose co-ordinates are also in the field $\mathbb{F}_p$. You may worry about your geometric intuitions in situations like this. How can one visualize points and curves and directions in $\mathbb{A}^2$ when the points of $\mathbb{A}^2$ are pairs (x, y) with $x, y \in \mathbb{F}_p$? There are two answers to this question. The first and most reassuring

> is that you can continue to think of the usual Euclidean plane, i.e., $\mathbb{R}^2$, and most of your geometric intuitions concerning points and curves will still be true when you switch to co-ordinates in $\mathbb{F}_p$. The second and more practical answer is that the affine and projective planes and affine and projective curves are defined *algebraically* in terms of ordered pairs (r, s) or homogeneous triples $[a, b, c]$ without any reference to geometry. So in proving things one can work algebraically using co-ordinates, without worrying at all about geometrical intuitions. We might summarize this general philosophy as: *Think Geometrically, Prove Algebraically.* (Silverman and Tate 2015, p. 277)

Affine and projective planes and curves can in principle be defined without any reference to our phenomenal intuition, which grounds the geometrical thinking referred to by Silverman and Tate, rather than spatial algebra, which defines them rigorously. Even in these more intuitively accessible cases, we think algebraically, too, by using spatial algebra, except possibly in dealing with low-dimensional topological and geometrical objects. I would argue, however, that spatial algebra is still irreducible there because one commonly converts topological operations into algebraic ones. This conversion, in low dimensions, was essential to the origin of algebraic topology, even in dealing with two-dimensional manifolds, such as Riemann surfaces. Most three-dimensional manifolds, such as three-dimensional surfaces, are beyond visualization. Certain more recent developments of low-dimensional topology, following, among others, William Thurston's pioneering work from the 1970s on, or knot theory, may be seen as a more purely geometrically or topologically oriented trend that counters modernist algebraization, because our phenomenal geometrical intuitions could be used directly and even with fuller mathematical rigor, analogously to Euclidean geometry. Riemann used them in this way, as diagrams in his papers indicate. The use of such diagrams as mathematical objects (rather than auxiliary visual tools) is part of this trend, as are, by now ubiquitous, digital topological and geometrical objects or videos. Along with the use of digital technology in other areas of mathematics (chaos theory pictures, computerized mathematical proofs, and so forth), this use may be seen as belonging to a new stage of mathematics, which some might want to see as "postmodern."[6] They may also be seen as a digital or digitally visualized return of geometry. Thurston was one of the pioneers of the use of digital technology in low-dimensional topology. Still, this turn to more immediate visualizable thinking in low dimensional topology and geometry, however digitally helped, is only partial because the algebraic structures, such as homotopy and cohomology groups, associated with these topological objects remain crucial. Some of the most powerful tools of algebraic topology have been developed in this field. Low-dimensional topology and geometry have played an important role in string and brane theories in physics. One should be careful, however. The so-called "string theory in low dimensions" does deal with three-dimensional manifolds. But it also deals with four or even six-dimensional manifolds, objects far beyond any geometrical visualization, which string theory uses, helped by digitally created two-dimensional images of such objects as Calabi–Yau manifolds, the actual dimension

---

[6] The reason for using this denomination is that, as noted, during the last 50 years or so, modernity entered a new stage, known as postmodernity, defined by the rise of digital information technology (Lyotard 1984).

of which is six. The subject, which can only be mentioned in passing here, is a powerful illustration of the modernist complexity of the interplay of algebra, geometry, and topology, and of mathematics and physics, regardless of the controversies accompanying string theory as a viable physical theory.

It is also true that a mathematician can develop and use intuition in dealing with discrete geometries, say, that of the Fano plane of order 2, which has the smallest number of points and lines (seven each). However, beyond the fact that they are made in the two-dimensional regular plane, the diagrammatic representations of the Fano plane are still difficult to think of as other than spatial algebra, in this case combinatorial. Euclidean intuitions are limited even when we deal with algebraic curves in the Euclidean plane, let alone in considering curves and other objects of finite or projective geometries, abstract algebraic varieties, Hilbert spaces, the spaces of noncommutative geometry, or geometric groups, a great example of the reversed extension of spatial algebra to algebraic objects themselves.

That said, however, these geometrical intuitions remain indispensable, thus giving the situation a further complexity. Tate's "*think* geometrically" may even be read as giving geometry a greater creative significance, although proving anything requires thinking, too. Our geometrical thinking shapes our practice of spatial algebra and other areas of abstract algebra, for example, when dealing with such objects as ideals of rings, even apart from algebraic geometry where manifolds or schemes are constructed of ideals (maximal or prime respectively) of commutative rings. For the moment, however, I would like to emphasize the creative role of geometrical *thinking*, in its Euclidean intuition and in spatial algebra. This thinking may need algebra to achieve the necessary mathematical rigor—Prove algebraically!—but it is crucial to mathematical thought whenever spatial algebra enters—Think geometrically!

Dirac, one of the greatest algebraic virtuosi of quantum theory, was, nevertheless, fond of referring to geometrical thinking in QM and QFT (e.g., Farmelo 2005). It is difficult to surmise what Dirac, famous for his laconic style, exactly had in mind. If, however, one is to judge by his writings, they appear to suggest that at stake are the algebraic properties and relations modeled on those found in geometrical objects, defined by algebraic structures, in short, spatial algebra. His often-noted fondness for projective geometry, which appears to have helped him to appreciate the algebraic noncommutativity of QM, discovered by Heisenberg, supports this argument. Notwithstanding his insistence on the role of geometry in Dirac's thinking, Oliver Darrigol's analysis of this thinking shows the significance of spatial algebra there (Darrigol 1993). This is, however, more indicative of the complexity of the relationships between geometry and algebra, and new ways in which geometry and geometrical thinking work, or return, in modernism, under discussion in this chapter. Thus, Darrigol says: "Roughly, Dirac's quantum mechanics could be said to be to ordinary mechanics what noncommutative geometry is to intuitive geometry" (Darrigol 1993, p. 307). Noncommutative geometry, however, the invention of which was in part inspired by QM, is made a geometry by its spatial algebra (Connes 1994, p. 38; Plotnitsky 2009, pp. 112–113). One encounters similar appeals to geometrical thinking in referring to transfers of geometrical methods to spatial algebraic or just algebraic objects (thus making them spatially algebraic), such as in dealing with groups and group representations in QM and QFT, initially developed in a more geometrical context beginning with Sophus Lie and Felix Klein or in using the idea of metrics in geometric group theory.

And yet, it would be misleading to think of both the mathematics of quantum theory or noncommutative geometry as only algebra, even as only spatial algebra, which, again, is never only algebra but is also geometry in its structure and in the thinking it requires alike. It is possible that Dirac had in mind this aspect of the mathematics of QM and QFT. A Hilbert space is still, mathematically, a space and has its geometry, which we must *think*, again, by using *both* our Euclidean geometrical intuition as a heuristic guidance and our technical geometrical thinking by using spatial *algebra*, but as *spatial* algebra, in quantum theory, or of course in mathematics. At least, this appears how Dirac himself was thinking in his work, often proceeding from the mathematics of the formalism to physical discovery, most famously, in his discovery of Dirac's equation, invented from formal (spatial-algebraic) considerations (e.g., Plotnitsky 2016, pp. 214–226). As became clear with Heisenberg's invention of QM, the Hilbert spaces of quantum-mechanical formalism are both the product and, once adopted, the mathematical technology of this thinking. Dirac arguably represents the greatest example of this thinking when it proceeds from mathematics to physics, more so than Heisenberg himself, who, while an inspiration for Dirac, was proceeding more from physics to mathematics.

Feynman, who was awarded a Nobel prize for his work on the renormalization of QED, instructively commented on the decisive role of visual intuition in thinking about quantum objects and behavior, confirming the significance of intuitive geometrical thinking in physics (Schweber 1994, pp. 465–466). Feynman invokes "a half-assedly thought out pictorial semi-vision thing," to which is he is "trying … to bring birth and clarity" in his creative process, thus thinking spatially and even geometrically (Schweber 1994, p. 466). Think geometrically! The proof in physics is not the same as in mathematics, but still: Prove algebraically! Obviously, apart from the fact that one's creative process could be different, such anecdotal evidence hardly suffices for any definitive claim. Feynman's account, however, appears to be in accord with current neurological and cognitive-psychological research, as just mentioned, which suggests the dependence of our spatial intuition, including visualization, on two- and three-dimensional phenomenality. This was in part why Kant thought of this intuition, which he saw as that of Euclidean three-dimensional spatiality, as given to us a priori. The claim that this intuition is entirely Euclidean or that it is a priori has been challenged. On the other hand, its three-dimensional character appears to be reasonably certain. This is, however, a separate subject.

Be it is it may on this score, my argument here is different. It concerns the joint role of visual geometrical intuition (regardless on its general cognitive or neurological status) and formal geometrical thinking in working with the *mathematical formalism* of QM or QFT, and not imagining or imaging quantum objects and behavior, which, in the present view, are beyond mathematical representation. On the other hand, Feynman's path integral formulation of QM and then QFT would illustrate my point because of its manifested geometrical aspects, of both types, spatial-algebraic and intuitive. This formulation was an alternative, although mathematically equivalent, way of calculating probabilities of quantum events by the standard procedures explained earlier (using the so-called probability amplitudes), without necessarily representing or even forming a conception of quantum objects and behavior. The geometry of this formulation is also more expressly linked to analysis. It is possible that Feynman himself connected his formulation to his heuristic visualization of quantum objects and behavior described above, regardless of which type of interpretation he had in mind, which is a complex issue that need not be addressed here. One must, however, be careful in avoiding loose but not uncommon formulations that a particle found first at point X and then at point B takes all possible paths, sometimes seen even as a manifestation of the principle "anything that can happen does happen," unless one qualifies that this does not refer to what happens in the same experiment. If one still uses this classical (and ultimately inapplicable) language, a particle can "jump" quantumly from one such paths to another in the way it cannot happen if one deals with a classical motion, which is one of the (many) reasons why such pictures or ultimately any visualizations are inapplicable here. Besides, in high-energy (QFT) regimes a particle of a given type, say, an electron can "jump" to become a positron or a photon, which will then be registered at point Y. Accordingly, in these regimes, even speaking of registering the same electron at point Y is meaningless, which support the present interpretation, according to which any quantum object is an idealization applicable only at the time of a measurement. QM or QFT make these predictions by using the mathematics of Hilbert spaces, a mathematics that is *geometrical*, beginning with the fact that the concept of distance can be defined there, via the concept of norm, even though it needs algebra as well. But then, so does, in one way or another, all geometry.

This joint, spatial-algebraic and intuitive, geometrical thinking is found in mathematics itself, to one degree or another, when one deals with geometry or topology. Consider Weyl's argument leading him to his definition of a Riemann surface as a manifold (a definition not expressly given by Riemann, who was, however, undoubtedly aware that Riemann surfaces were manifolds):

> [O]ne's intuitive grasp of an analytic form [an analytic function to which a countable number of irregular elements have been added] is greatly enhanced if one represents each element of the form by a point on a surface F in space in such a way that the representative points cover F simply and so that every analytic chain of elements of the form becomes a continuous curve on F. To be sure, from a purely objective point of view, the problem of finding a surface to represent the analytic form in this visual way may be rejected as nonpertinent; for in essence, three-dimensional space has nothing to do with analytic forms, and one appeals to it not on logical-mathematical grounds, but because it is closely associated with our sense perception. To satisfy our desire for pictures and analogies in this fashion by forcing inessential representation of objects instead of taking them as they are could be called an anthropomorphism contrary to scientific principles. However, these reproaches of the pure logicians are no longer pertinent if we pursue the other approach, already hinted at, in which the *analytic form is a two dimensional manifold* to which all the ideas of continuity that we meet in ordinary geometry may be applied. To the contrary, not to use this approach is to overlook one of the most essential aspects of the topic. (Weyl 2013, pp. 16–17)

Weyl's own approach is grounded in the joint way of thinking geometrically, but using *both* our geometrical intuition (ordinary geometry, including its ideas of continuity) and the concepts and techniques of geometry defined by spatial algebra, which is, again, still geometry. The second is represented by the fact that "the concept 'two dimensional manifold' or 'surface' will not be associated with points in three-dimensional space; rather it will be a much more general abstract idea," a spatial-algebraic one (Weyl 2013, p. 17). In this regard Weyl follows Riemann. Admittedly, Weyl deals with (topologically) two-dimensional objects, Riemann surfaces. The point is, however, more general and applies to all manifolds, over $\mathbb{R}$ and $\mathbb{C}$, or to algebraic varieties over finite fields, $\mathbb{F}_p$, which are represented strictly in terms of spatial algebra. Weyl's argumentation would of course also apply to Hilbert spaces over $\mathbb{C}$.

The question of continuity in modernist mathematics and physics naturally fits into this conceptual and epistemological architecture as well. According to Gray, one of the characteristic features, and for him the most important one, of modernist mathematics, is its separation from its connections to nature and physics, emerging from what he calls "the crisis of continuity," most especially in analysis, but clearly also elsewhere in modernist mathematics. As he says: "This is the widespread feeling among mathematicians around 1900, documented in many sources, that the basic topic of analysis, continuity, was profoundly counterintuitive. This realization marks a break with all philosophy of mathematics that present mathematical objects as idealizations from natural ones: it is characteristic of modernism" (Gray 2008, p. 20). These claims require qualifications. First of all, mathematics separated, abstracted, itself from representing natural objects and thus also from physics much earlier, already with the Pythagoreans, or during modern times, with the rise of algebra, especially the study of algebraic equations, pursued largely apart from physics. On the other hand, it is true that geometry and analysis, Gray's main context in this statement, had kept close relationships to representing nature and physics. It is also easier to argue this type of case in its disciplinary sense, insofar as the possibility of this separation was used to ground the disciplinary field of mathematics. The drive toward this independence had been emerging in modern mathematics from the 1800s on. With mathematical modernism, however, this drive reached the stage of breaking with representing (mathematically idealizing) natural objects in most areas of mathematics, including geometry and analysis, especially following the rise of Cantor's set theory, perhaps the most significant event leading to "the crisis of continuity" invoked by Gray, not the least as concerns the separation of the *concept* of continuity in mathematics from our phenomenal intuition of continuity.

It might be added that the continuum is a mathematical concept that poses major and even insoluble problems topologically. These problems emerged with Cantor's discovery of the multiplicity of infinities, the infinity of infinities, and his continuum hypothesis, and became especially dramatic with Gödel's incompleteness theorems and Cohen's proof of the undecidability of the continuum hypothesis. These findings made us realize that we do not and cannot know how a continuous line, straight or curved (which does not matter topologically), is constituted by its points. Quantum theory, beginning with Planck's discovery of it in 1900, contributed to this crisis, in

QM through complex relationships between mathematical continuity and physical discontinuity, as discussed above.

Weyl responded to this crisis of continuity in his 1918 *The Continuum* (Weyl 1928), followed closely by his 1918 classic *Time Space Matter* (Weyl 1952), shaped by modernist mathematics as well, by which point quantum theory (not yet QM) enters his thinking concerning the subject and its role in physics (Weyl 1952, p. 23). According to Weyl: "the conceptual world of mathematics is so foreign to what the intuitive continuum presents to us that the demand for coincidence between the two must be dismissed as absurd" (Weyl 1928, p. 108). "Coincidence" is, however, not the same as "relations," which are unavoidable, at least insofar as it is difficult to think of continuity spatially apart from our phenomenal intuition. Weyl's point concerning the world of mathematics as foreign to our ordinary phenomenal intuition applies far beyond the idea of continuity. On the other hand, modernist physics, such as relativity and quantum theory, was able to take advantage of this divorce of mathematics from ordinary phenomenal intuition. Weyl indeed added to his statement just cited: "Nevertheless, those abstract schemata supplied us by mathematics must underlie the exact sciences of domains of objects in which continua play a role" (Weyl 1928, p. 108).

This statement was most likely made with Einstein's relativity, a fundamentally continuous theory, in mind, as Weyl's next book, *Time Space Matter*, was already in the works. QM was still a few years away, although the old quantum theory already broke with classical physics on the account of continuity, a point Weyl did not fail to note in *Space Time Matter* (Weyl 1952, pp. 97–98). Eventually, Weyl came to play a major role in the development of QM, first, assisting Schrödinger's work on his equation. Weyl also made several important contributions of his own, especially as concerns the role of group theory in QM and QFT, such as the concept of gauge symmetry, central for particle physics and the standard model. His 1931 book, *The Theory of Groups and Quantum Mechanics* (Weyl 1931), had a shaping influence in this area. Weyl's view under discussion at the moment retains a modernist twist given that the *abstract* schemata used in relativity, special and general, and QM are divorced from our phenomenal intuition. Both used modernist geometry, respectively, that of Riemannian geometry (Minkowski's spacetime of special relativity is a pseudo-Riemannian manifold) and that of infinite-dimensional Hilbert spaces, and group theory plays a key role in both. There was, however, a crucial difference.

Relativity still did this in a representational (realist) way, which was, again, a permissible idealization, even in the U-RWR view, but in this view, under the assumption that it is still underlain by the ultimate constitution of the reality responsible for the phenomena considered which is beyond conception. This was possible because, as discussed in Chap. 2, in this case, the break from our phenomenal intuition need not entail a divorce from realism by means of mathematized physical concepts. As indicated earlier, the discovery of singularities in general relativity introduced new complexities as early as 1916, but these complexities did not appear threatening to the representational nature of the theory and were expected to be resolved in one way or another. Einstein, again, believed that such solutions could be avoided physically, a view that became progressively more difficult to sustain given the development of the physics of collapsing stars, in the extreme case, black holes, the existence of which is pretty much universally accepted now. On the other hand, black holes are quantum objects.

In any event, in contrast to relativity, QM, at least in RWR-type interpretations, such as the one, of the weak RWR type, implicitly adopted by Heisenberg (as he did not offer an interpretation of QM as such), used modernist mathematics, that of Hilbert spaces over $\mathbb{C}$, for providing probabilistic predictions of the outcomes of quantum experiments, without providing a representation of the ultimate nature of reality responsible for these events, even a purely mathematical one in the absence of a physical one, by means of physical concepts, such as motion. The latter possibility has been entertained, including by Heisenberg in his later work or in certain ("ontic") forms of structural realism (Ladyman 2016). QM, in RWR-type interpretations, used "abstract schemata" of modernist mathematics in a radically new way and by doing so established a new type of relationship between the mathematics of a physical theory and the experimental data considered. By using the *geometry* of Hilbert spaces over $\mathbb{C}$, and the algebra of the operators there, QM was able to connect, on RWR-lines, to a reality that is beyond representation and even thought, and thus cannot be given a geometry, any more than algebra. This was the mathematical nature of the Heisenberg method. It only enabled him to establish such connections in terms of probabilistically or statistically predicting effects of the ultimate constitution of reality (or part of it) upon the world we can observe and represent, effects manifested in quantum phenomena. But this was no small achievement. Indeed, as I argue, it changed the nature of fundamental physics, theoretical and experimental. It made theoretical physics the creation of abstract mathematical concepts, predicting physical phenomena without representing how they comes about, and it made experimental physics the creation of new technology through which such phenomena could be constructed, without an aim to reach the ultimate nature of physical reality responsible for them either, as in classical physics.

4.6 Conclusion

While continuing the project of *modern* post-Galilean physics as a mathematical-experimental science, Heisenberg's thinking that led him to his discovery of QM was a radically new way of relating mathematics, moreover, a mathematics of an entirely new type in physics, and the experiment. To better illustrate the mathematical nature of his thinking, let us imagine for the moment that the data that led Heisenberg to his invention of QM, were given to a mathematician on the cutting edge of mathematics at the time, say, a doctoral student of Hilbert, somebody like Emmy Noether or von Neumann, who would, however, be unfamiliar with the old quantum theory or even classical physics. This is not true historically as concerns either von Neumann or Noether, both of whom by that time had made major mathematical contributions to general relativity and proved her celebrated theorems, discussed in Chap. 8, or arguably most students of Hilbert, given his own interest in contemporary physics, including the old quantum theory, which was widely discussed in Göttingen at the time. Born was a major presence there, and Heisenberg (at the time of his invention of QM), Pauli, and Jordan were Born's postdoctoral students. Accordingly, the relationships between mathematics and physics, including in their modernist manifestations, were at the center stage at Göttingen, which was the center stage of mathematical modernism. Nevertheless, let us imagine such a mathematician.

This mathematician would then have been asked to develop a mathematics that would enable one to predict, probabilistically, these data, a mathematics it follows independent of physics, just a technology of human predictions of a certain type of data. This would have been a formidable problem, especially in the absence of physics, although, as things actually happened, physics (as it existed then) was almost more inhibiting than helpful in solving it. The mathematician needed not to think in terms of describing mathematically some classical-like objects and motion, as a physicist would be likely to do, but only to predict, again, probabilistically, the patterns in question, observed in spectra. The mathematician only needed to find the mathematics that predicts them in terms of their probability distribution. Assuming the invented mathematics would be the same of that of QM (for it could in principle be something else), this mathematician would need to make two extraordinary guesses—one truly extraordinary and, with the first guess in hand, the other, somewhat less so. The first guess is that one needs a Hilbert space over $\mathbb{C}$, which is difficult, but for the like of Noether or von Neumann, who would, for example, have been familiar with Fourier's analysis (again, separating it from physics), is not unimaginable. As Schwinger noted in the passage cited above, given the data in question, "the mathematical scheme [of QM] can certainly not be the assignment, the association, or the representation of physical properties by numbers," especially by real numbers (Cited in Schweber 1994, p. 361). But even complex numbers would not represent the physical properties of quantum objects, except that, in my fable, no physical properties of anything connected to the date in question would be at stake, only a mathematical connections of two sets of data or information. The second guess would have been Born's rule, and while still not easy, would be almost natural because probabilities are real numbers, and when moving from complex to real numbers, the square moduli of complex numbers is the most obvious way to do so. Such a mathematician could have thought of von Neumann's projection postulate as well.

The reason that I brought up this fable is that it is not that far from how Heisenberg made his discovery of QM. In effect, he had to suspend, to "forget," classical physics or the old quantum theory to arrive at his new mathematical scheme. His physics was defined by finding the mathematics probabilistically predicting the data or information in question, that of hydrogen spectra. The correspondence postulate enabled him to use the already available equations of classical mechanics. Still, he reinvented, along the lines of my fable, matrix algebra used as variables in these equations, which enables these predictions. This was a radical transformation of fundamental physics, enabled by a new, essentially mathematical, way of thinking, defined by one's invention of a new predictive mathematical scheme, rather than by finding the mathematics that would represent the physical reality considered, which could be nontrivial as well, as in Einstein's general relativity. This was Heisenberg's revolution, with which, in Bohr's words, "a new epoch of mutual stimulation of mechanics and mathematics has commenced:"

> It will interest mathematical circles that the mathematical instruments created by the higher algebra play an essential part in the rational formulation of the new quantum mechanics. Thus, the general proofs of the conservation theorems in Heisenberg's theory carried out by Born and Jordan are based on the use of the theory of matrices, which go back to Cayley and were developed especially by Hermite. It is to be hoped that a new era of mutual stimulation of mechanics and mathematics commenced. To the physicists it will at first seem deplorable that in atomic problems we have apparently met with such a limitation of our usual means of visualization. This regret will, however, have to give way to thankfulness that mathematics in this field, too, presents us with the tools to prepare the way for further progress. (Bohr 1987 v. 1, p. 51)

"Visualization" here is, again, also a translation of German *Anschaulichkeit,* which has a general meaning of phenomenal intuition, beyond which quantum objects and behavior were placed as a result. While a new era of mutual stimulation of, and the new, RWR-type, relationships between physics and mathematics are clearly at stake, Bohr speaks a "mutual stimulation of *mechanics and mathematics*" because the difference between classical and quantum physics concerned most essentially individual quantum objects and their behavior. The corresponding mathematical theory in classical physics, a theory mathematically representing individual classical objects and their behavior, is classical mechanics. In quantum *mechanics*, which considers individual quantum objects and behavior, mathematics plays an essentially different role, at least in RWR-type interpretations, such as the one assumed by Bohr here, because, to return to Bohr's formulation in the same article, "Atomic Theory and Mechanics," "in contrast to ordinary [classical] mechanics, the new quantum mechanics does not deal with a space–time description of the motion of atomic particles" and thus does not offer a mathematical representation of the ultimate constitution of the reality responsible for quantum phenomena (Bohr 1987, v. 1, p. 48). QM, in RWR-type interpretations, introduced a new form of relationships between the mathematics of QM and the data considered, which compelled Bohr to speak of this mechanics as symbolic, while confirming the fundamental role of mathematics in this new physics.

As far as the physicists' attitude is concerned, the subsequent history has proven that Bohr was too optimistic. Discontent with "the limitation" in question has never subsided. Einstein, again, was in the forefront of this resistance. He did not find satisfactory or even acceptable this state of affairs as concerns physics or this type of the relationships between mechanics, or physics in general, and mathematics, and he never stopped seeing this limitation as anything less than deplorable. Schrödinger was quick to join Einstein, followed by most physicists and philosophers, all guided, as was Einstein, by the hope, the Einsteinian hope, that a future theory will return us to the realist form of the relationships and mutual stimulation (which cannot be denied) of physics, including mechanics, and mathematics. This is still our own situation, one hundred years since Bohr's comments, and whether the future will lead to a fulfillment of the Einsteinian hope or will leave the Heisenberg method in place remains as uncertain as ever.

Whatever the future holds, as stated from the outset of this study, the Heisenberg method, its algebra and geometry, made quantum *mechanics* the most *mathematical* physical theory ever, because it provided *no mechanics* for the ultimate constitution of the reality responsible for quantum phenomena. In the RWR view, there is no such mechanics, which, however, makes mathematics all the more necessary in dealing with quantum phenomena. There are the experimental technology of quantum physics and the mathematics of quantum theory, QM and QFT, and the rules through which we relate this mathematics to quantum phenomena, created by this technology, in predicting, probabilistically or statistically, the outcome of our experiments manifested in these phenomena. No other predictions are experimentally possible, at least as things stand now. By bringing together geometry, algebra, and the unthinkable, the Heisenberg method is a Pythagorean method, which, as will be discussed in Chap. 6, contains the unthinkable as well, in the form of "alogon," as the ancient Greeks called it. The RWR unthinkable is, however, a new and radical form of the unthinkable, unlike anything ever contemplated by the Pythagoreans or by anyone else before QM, but Pythagorean nevertheless.

References

Aristotle: The Complete Works of Aristotle (ed. J. Barnes). Princeton University Press, Princeton, NJ, USA (1984)

Bohr, N.: On the constitution of atoms and molecules (Part 1). Phil. Mag. **26**(151), 1–25 (1913)

Bohr, N.: Can quantum-mechanical description of physical reality be considered complete? Phys. Rev. **48**, 696–702 (1935)

Bohr, N.: The Philosophical Writings of Niels Bohr, vol. 3, Ox Bow Press, Woodbridge, CT, USA (1987)

Born, M.: Zur Quantenmechanik der Stoßvorgänge. Z. Phys. **37**, 863–867 (1926a)

Born, M.: Quantenmechanik der Stoßvorgänge. Z. Phys. **38**, 803–827 (1926b)

Born, M.: Natural Philosophy of Cause and Chance. Oxford University Press, Oxford, UK (1949)

Born, M.: The Einstein-Born Letters (tr. Born, I.), Walker, New York, NY, USA (2005)

Born, M., Jordan, P.: Zur Quantenmechanik. Z. Phys. **34**, 858–888 (1925)

Born, M., Heisenberg, W., Jordan, P.: On quantum mechanics. In: Van der Waerden, B.L. (ed.) Sources of Quantum Mechanics, 1968, pp. 321–385. Dover, New York, NY, USA (1926)

Bose, S.N.: Plancks Gesetz und Lichtquantenhypothese. Z. Phys. **26**, 178–181 (1924)

Bub, J.: Quantum mechanics as a principle theory. Studies in the History and Philosophy of Modern. Physics **31**(1), 75–94 (2000)

Connes, A.: Noncommutative Geometry (tr. Berberian, S. K.). Academic Press, San Diego, CA, USA (1994)

D'Ariano, G.M.: No purification ontology, no quantum paradoxes. Found. Phys. **50**, 1921–1933 (2020). https://doi.org/10.1007/s10701-020-00398-6

D'Ariano, G.M., Chiribella, G., Perinotti, P.: Quantum Theory from First Principles: An Informational Approach. Cambridge University Press, Cambridge, UK (2017)

Darrigol, O.: From c-Numbers to q-Numbers: The Classical Analogy in the History of Quantum Theory. University of California Press, Berkeley, CA, USA (1993)

Dirac, P.A.M.: The fundamental equations of quantum mechanics. In: van der Warden, B.L. (ed.) Sources of Quantum Mechanics New York, 1968, pp. 307–320. Dover, NY, USA (1925)

Dirac, P.A.M.: The Principles of Quantum Mechanics. Clarendon, Oxford, UK (1930)

Dirac, P.A.M.: The Principles of Quantum Mechanics (4th edn), rpt. 1995. Clarendon, Oxford, UK (1958)
Einstein, A.: Theorie der Lichterzeugung und Lichtabsorption.Annalen der Physik **20**, 206 (1906)
Einstein, A.: What is the theory of relativity? In: Einstein, A. (ed.) Ideas and Opinions, 1954, pp. 227–231. Bonanza Books, New York, NY, USA (1919)
Einstein, A.: Physics and reality. J. Franklin Inst. **221**, 349–382 (1936)
Einstein, A.: Autobiographical Notes (tr. Schilpp, P. A.), Open Court, La Salle, IL, USA (1949)
Farmelo, G.: Dirac's hidden geometry. Nature **437**, 323 (2005)
Gleason, A.: Measures on the closed subspaces of a Hilbert Space. Indiana Univ. Math. J. **6**(4), 885–893 (1957). https://doi.org/10.1512/iumj.1957.6.56050.MR0096113
Gray, J.: Plato's Ghost: The Modernist Transformation of Mathematics. Princeton University Press, Princeton, NY, USA (2008)
Hacking, I.: The Emergence of Probability: A Philosophical Study of Early Ideas about Probability, Induction and Statistical Inference, 2nd edn. Cambridge University Press, Cambridge, UK (2006)
Hardy, L.: Quantum mechanics from five reasonable axioms, arXiv:0101012v4 [quant-ph] (2001)
Heisenberg, W.: Quantum-theoretical re-interpretation of kinematical and mechanical relations. In Van der Waerden, B.L. (ed.) Sources of Quantum Mechanics, rpt. 1968, pp. 261–277. Dover, New York, NY, USA (1925)
Heisenberg, W.: The physical content of quantum kinematics and mechanics. In: Wheeler, J.A., Zurek, W.H. (eds.) Quantum Theory and Measurement, 1983, pp. 62–84. Princeton University Press, Princeton, NJ, USA (1927)
Heisenberg, W.: The Physical Principles of the Quantum Theory (tr. Eckhart, K., Hoyt, F.C.), rpt. 1949. New York, NY, USA, Dover (1930)
Heisenberg, W.: Physics and Philosophy: The Revolution in Modern Science. Harper & Row, New York, NY, USA (1962)
Hilbert, D.: Foundations of Geometry (tr. Unger L., Bernays, P.), Open Court, La Salle, IL, USA (1999)
Jaeger, G.: Grounding the randomness of quantum measurement. Philosoph. Trans. Royal So. A (2016). https://doi.org/10.1098/rsta.2015.0238
James, W.: The Meaning of Truth. MA, USA, Harvard University Press, Cambridge (1978)
Kolmogorov, A.: Foundations of the Theory of Probability. Chelsea, New York, NY, USA (1956)
Ladyman, J.: Structural realism. In: Zalta, E.N. (ed.) Stanford Encyclopedia of Philosophy (2016). https://plato.stanford.edu/archives/win2016/entries/structural-realism/
Lyotard, J-F.: The Postmodern Condition: A Report on Knowledge (tr. Bennington, G., Massumi, B.). University of Minnesota Press, Minneapolis, MN, USA (1984)
Mehra, J., Rechenberg, H.: The Historical Development of Quantum Theory, 6 vols. Springer, Berlin, Germany (2001)
Papadopoulos, A.: Topology and biology: from Aristotle to Thom. In: Dani, S.G., Papadopoulos, A. (eds.) Geometry in History, pp. 89–128. Springer/Nature, Berlin, Germany (2019)
Penrose, R.: Gravitational collapse and space-time singularities. Phys. Rev. Lett. **14**(3), 57–59 (1965). https://doi.org/10.1103/PhysRevLett.14.57
Plotnitsky, A.: Quantum atomicity and quantum information: Bohr, Heisenberg, and quantum mechanics as an information theory. In: Khrennikov, A. (ed.) Quantum Theory: Reexamination of Foundations 2, pp. 309–342. Växjö University Press, Växjö, Sweden (2002)
Plotnitsky, A.: Epistemology and Probability: Bohr, Heisenberg, Schrödinger and the Nature of Quantum-Theoretical Thinking. Springer, New York, NY, USA (2009)
Plotnitsky, A.: The Principles of Quantum Theory, from Planck's Quanta to the Higgs Boson: The Nature of Quantum Reality and the Spirit of Copenhagen, New York, NY, USA: Springer/Nature (2016)
Plotnitsky, A.: On the concept of curve: Geometry and algebra, from mathematical modernity to mathematical modernism. In Dani, S.G., Papadopoulos, A. (eds), Geometry in History, pp. 153–212. Springer/Nature, Berlin, Germany (2019)

Plotnitsky, A.: The ghost and the spirit of Pythagoras: the twentieth and twenty-first century mathematics between and beyond geometry and algebra. In: Sriraman, B. (ed), Handbook in the History and Philosophy of Mathematics. Springer/Nature, Berlin, Germany (2020). https://doi.org/10.1007/978-3-030-19071-2_7-1

Schrödinger, E.: The present situation in quantum mechanics. In: Wheeler, J.A., Zurek, W.H. (eds.) Quantum Theory and Measurement, 1983, pp. 152–167. Princeton University Press, Princeton, NJ, USA. Princeton (1935a)

Schrödinger, E.: Discussion of probability relations between separated systems. Proc. Cambridge Philos. Soc. **31**, 555–563 (1935b)

Schrödinger, E.: Discussion of probability relations between separated systems. Proc. Cambridge Philos. Soc. **32**, 446–452 (1936)

Schweber, S.S.: QED and the Men Who Made It: Dyson, Feynman, Schwinger, and Tomonaga. Princeton University Press, Princeton, NJ, USA (1994)

Schwinger, J.: Quantum Mechanics: Symbolism of Atomic Measurement. NY, USA, Springer, New York (2001)

Silverman, J., Tate, J.: Rational Points on Elliptic Curves. Springer, Heidelberg, Germany (2015)

Thom, R.: Les intuitions topologiques primordiales de l'aristot´elisme. Revue Thomiste **3**(88), 393–409 (1988)

Von Neumann, J.: Mathematical Foundations of Quantum Mechanics (tr. Beyer, R. T.) 1983. Princeton University Press, Princeton, NJ, USA. Princeton, NJ, rpt. (1932)

Weyl, H.: The Continuum: A Critical Examination of the Foundation of Analysis, (tr. Pollard, S., Bole, T.) 1994. Mineola, NY, USA: Dover, rpt. (1928)

Weyl, H.: Theory of Groups and Quantum Mechanics (tr. Robertson, H.). Dover, rpt., Mineola, NY, USA 1984 (1931)

Weyl, H.: Space Time Matter (tr. Brose, H. L.). Dover, Mineola, NY, USA (1952)

Weyl, H.: The Concept of a Riemann Surface (tr. MacLane, G. L.). Dover, Mineola, NY, USA (2013)

Chapter 5
Schrödinger's Great Guess: The Time-Dependent Wave Equation

Schrödinger: If all this damned quantum jumping were really here to stay, I should be very sorry that I ever got involved with quantum theory.
Bohr: But the rest of us are extremely grateful that you did. Your wave mechanics has contributed so much to mathematical clarity and simplicity that it represents a gigantic advance over all previous forms of quantum mechanics.
—From the conversation between Bohr and Schrödinger (1926), reported in Werner Heisenberg, Physics and Beyond: Encounters and Conversations (Heisenberg 1971, p. 76)

Abstract This chapter discusses Schrödinger's time-dependent equation and its mathematical features. The time-dependent equation and its derivation have been given less attention in the literature, in contrast to his time-independent equation and its derivation, which, along with Schrödinger's program for wave mechanics, have been discussed extensively. The present approach to his time-dependent equation is new because it focuses on those features of the equation that emerge against Schrödinger's own grain. Section 5.2 revisits Schrödinger's derivation of his time-independent equation. Section 5.3 considers Schrödinger's derivation of his time-dependent equation. Section 5.4 comments on Dirac's equation in connection with Schrödinger's time-dependent equation, which is the nonrelativistic limit of Dirac's equation.

Keywords Complex wave function · Dirac's equation · Eigenvalue (proper value) problem · Quantum jumps · Schrödinger's time-dependent equation · Schrödinger's time-independent equation · Wave mechanics

5.1 Introduction

This chapter considers Schrödinger's time-dependent equation and its mathematical features, in particular the first-order derivative in time and the complex wave function, and their significance and implications, especially those that, against Schrödinger's own grain, invite an RWR-type interpretation. Schrödinger's derivation of this equation or, given that, just as that of Heisenberg's matrix mechanics, it was not strictly

A. Plotnitsky, *Reality Without Realism*,
https://doi.org/10.1007/978-3-030-84578-0_5

derived but partially guessed (as was also the case with his time-independent equation), the emergence of this equation no less remarkable than that of Heisenberg's matrix mechanics. Schrödinger's time-dependent equation and its discovery have been given less attention in literature, in contrast to his time-independent equation and its derivation, which, along with Schrödinger's program for wave mechanics, were discussed more extensively, and were previously considered for the RWR perspective by the present author (Plotnitsky 2016, pp. 84–99).[1] Nevertheless, it is helpful to revisit, first, this program, which I do in Sect. 5.2. Schrödinger's derivation of his time-dependent equation is considered in Sect. 5.3. Section 5.4 comments on Dirac's equation in connection with Schrödinger's time-dependent equation, which is the nonrelativistic limit of Dirac's equation.

5.2 "The Wave Radiation Forming the Basis of the Universe" Versus Quantum Discontinuity

About half a year after its discovery by Heisenberg, QM was co-discovered by Schrödinger in early 1926, in a different mathematical form, or so it appeared before the mathematical equivalence of both mechanics was established. In contrast to that of Heisenberg, Schrödinger's approach was based on realist principles, with the aim of finding a physical description and mathematical representation of the reality responsible for quantum phenomena in undulatory or wave terms. Hence, he characterized his theory as a wave mechanics. Indeed, initially Schrödinger did not use the term "quantum mechanics." This use was adopted later, after both theories were shown, by, among others, Schrödinger himself, to be mathematically equivalent. While, as noted earlier, the initial demonstrations of this equivalence, that by Schrödinger included, did not have the kind of mathematical rigor as had that by von Neumann in the 1930s, which finalized the matter, this equivalence was pretty much taken for granted a few months after Schrödinger's discovery in 1926. Initially, however, his theory had appeared to offer an alternative account of the same phenomena. Schrödinger made a revealing comment in his paper on the Bose–Einstein theory, written just before he commenced his work on his wave mechanics and inspired,

[1] Thus, while J. Mehra and H. Rechenberg, in their classic study, in which Schrödinger and wave mechanics was given more space than any other subject (close to a thousand pages!), discuss his time-dependent equation and note its significance, especially its introduction of the complex wave function, their analysis is brief and does not address the deeper physical and philosophical aspects of this equation (Mehra and Rechenberg 2001, v. 5, pp, 786–88, 796–797). Schrödinger's paper "*Quantisierung als Eigenwertproblem* 4" [Quantization as a Problem of Proper Value 4], which offers his derivations of his time-dependent equation, was recently addressed in more detail in (Karam 2020), focusing primarily on the role of the complex wave function there and Schrödinger's attitude toward it. The paper does not, however, consider the role of the first-order time derivative in Schrödinger's equation either, which is equally important and is essentially linked to the complex wave function. Even more crucial is the nature of this derivation, as discussed here (but not addressed by Ricardo Karam or others who commented on the subject), as essentially a guess, rather than a logical deduction.

as was his approach to wave mechanics, by De Broglie's theory of matter waves. Schrödinger envisioned there a picture of the ultimate physical reality in which a moving particle would appear "as nothing more than a kind of 'white crest' on the wave radiation forming the basis of the universe" (Schrödinger 1926, p. 95; cited in Mehra and Rechenberg 2001, p. 435).

The phrase is remarkable in reflecting a major philosophical ambition, *a philosophical ambition in physics*. It is true that Schrödinger was greatly interested in philosophy, both European, in particular that of Baruch Spinoza and Arthur Schopenhauer, and Indian, and engaged in extensive reading in philosophy before and during his work on his wave mechanics. The connections between these philosophical systems or Schrödinger's philosophical interest and his work on wave mechanics is a complex issue, which will be put aside because it is tangential to my argument.[2] By a philosophical ambition, I refer to the ontological nature of his project in its specific physical and mathematical character, in particular his equations (time-independent and time-dependent) as, in his words, "a *continuum*-theory," as against matrix mechanics as "a theory of a *discontinuum*" (Schrödinger 1982, p. 45). Both of Schrödinger's assessments would require qualifications of the type considered in Chapter 4 in assessing Einstein's view of the Heisenberg method as fundamentally algebraic. These qualifications would, however, not undermine Schrödinger's point concerning the nature of wave mechanics as an attempt at an (idealized) representation of the *continuous* physical reality responsible for quantum phenomena. One might, it is true, argue that the main shared trend of his philosophical interests and his philosophy of wave mechanics is the question of continuity, including that of consciousness (e.g., Mehra and Rechenberg, v. 5, pp. 405–410). It does not appear to me, however, that this fact is especially helpful in understanding his work on wave mechanics, which, "the *continuum*-theory" as it might be, was as inextricably tied to its mathematical structure as was, as "a theory of *discontinuum*," matrix mechanics. Accordingly, I would argue that, just as in the case of Heisenberg's work on QM, as considered in Chap. 4, one can explore more directly the relationships between the philosophical nature of Schrödinger's wave mechanics and its physical and mathematical structure. This is what I shall attempt to do here.

The essentially discrete nature of quantum phenomena was in the way of fulfilling Schrödinger's ambition. De Broglie's theory appeared to offer a pathway, and it was both a major inspiration for wave mechanics and a major technical help in inventing it. As is clear from his notebooks, Schrödinger initially derived his (time-independent) equation more directly by using de Broglie's formulas for matter waves. His first published paper replaced this derivation with the one based on the concept of eigenvalue, announced by its title "Quantization and a Problem of Proper Values"

[2] Michel Bitbol's book offers a helpful discussion of connections between modern philosophy (of Hume, Kant, and Schopenhauer) and Schrödinger's philosophy of QM, including as against that of the main representatives of the Göttingen–Copenhagen circle, Bohr and Heisenberg in particular, whose views are given a subtler treatment than is customary, even if bypassing the more radical aspects of their thinking, such as those of the RWR type (Bitbol 1996). Bitbol also notes the stratifications and changes in Bohr's approach to complementarity, which I shall discuss in Chaps. 5 and 6 (Bitbol 1996, pp. 213–219).

(Schrödinger 1982, pp. 1–12).[3] As he said in his note in his article which considered both theories and their mathematical equivalence: "My theory was inspired by L. de Broglie ... and by brief, yet infinitely far seeing remarks by A. Einstein [in his work on the Bose–Einstein theory, which used de Broglie's theory]. I did not at all suspect any relations to Heisenberg's theory at the beginning, by what appeared to me as a very difficult method of transcendental algebra, and by the want of intuitive visualization (*Anschaulichkeit*)" (Schrödinger 1982, p. 46, n.1; translation modified). Schrödinger was right on both points, in particular, on the second, because, as discussed in Chap. 4, Heisenberg's theory, accompanied by the (weak) RWR view of the reality ultimately responsible for quantum phenomena, offered no representation of the behavior of quantum objects in space and time. Schrödinger's ambition was to remedy this deficiency, and to do so by means of a continuous, wave, mechanics. The situation may be seen in the following terms, taking Bohr's 1913 theory and his concept of "quantum jumps," the inaugural event of both quantum theory as leading to QM and of the RWR view, as the grounding event of both approaches, proceeding in the opposite direction from it. Heisenberg extended the idea of quantum jump to the relationships (which one could predict probabilistically, in accord with what was actually observed), between all quantum states or events: in his theory there were, in quantum physics, no other connections, especially, by definition, continuous and classical causal, between them. Schrödinger, by contrast, aimed to eliminate quantum jumps through a wave process, continuously and classically causally (thus without the recourse to probability) connecting all quantum events. Hence, as he said in his famous conversation with Bohr, discussed in closing this chapter: "If all this damned quantum jumping were really here to stay, I should be very sorry that I ever got involved with quantum theory" (Heisenberg 1971, p. 76).

As often happens, the task, hardly assumed to be simple by Schrödinger, proved to be even more difficult than it appeared and ultimately became insurmountable. "This damned quantum jumping" was there to stay and still are with us. Schrödinger never managed to reconcile his program with the observed discreteness of quantum phenomena, which is, again, not the same as discreteness of quantum objects or the ultimate constitution of the reality responsible for quantum phenomena, assumed to be continuous by Schrödinger. The fact that the wave function had to be considered not in three-dimensional physical space but in the configuration space of higher dimensions, meant that it had to have an indirect relation to whatever continuous physical processes were there. Aware of the difficulties of attributing physical reality to the waves in the configuration space, Schrödinger thought of such waves "as something real in a sense, and the constant h universally determined their frequencies or their wave lengths" (Schrödinger to Wilhelm Wien, February 22, 1926, cited in Mehra and Rechenberg 2001, v. 5, p. 536). In other words, one needed to find a relationship and, hopefully, a representational correspondence between these waves in configuration space and some physical vibrations—correspondence, but not identification, for which one could not hope.

[3] Schrödinger's papers of wave mechanics were reprinted in this collection, originally published in 1928, which I shall cite in this chapter.

The waves in configuration space alarmed Einstein, who initially welcomed Schrödinger's theory, especially as a constructive geometrical theory, as against Heisenberg's principle algebraic one. This discontent was not helped by the mathematical equivalence of both schemes, which further suggested that Schrödinger's mechanics might not have been the alternative Einstein hoped for. These features of Schrödinger's wave mechanics added to Einstein's discontent with QM in general, a discontent ultimately shared by Schrödinger after his hopes for a proper wave mechanics as a physical theory failed to materialize. Born's probabilistic interpretation of the wave function later in 1926 resolved the problem of waves in the configuration space in QM, but hardly in accord with Schrödinger's aims for wave mechanics (Born 1926a, b). The probabilistic or statistical nature of QM, not initially a feature of Schrödinger's theory, proved to be another major obstacle to his agenda. This feature was never accepted as fundamental by Schrödinger either, any more than by Einstein. Both saw this feature as an indication that a better theory should be found, possibly, as discussed in Chap. 2, by means of new representational concepts. As noted earlier, the factors just mentioned do not in principle exclude the possibility of a continuous and classically causal representation of the ultimate constitution of the reality responsible for quantum phenomena. Schrödinger's theory, however, proved not to be such a theory that was able to do so. Schrödinger was aware of the problems his program faced and he commented on them throughout his papers of wave mechanics and his correspondence, still for a while hoping to resolve them.

Proving, by, among others, Schrödinger himself, that both matrix and wave-mechanical formalism were mathematically equivalent or transformable into each other (and were then unified in the transformation theory of Dirac and Jordan in 1927 and especially by von Neumann's Hilbert space formalism) resolved the puzzle of why these, seemingly so different, schemes were in complete agreement as concerns their predictions. This resolution was, for those of the Göttingen–Copenhagen persuasion, a further sign of difficulties for Schrödinger's physical program. Heisenberg's approach naturally accommodated both quantum discreteness and probability, indeed, as discussed in Chap. 4, assuming both as postulates (the QD and the QP/QS postulates), without making underlying physical assumptions of the type made by Schrödinger. Schrödinger was, again, well aware of this difference, and instructively commented on it in his 1926 article on the mathematical equivalence of both mechanics, while still retaining his hopes for wave mechanics at the time. As he said:

> Considering the extraordinary differences between the starting-points and the concepts of Heisenberg's quantum mechanics and of the theory which has been designated "undulatory" or "physical" mechanics and has been lately describe here [in *Annalen der Physik*], it is very strange that these two new theories agree *with one another* with regard to the known facts, where they differ from the old quantum theory. I refer, in particular, to the peculiar "half-integralness" which arises in connection with the oscillator and the rotator. That is really very remarkable, because starting-points, presentations, methods, and in fact the whole mathematical apparatus appear fundamentally different. Above all, however, the departure from classical mechanics in the two theories seems to occur in diametrically opposing directions. In Heisenberg's work the classical continuous variables are replaced by systems of discrete numerical quantities (matrices), which depend on a pair of integral indices, and are defined by *algebraic* equations. The authors themselves [Heisenberg, Born, and Jordan] describe the

> theory as a "true theory of a dis-continuum." On the other hand, wave mechanics shows just the reverse tendency; it is a step from classical point-mechanics toward a *continuum-theory*. In place of a process described in terms of a finite number of dependent variables occurring in a finite number of total differential equations, we have a continuous *field-like* process in configuration, which is governed by a single *partial* differential equation, derived from the principle of action. This principle and this differential equation replace the equations of motion *and* the quantum conditions of the older "classical quantum theory." (Schrödinger 1982, pp. 45–61)

Schrödinger was about to show that the mathematical apparatus of each theory is in fact translatable into that of the other, which makes physical and conceptual difference all the more striking, ultimately not in favor of his own program, although that did not appear to Schrödinger to be the case at the time. In Heisenberg's theory, in the language of Hilbert-space formalism, operators were time-dependent, while state vectors were not, as they were to become in Schrödinger's time-dependent equation in a few months, a major advance, crucial to all subsequent quantum theory. For the reasons explained in Chap. 4, Schrödinger's assessment of the matrix mechanics requires qualification, both as concerns the mathematical continuities the theory involves and, most crucially, the fact that the "dis-continuum" in question refers only to the discreteness of quantum phenomena, rather than to any description of the behavior of quantum objects, which behavior Schrödinger wanted to replace with undulatory processes. One of the problems of his project, perhaps its greatest problem, was that he could not account for this discreteness in terms of wave propagation, but could only predict them by using the waves, the "symbolic waves" (as Born and then Bohr came to seem them), in the configuration space of a quantum system. This space in general has a larger number of dimensions (in the case of continuous variables considered by Schrödinger, an infinite number of dimensions) than the three-dimensional physical space in which quantum behavior was assumed to occur.

Schrödinger's appeal to the difference between the concepts, physical and mathematical, of the two theories merits registering here, given this study's emphasis on the roles of both concepts and mathematics in quantum theory. Their mathematical concepts proved to be essentially linked, indeed were parts of the same overall mathematical framework, eventually given a Hilbert-space version by von Neumann, who as noted earlier, also gave a fully rigorous proof of the mathematical equivalence of both formalisms. Their physical concepts were ultimately irreconcilable, which was one of Schrödinger's points. His description of his mechanics as a "physical" mechanics is worth registering, too. It reflects an accord with Einstein's view that Heisenberg's theory was not even properly mechanics, insofar as, while providing correct probabilistic predictions, it did not, and did not aim to, represent the ultimate nature of reality responsible for quantum phenomena. This, in Einstein's view, shared by Schrödinger, should be the task of a fundamental theory, the task that wave mechanics pursued. Schrödinger's article, however, clearly indicated that, despite the problems yet to be resolved, his hopes for his wave mechanics did not diminish, now allowing for working with either or both "very similar analytical mechanisms," that of Heisenberg or his own (Schrödinger 1982, p. 61). This was early in the game, and wave mechanics was still a work in progress. In particular, his time-dependent equation was yet to be introduced (Schrödinger 1982, p. 103).

The future proved to be less charitable to Schrödinger's physical program than he had hoped, in a few years making him abandon his project for wave mechanics, and eventually to assess QM itself, especially as interpreted on RWR lines, as "a doctrine born of distress" in his cat-paradox paper (Schrödinger 1935, p. 154). Schrödinger saw QM, if not as necessarily incomplete, as was argued in EPR's paper, to which his cat-paradox paper responded, but then as "perhaps after all only a convenient calculational trick" (Schrödinger 1935, p. 167). EPR's paper gave Schrödinger, just as it did to Einstein, new hopes that an alternative theory might be necessary, although, as indicated in Chap. 2, Schrödinger returned to his initial approach in the 1950s (Schrödinger 1995). I shall discuss EPR's paper and Bohr's reply (not much appreciated and perhaps not fully understood by either Einstein or Schrödinger) in Chap. 7. Assuming QM to be only a convenient calculational trick requires assuming that any theory of quantum phenomena can provide more than probabilistic calculations, which remains an open question. Besides, as I argue here, calculations are not to be disparaged, as there is no theoretical physics apart from calculations. I am not saying that Schrödinger aimed to do so, his appeal to a trick notwithstanding. Furthermore, calculations, while they can be divorced from realism, need not be divorced from concepts, as Schrödinger's derivation of both of his equations demonstrates.

In deriving them, he made, against his own grain, several mathematical adjustments, either in order to bring his calculation in accord with the experiment or for purely mathematical reasons, that were in effect in conflict with or at least posed difficulties for his agenda, and brought his mathematics closer to Heisenberg's physics, including as concerns discreteness and probability. This was, again, further confirmed by Born's probabilistic interpretation of the wave function, again, never welcomed by Schrödinger. Indeed, it became quickly apparent that one could establish from within Schrödinger's scheme all the key features that grounded Heisenberg's matrix mechanics and helped, via the correspondence principle, to derive it, as discussed in Chap. 4. Bohr, as noted, spoke of "the symbolic character of Schrödinger's method," shared by it with the matrix theory, referring by "method" to Schrödinger's formalism, interpreted accordingly, rather than to the wave-like physical ontology aimed by Schrödinger. Bohr said:

> The symbolic character of Schrödinger's method appears not only from the circumstance that its simplicity, similarly to that of the matrix theory, depends essentially upon the use of imaginary arithmetic quantities. But above all there can be no question of an immediate connection with our ordinary conceptions because the 'geometrical' problem represented by the wave equation is associated with the so-called co-ordinate space, the number of dimensions of which is equal to the number of the degrees of freedom of the system, and, hence in general greater than the numbers of dimensions of ordinary space. Further, Schrödinger's formulation of the interaction problem, just as the formulation offered by matrix theory, involves a neglect of the finite velocity of propagation of the forces claimed by relativity theory. (Bohr 1987, v. 1, pp. 76–77)

The use of complex numbers never quite satisfied Schrödinger either, any more than did Born's probabilistic interpretation, tied to this use (e.g., Schrödinger 1982, pp. 61, 123, 206). As will be seen, however, just as in Heisenberg's scheme, their role is irreducible given how the ultimate mathematical structure of Schrödinger's scheme turned out. Schrödinger came to accept this role along with the complex wave function, at least at certain stages of his thinking, but never without ambivalence (e.g., Schrödinger 1928). While this role was related to the nonrelativistic nature of his mechanics, it was also to affect the relativistic theory, QED, beginning with Dirac's equation.

In retrospect, this view of Schrödinger's theory as "symbolic" and, as a result, RWR-type interpretations of Schrödinger's mathematics are not surprising, given the mathematical equivalence of both theories. However, just as in Heisenberg's case, considered in Chap. 4, it is Schrödinger's thinking in deriving his formalism and its mathematical features in question that is my main interest in this chapter. Unlike in Heisenberg's derivation, where these features were in accord with physical principles and assumed by Heisenberg from the start, in Schrödinger these features proved to undermine or even ultimately defeat his realist physical program, not the least as a fundamentally continuous theory. It is from this perspective of Schrödinger's mathematics against Schrödinger's physics, or that part of his physics that was defined by his program for wave mechanics, that I would like to consider Schrödinger's time-dependent equation, and its mathematical features, in particular the first-order derivative in time and the complex wave function defining its structure, and their physical and philosophical implications.

5.3 From "The Amplitude Equation" to "The Real Wave Equation" to the Time-Dependent Equation

Schrödinger's time-dependent equation was introduced in the fourth installment of his "*Quantisierung als Eigenwertproblem*" [Quantization as a Problem of Proper Value]. Schrödinger opens by explaining the necessity of a time-dependent wave equation. The original wave equation, the first of which (5.1) contains *the second-order derivative in time* and the second one (5.1′) contains no time at all, while being no less general that the first.

He writes:

$$\nabla^2\psi - \frac{2(E-V)}{E^2}\frac{\partial^2\psi}{\partial t^2} = 0 \tag{5.1}$$

or

$$\nabla^2\psi + \frac{8\pi^2}{h^2}(E-V)\psi = 0, \tag{5.1′}$$

which forms the *basis* for the re-establishment of mechanics attempted in this series of papers, suffers from the disadvantage that it expresses the law of variations of the "mechanical field scholar" ψ, neither *uniformly* nor *generally*, and is valid … with a *definite* E-value inserted, for processes which depend on the time exclusively through a *definite* periodic factor:

$$\psi \sim real\ part\ of\ \left(e^{\pm\frac{2\pi iEt}{h}}\right) \tag{5.2}$$

Equation (5.1) is thus not really any more general than Eq. (5.1′), which takes account of the circumstances just mentioned and does not contain time at all.

Thus, when we designated Eq. (5.1) or (5.1′), on various occasions, as "the wave equation", we were really wrong and would have been more correct if we had called it a "vibration-" or "amplitude-" equation. However, we found it sufficient, because to it is linked the Sturm–Liouville proper value problem—just as in the mathematically strictly analogous problem of the free vibrations of strings and membranes—and not to the *real* wave equation.

As to this, we have always postulated up till now that the potential energy V is a pure function of co-ordinate s and does *not* depend explicitly on time. There arises, however, an urgent need for the extension of the theory to *nonconservative* systems, because it is only in this way that we can study the behavior of a system under the influence of prescribed physical forces, *e.g.,* a light wave, or a strange atom flying past. Whenever V contains the time explicitly, it is manifestly *impossible* that Eq. (5.1) or (5.1′) should be satisfied by a function ψ, the method of dependence of which on time is given by (5.2). We then find that the amplitude equation is no 1 onger sufficient and that we must search for the real wave equation. (Schrödinger 1982, pp. 102–103)

Schrödinger then derives *the real wave equation* for a conservative system by first replacing (5.2) with:

$$\frac{\partial^2\psi}{\partial t^2} = -\frac{4\pi^2E^2}{h^2}\psi \tag{5.3}$$

and then by eliminating E from (5.1′) and (5.3) by differentiation, obtains the equation:

$$\left(\nabla^2 - \frac{8\pi^2}{h^2}\mathrm{V}\right)^2\psi + \frac{16\pi^2}{h^2}\frac{\partial^2\psi}{\partial t^2} = 0 \tag{5.4}$$

which would be "satisfied by every ψ which depends on the time as in (5.2), though with E arbitrary, and consequently also by every ψ which can be expanded in a Fourier series with respect to the time (naturally with functions of the co-ordinates as coefficients)" and which is "thus evidently *the uniform and general wave equation for the field scalar* ψ" (Schrödinger 1982, p. 103). The fourth order of the equation does not pose serious mathematical problems in itself. Schrödinger also notes that "if V does *not* contain time, one can, proceeding from (5.4), apply (5.2), and then split up the operator as follows:

$$\left(\nabla^2 - \frac{8\pi^2}{h^2}V + \frac{8\pi^2}{h^2}E\right)\left(\nabla^2 - \frac{8\pi^2}{h^2}V - \frac{8\pi^2}{h^2}E\right)\psi = 0'' \tag{5.4′}$$

Schrödinger (1982, p. 103).

This means that the decomposition into the original second-order Eq. (5.1′) is obtainable. "Thus," Schrödinger says, "we see that the wave Eq. (5.4), which contains in itself the law of dispersion, can really stand as the basis of the theory previously developed for conservative systems" (Schrödinger 1982, p. 103). Schrödinger, however, does not proceed to developing such a theory, by "generaliz[ing] it for the case of a time varying potential function," for the following reason:

> [This generalization] demands caution, because terms with a time derivative of V might appear, about which no information can be given to us by Eq. (5.4), owing to the way we obtained it. In actual fact, if we attempt to apply Eq. (5.4) as it stands to non-conservative systems, we meet with complications, which seem to arise from the term $\partial V/\partial t$. (Schrödinger 1982, p. 104).

At this point Schrödinger changes course:

> Therefore, in the following discussion, I have taken a somewhat different route, which is much easier for calculations, and which I consider justified in principle.
>
> We *need* not raise the order of the wave function to four, in order to get rid of the energy parameter. The dependence of ψ on the time, which must exist if (5.1′) is to hold, can be expressed by
>
> $$\frac{\partial \psi}{\partial t} = \pm \frac{2\pi i}{h} E\psi \tag{5.3'}$$
>
> as well as by (5.3). We thus arrive at one of the two equations:
>
> $$\nabla^2 \psi - \frac{8\pi^2}{h^2} V\psi \mp \frac{4\pi i}{h}\frac{\partial \psi}{\partial t} = 0 \tag{5.4''}$$
>
> *We will require the complex wave function ψ to satisfy one of these two equations.* Since the conjugate complex function $\overline{\psi}$ will then satisfy the *other* equation, we may take the real part of ψ as the real wave function (if we require it). In the case of a conservative system, (5.4″) is essentially equivalent to (5.4) as the real operator may be split into the product of the two conjugate complex operators if V does not contain time. (Schrödinger 1982, p. 104)

This is hardly only "a *somewhat* different route." It is a radically new way of dealing with the problem at hand and quantum theory. This is a remarkable moment in the history of QM, although rarely, if ever, seen as such, or even discussed at all. Thus, Mehra and Rechenberg merely say that "it was possible to write, for the time-dependent equations by considering the fact that the time dependence of the wave function might be expressed not by the second-order differential equation (5.3) but by a linear relation (5.3′). … The price one had to pay was that the wave function ψ … definitively became a *complex* object" (Mehra and Rechenberg 2001, v. 5, p. 789). Is it a price to pay or a major benefit for the future of QM? Schrödinger clearly thought the first. He speaks of "a certain crudeness in the use of a *complex* wave function" and, again, sees it merely as a calculational tool, with a proper theory requiring a real wave function of "probably the fourth order" (Schrödinger 1982, p. 122–123). On the other hand, subsequent history proves the second for both the complex wave function and the first-order derivative in time (on which Mehra and Rechenberg do not comment at all), and not only for calculations enabling correct predictions but also conceptually. For this might well be how our interactions with nature via mathematics and experimental technology work in the case of quantum phenomena.

The first-order derivative in time and the complex wave function are fundamentally related to probability, as was discovered by Born, which was, again, not something that Schrödinger wanted, as he often said. Schrödinger's time-dependent equation preserves the probability current, just as does Dirac's equation, whose nonrelativistic limit is Schrödinger's time-dependent equation.

Mehra and Rechenberg do not consider how this relation, "for computational purposes," comes about. Nor do they ask what does it mean to be "correct *in principle*," as Schrödinger says. Rather than a deduction, this relation may be seen as a *guess* demanded by calculational purposes, not unlike Heisenberg's guess of his matrix variables. Is there a more rigorous derivation, or does a right guess suffice? There is no simple answer to this question. It is worth nothing, however, that, although we have different derivations of Schrödinger's time-dependent equation, we do not appear to have a fully rigorous derivation of it, or his time-independent equation or of QM for continuous variables in general.[4] As it happened, mathematical and specifically calculational demands led to a major discovery that was consistent with the mathematical structure of the theory being developed but was not derived from this structure, while decisively advancing the theory itself. That was already true in the case of Schrödinger's time-independent equation, which, too, was mainly guessed, certainly initially, as is clear from Schrödinger's notebooks. It is, however, difficult to see his derivation of it in his published articles as a fully rigorous deduction from first principles either.

He appears to have proceeded roughly as follows. De Broglie's formula for the speed of the phase wave of an electron, adjusted for the speed of the electron in the electric field of a hydrogen nucleus, is inserted into the classical relativistic wave equation. Since de Broglie's formula conveys both the particle and the wave aspects of the behavior of quantum objects, the nature of the equation changes. (One must of course keep in mind that de Broglie's waves are not the same as Schrödinger's standing wave, which his time-independent equation describes.) Unfortunately, the resulting equation, now known as the Klein–Gordon equation (which Schrödinger appears to be the first to have written), does not work for a relativistic electron because the predictions one makes by using it are in conflict with experimental results. One needs Dirac's equation to make correct relativistic predictions. However, if one drops terms that are small in the nonrelativistic limit, one arrives, yet again, (essentially by way of a guess!) at a different equation, (5.1) or (5.1′) above. The resulting equation happens to offer correct predictions in the nonrelativistic case. Accordingly, Schrödinger's time-independent equation becomes even more of a guess, albeit a correct guess, which would need to be justified otherwise or even require that the equation be derived otherwise. This is what Schrödinger attempted to do in his first published paper on the subject, which derives his equation differently. Eventually, Dirac showed Schrödinger's (time-dependent) equation to be the nonrelativistic limit of his equation, which straightened out the difficulty of connecting a wrong relativistic equation to the right nonrelativistic one. As discussed below, this was an important

[4] The case of deriving QM for discrete variables (which may be considered more rigorous but which has its complexities as well) will be considered in Chap. 9.

point of Dirac's argument concerning his equation. By that time, Schrödinger's time-dependent and time-independent equations were firmly established as the correct nonrelativistic equations for the electron, essentially, again, as mathematical postulates of QM, rather than deductions from first principles, as Heisenberg said about the equations of QM in general at the time (Heisenberg 1930, p. 108).

Schrödinger's new derivation was also more representative of his program for a wave mechanics. Schrödinger now viewed the problem of quantization as an eigenvalue problem, treated by variational methods, with atomic spectra derived accordingly from his equation. As is clear from his notebooks, Schrödinger's alternative derivation was linked to and possibly motivated by the fact that some of the predictions based on his wave equation coincided with those of Bohr's and Sommerfeld's theory, which used a classical-like Hamiltonian approach. Schrödinger explained these relationships in a way that led him to the derivation of his equation found in his first paper. He did so by replacing—again, without a rigorous theoretical justification from first principles—the mechanical Hamilton–Jacobi equation $H(q, (q, \partial S/\partial q) = E$ with a wave equation by substituting $S = K\ ln\ c$ (where K is a constant that has a dimension of action). This step led him, via a mechanical-optical analogy, to the equation and then to the right time-independent equation for the nonrelativistic hydrogen atom (5.1) or (5.1′). Schrödinger's key step—made via the mechanical-optical analogy and the connections between the principle of least action in mechanics and Fermat's principle in optics—was to give to S the wave form by putting in $S = K\ ln\ c$. One thus also replaces the deterministic mechanics of the particle motion with amplitudes and then probabilities, although Schrödinger did not realize this at the time.

Schrödinger extended his program in his second and third papers, which, as Mehra and Rechenberg note, "establishe[s] the foundations and the definite outlines of what was later [in his paper] called 'wave mechanics'" (Mehra and Rechenberg 2001, v. 5, p. 533). I shall only comment on a few key points here. The program was still to be enacted through the mechanical-optical analogy, which accompanied the Hamilton–Jacobi framework for classical mechanics since Hamilton's work. Schrödinger wanted to "throw more light on the general correspondence which exists between the Hamilton–Jacobi differential equation of a mechanical problem and the 'allied' *wave equation*," Schrödinger's time-independent equation (Schrödinger 1982, p. 16). The mathematical procedures used in the first paper were declared "unintelligible" and "incomprehensible." It becomes clear from his papers on wave mechanics and from his correspondence that Schrödinger continued to adjust his theory and expected that some changes of the physical concepts involved were likely to be necessary in order to preserve the character of the theory as he envisioned it. As his conclusion, cited below, to the fourth installment of his program indicates, he thought that such further adjustments will still be necessary even after the formalism of his wave mechanics was essentially established with his time-dependent equation. It would perhaps be more accurate to say that the subsequent history proved that this formalism was, thus, essentially established, because Schrödinger himself did not quite see it as finalized, especially in view of his use of the complex wave function. The situation was not destined to change, however, including as concerns this use, unhappily for Schrödinger but with great benefits for QM. Schrödinger aimed at a wave-like deterministic picture, but arrived elsewhere.

This "arriving elsewhere" from his physical program of capturing, by means of his wave-mechanics, "the wave radiation forming the basis of the universe," is the story of Schrödinger's project. This story was manifested in particular in his moves made from within his mathematics without a proper physical justification from within his representational program. That is not to say that these moves were not justified otherwise. The ultimate justification is in their mathematical structures that lead to the agreement of the predictions with the experiment enabled by these structures, as Heisenberg, again, said concerning the derivation of the formalism, by that time (1930), including Schrödinger's time-dependent equation (Heisenberg 1930, p. 108). In other words, the ultimate justification of such moves is in the possibilities of calculations created by them. Some of these moves are dictated by the inner mathematical demands but are not rigorously derived from the formalism, as in the case of Schrödinger's introduction of his first-order derivative in time and the complex wave function in his time-dependent equation. The necessary calculations would not be possible without getting rid of the energy parameter in the time-independent equation. However, doing so in a way that would be justified from within his program, would require an equation (likely that of the fourth order) that Schrödinger could not properly establish, in in view the difficulties mentioned above. As he said in closing his article:

> [T]here is no doubt a certain crudeness in the use of a *complex* wave function. If it were unavoidable *in principle*, and not merely a facilitation of calculations, this would mean that there are in principle *two* wave functions, which must be used together to obtain information on the state of the system. This somewhat unacceptable inference admits, I believe, of the very much more congenial interpretation that the state of the system is given by a real function and its time derivative. Our inability to give more accurate information about this is ultimately connected to the fact that, in the pair of equations (5.4″), we have before us only the *substitute*—extraordinarily convenient for the calculations, to be sure—for the real wave equation of probably the fourth order, which, however, I have not succeeded in forming for the non-conservative [time-dependent] case. (Schrödinger 1982, p. 123)

There might be some doubt, however, given that one might find, and not a few, Bohr, for one, had found, in this use a form of refinement of the mathematics of quantum theory. This use one has "a certain crudeness" only if one assumes, as Schrödinger did, that it is merely a substitute for a proper mathematical representation of the physical reality responsible for quantum phenomena. Otherwise, the predictions made by means of this function were, and still are, in exact correspondence with what is observed. There was not yet a probabilistic interpretation of the wave function, which gives one less, if any, hope that the same function could be used for representing and predicting the behavior of electrons ideally exactly, as Schrödinger wanted in setting up his program. Was "a *real* wave equation," able to provide such representation and predictions, possible? Perhaps. Is it still possible? Perhaps, although at present it appears unlikely in the framework of QM or QFT. In any event, Schrödinger never found such an equation. His substitute, extraordinarily convenient for calculations, derived by a "somewhat unacceptable inference" may in fact be, against Schrödinger's hope expressed here, "unavoidable *in principle*." Moreover, by virtue of its mathematical features, taken beyond their calculational function, it offered new pathways for conceptual thinking concerning QM and QFT, which retains these features. Mathematics led our thinking in the *mathematical*-experimental project of modern physics to a new theory, QM, but in a different way, along a different path.

This unexpected destination did not satisfy Schrödinger himself, who, while accepting QM as an effective calculational technology, came to repudiate it conceptually. In his cat-paradox paper, he described it as "a doctrine born of distress," in part, again, as "perhaps [he was wise to qualify] after all only a convenient calculation trick," which, however, "today … has attained influence of unprecedented scope over our basic attitude toward nature" (Schrödinger 1935, pp. 154, 167). As discussed earlier, one might doubt Schrödinger's claim concerning the scope of this influence. Most physicists and philosophers then and since took Einstein's and Schrödinger's view of the theory, rather than that of Bohr, as Bohr acknowledged (Bohr 1987, v. 2, p. 65). Schrödinger's assessment, however, would have applied to the Göttingen–Copenhagen circle, clearly on his mind.

Be that as it may on this score, calculations and the ways of calculations are, as I argue in this study, not separate from concepts, mathematical, physical, and philosophical, and may give rise to these concepts, just as reciprocally, concepts can lead to new ways of calculations. And sometimes either can happen against the grain of our aims and agendas. This was the case in Schrödinger's work. His calculations were continually bringing him to other places than he wanted to reach, ultimately landing his mathematics far from its intended destination. Schrödinger discovered new territories in his journey, sometimes by trying to escape them and yet landing somewhere else than he aimed to land.

It is true, that the mathematics of QM leaves open the question concerning the ultimate nature of physical reality and our interaction with it by means of mathematics and experimental technology, the interactions that ground the concepts and calculations of any theory, and hence also the possibility of alternative theories of this reality. Schrödinger came to strongly doubt, as did Einstein, that QM could serve as a starting point for developing such as a theory, or be interpreted in conformity with this imperative, although such interpretations have been proposed. These considerations, however, do not affect my argument that new theories, and their new concepts *may* (it need not always happen in this way) emerge independently, or against the grain, of their intended physical or mathematical aims and procedures. This was the case in Heisenberg's invention of his new variables and (in this case, against his own grain) in Schrödinger's invention of both of his equations. Bohr's concept of "quantum jumps" was even more radical in this respect, going against all preceding assumptions concerning how nature works. It brought physics to an entirely new and even previously unimaginable place, although in this case mathematics did not play the kind of role it did in the case of Heisenberg's, Schrödinger's, and Dirac's work. That does not mean that these inventions come from nowhere, but only that they come not only from within a given theory (established or being developed) but also from other trajectories of thought, some of which are difficult and even impossible to track. "Their ultimate justification," as Heisenberg said, "lies in the agreement of [their] predictions with the experiment," and thus in the calculations they enable (Heisenberg 1930, p. 108). Apart from these calculations, these predictions are not possible. But calculations also lead to new concepts, theories, and interpretations, to new ways of thinking.

5.4 From Schrödinger to Dirac

Dirac's thinking concerning the free relativistic electron was based in the fundamental principles of both relativity and quantum theory, with those of quantum theory being represented Schrödinger's time-dependent equation. The first-order derivative in time was to play a key role in Dirac's theory. Dirac's starting point, shaped by the mathematical character of his thinking, was that, in order to satisfy, and to mathematically express, the physical principles of both theories, the equation he needed had to have certain specific mathematical features. One of them, in accordance with the principles of QM, as manifested in Schrödinger's time-dependent equation, was the first-order derivative in time. By symmetry, required by special relativity, this equation would then, now unlike Schrödinger's time-dependent equation, also have to have the first-order derivative in spatial coordinates. Then the right variables that satisfy this equation needed to be found. Dirac accomplished both tasks.

Dirac followed Heisenberg's first paper on QM (Heisenberg 1925), discussed in Chap. 4, insofar as his aim was the invention of a consistent mathematical scheme by means of which one could predict, probabilistically or statistically, the outcomes of relevant experiments. Dirac, as is well known, read Heisenberg's paper very carefully earlier and used its key ideas, in particular, the noncommutative nature of Heisenberg's formalism, in developing his own version of QM (Dirac 1925). Dirac's theory, thus, conformed to the QD postulate, applied to observed high-energy quantum phenomena, the QP/QS postulate. Dirac also adopted a form of the correspondence postulate, defined by the fact that his equation should convert into Schrödinger's equation in the nonrelativistic limit. In Heisenberg, this correspondence was between the equations of QM and those of classical mechanics, and it was more automatic because his equations were formally the same as those of classical mechanics. Only his variables were different. By contrast, Dirac's equation not only used different variables but was also not formally the same as either any classical equation or Schrödinger's equation. Dirac used the mathematical correspondence postulate in his earlier work on quantum mechanics, inspired by Heisenberg's paper. Indeed, while Heisenberg was the first to convert Bohr's correspondence *principle* into a mathematical *postulate*, it was Dirac who appears to have been the first to expressly formulate this postulate. In his first paper on quantum mechanics, he said: "The correspondence between the quantum and classical theories lies not so much in the limiting agreement when $h = 0$ as in the fact that the mathematical operations on the two theories obey in many cases the same [formal] laws" (Dirac 1925, p. 315). Dirac's equation was a new equation even formally. Both the equations and the variables in it were new. Accordingly, proving that his equation would convert into Schrödinger's time-dependent equation in the nonrelativistic limit was mathematically more difficult to prove than in Heisenberg's case.

As against Heisenberg, who, at least in his paper introducing QM, adopted a form of RWR view, Dirac, as mentioned, appears to have always retained the view that he introduced in his paper on transformation theory (Dirac 1927). According to this view, the probabilistic or statistical nature of quantum predictions is due to our interference into quantum processes by measurement, while the undisturbed behavior of quantum objects is classically causal and is represented by the formalism of quantum theory. Dirac never addressed the difficulties that this claim entails, for example, the role of complex variables in the formalism, which one cannot directly associate with what occurs in space and time. Whatever Dirac's own views might have been, however, his formalism is open to an RWR-type interpretation, as discussed in relation to QFT in Chap. 8.

Dirac's starting point, as that of others, in particular, Oscar Klein, who were pursuing a relativistic extension of QM at the time, was that, in accordance with the *principles of special relativity theory*, time and space must enter symmetrically and be interchangeable. This feature is not found in Schrödinger's time-dependent equation, which contained the first derivative of time and the second derivatives of coordinates, although the equation is consistent with relativity, because it satisfies the locality principle (which prohibits an action at a distance). On the other hand, by consistency with QM, the equation Dirac needed must have been *first-order* linear in time, just as Schrödinger's time-dependent equation was.

Remarkably, given the naturalness of this logic, it appears that only Dirac thought of the situation in this way at the time. His famous conversation with Bohr reflects this fact:

> Bohr: What are you working on?
>
> Dirac: I am trying to get a relativistic theory of the electron.
>
> Bohr: But Klein already solved that problem. (Dirac 1962)

Dirac disagreed, and, for the reasons just explained, it is clear why he did and why Bohr should have known better. If Schrödinger's time-dependent equation was correct, as it was surely assumed to be by Bohr, the Klein–Gordon equation, to which Bohr referred, was unlikely to be right because it contains the second-order derivative in time and thus would not to give Schrödinger's equation as its nonrelativistic limit. Bohr of course changed his mind after Dirac's discovery, and subsequently saw Dirac's theory of the electron as a confirmation of the principles of quantum theory that he advocated, including along RWR lines. The Klein–Gordon equation is relativistic and symmetrical in space and time with both variables entering via the second derivative, $\frac{\partial^2}{\partial t^2}, \frac{\partial^2}{\partial x_i^2}$. One can derive the continuity equation from it, but the probability density is not positive definite. Nor, again, does it give one Schrödinger's time-dependent equation in the nonrelativistic limit. Schrödinger, who, as noted, appears to be the first to have written down the Klein–Gordon equation in the process of his discovery of his wave mechanics, abandoned it in view of this latter fact. Schrödinger manipulated the equation to derive, essentially by a guess, his time-independent equation. Later on, he established his time-dependent equation by considerations, discussed above, which no longer had anything to do with the Klein–Gordon equation and which led Schrödinger to the first-order derivative in time and the complex wave function, both features found in Dirac's equation. Dirac's equation, a kind of a square root of the Klein–Gordon equation, converts into Schrödinger's equation in the nonrelativistic limit, which was a major factor in Dirac's thinking. Technically, at its immediate nonrelativistic limit, Dirac's equation, which contains spin automatically, converts into Pauli's spin-matrix theory, while Schrödinger's equation, which does not contain spin, is the limit of Pauli's theory, if one neglects spin (Pauli 1927).

As those of Heisenberg's matrix mechanics, Dirac's new variables, guessed or just about by Dirac as well, proved to be noncommuting matrix-type variables, but of a more complex character, involving the so-called spinors and the multicomponent wave functions, the concept discovered by Pauli in his nonrelativistic theory of spin (Pauli 1927). Just as Heisenberg's matrices, Dirac's spinors had never been used in physics previously, although they were introduced in mathematics by William C. Clifford about fifty years earlier (following the work of Hermann Grassmann on exterior algebras). And just as Heisenberg in the case of his matrix variables, Dirac, too, was unaware of the existence of spinors and reinvented them. His equation enabled him to answer, for the relativistic electron, the question, "What is the probability of any dynamical variable at any specified time having a value laying between any specified limits, when the system is represented by a given wave function ψ_{n}?" which the Klein–Gordon theory could only answer for "the position of the electron … but not [for] its momentum, or angular momentum, or any other dynamic variable" (Dirac 1928, pp. 611–612). Given my main concern here, that of the significance of the key mathematical features of Schrödinger's time-dependent equations, I shall put aside Dirac's derivations of his equation, which I discussed from the RWR perspective elsewhere (Plotnitsky 2016, pp. 201–225). I shall address the implications of Dirac's theory itself, in particular the discovery of antimatter as a consequence of his equation in considering QFT and the question of elementary particles in Chap. 8. It is, however, difficult to bypass one key aspect of this discovery. Dirac's equation contained a major difficulty, or what appeared as such when the equation was introduced, a difficulty inherited from the Klein–Gordon equation. According to Dirac:

> [Either equation] refers equally well to an electron with charge *e* as to one with charge – *e*. If one considers for definitiveness the limiting case of large quantum numbers one would find that some of the solutions of the wave equation are wave packets moving in the way a particle of – e would move on the classical theory, while others are wave packets moving in the way a particle with charge e would move classically. For this second class of solutions *W* has a negative value. One gets over the difficulty on the classical theory by arbitrarily excluding those solutions that have a negative *W*. One cannot do this on the quantum theory, since in general a perturbation will cause transitions from states with *W* positive to states with W negative. Such a transition would appear experimentally as the electron suddenly changes its charge from –*e* to *e*, a phenomenon which has not been observed. The true relativi[stic] wave equation should thus be such that its solutions split up into two non-combining sets, referring respectively to the charge –*e* and the charge *e*. …

> The resulting theory is therefore still only an approximation, but it appears to be good enough to account for all the duplexity phenomena without arbitrary assumptions. (Dirac 1928, p. 612)

Dirac's theory inherits this (second) problem of the Klein–Gordon theory because, as noted, mathematically Dirac's equation may be seen as a "square root" of the Klein–Gordon equation, which means that every solution of Dirac's equation is a solution of the Klein–Gordon equation. The opposite is not true, which reflects the fact that the Klein–Gordon equation ultimately does not work. The difficulty does not appear in the low-energy regime, or rather, it disappears at the low energy limit, because Dirac's equation converts into Schrödinger's equation, where this problem does not arise. In commenting on the problem in his Chicago lectures, Heisenberg amplified Dirac's assessment: "The classical theory could eliminate this difficulty by arbitrarily excluding the one sign, but this is not possible to do according to *the principles of quantum theory*. Here spontaneous transitions may occur to the states of negative value of energy E; as these have never been observed, *the theory is certainly wrong*. Under these conditions it is very remarkable that the positive energy-levels (at least in the case of one electron) coincide with those actually observed" (Heisenberg 1930, p. 102; emphasis added).

"Certainly right" would have been a better assessment, admittedly difficult initially. Dirac's theory proved to be better than it appeared at the time of its introduction, even to its creator. It has proven to be right (to the first order of approximation, with the well-known difficulties arising at higher orders of approximation where the infinities occur, requiring renormalization). What was wrong was the understanding of nature in high-energy quantum regimes at the time. That, in general, a perturbation will cause transitions from states with positive E to states with negative E, and that such a transition would appear experimentally as the electron suddenly changes its charge from $-e$ to e, is what actually happens, as was experimentally established in a year or so. (The energy itself is of course always positive.) This feature is inherent in all high-energy quantum processes, and is a reflection of the fact that particles are born and disappear, and transform into each other in these regimes. Antimatter was staring into the physicists' eyes. It took, however, a few years to realize that it was antimatter and that this type of transition, eventually understood in terms of the creation and annihilation of particles, defined all high-energy quantum physics.[5]

In spite of the elegant simplicity of its compact form,

$$i\gamma \cdot \partial\psi = m\psi,$$

reproduced on the plate in Westminster Abbey commemorating Dirac, Dirac's equation encodes a complex Hilbert-space machinery. The equation, as introduced

---

[5] Later on, Dirac lost this confidence in QED. This was not because he lost his faith in the fundamental principles of quantum theory, but rather because of the theory's inability to give these principles a proper (which for Dirac also meant elegant) mathematical expression. However, inelegant and even messy, and to some mathematically questionable (especially because of its dependence on renormalization), as it might be, QED and then QFT, proved to be a remarkable success, with QED being the best experimentally confirmed theory ever.

by Dirac, was.

$$\left(\beta mc^2 + \sum_{k=1}^{3} \alpha_k p_k c\right)\psi(x,t) = i\hbar\frac{\partial\psi(x,t)}{\partial t}$$

The new mathematical elements here are the 4 × 4 matrices α_k and β and the four-component wave function ψ. The Dirac matrices are all Hermitian,

$$\alpha_i^2 = \beta^2 = I_4$$

(I_4 is the identity matrix), and they mutually anticommute:

$$\alpha_i\beta + \beta\alpha_i = 0$$
$$\alpha_i\alpha_j + \alpha_j\alpha_i = 0$$

The above single symbolic equation unfolds into four coupled linear first-order partial differential equations for the four quantities that make up the wave function. The matrices form a basis of the corresponding Clifford algebra. One can think of Clifford algebras as quantizations of Grassmann's exterior algebras, in the same way that the Weyl algebra is a quantization of symmetric algebra. ***p*** is the momentum operator in Schrödinger's sense, but in a more complicated Hilbert space than in standard quantum mechanics. The wave function *ψ (t,* ***x****)* takes value in a Hilbert space $X = C^4$ (Dirac's spinors are elements of *X*). For each *t, y (t,* ***x****)* is an element of $H = L^2 (R^3; X) = L^2 (R^3) \otimes X = L^2 (R^3) \otimes C^4$. This mathematical architecture allows one to predict the probabilities of quantum-electrodynamical (high-energy) events, which, as explained in detail in Chap. 8, have a greater complexity than low-energy events.

Beginning with Heisenberg, inventing or finding new matrix-type variables or, more generally, Hilbert-space operators became the defining mathematical element of quantum theory. The current theories of weak forces, electroweak unifications, and strong forces (QCD) were all discovered by finding such variables. This is correlative to establishing the transformation group, a Lie group, of the theory and finding representations of this group in the corresponding Hilbert spaces. This is true for Heisenberg's matrix variables as well, as was discovered by Weyl and Eugene Wigner, who introduced the Heisenberg group. In modern elementary-particle theory, irreducible representations of groups correspond to elementary particles, the idea that was one of Wigner's major contributions to quantum physics (Wigner 1939; Newton and Wigner 1949). This was how Murrey Gell-Mann discovered quarks, because at the time there were no particles corresponding to the irreducible representations (initially there were three of those, corresponding to three quarks) of the symmetry group of the theory, the so-called SU (3). It is the group of all rotations around the origin in three-dimensional space, R^3, rotations represented by all three-by-three orthogonal matrices with their determinant equal to one. (This group is noncommutative.) The electroweak group that Gell-Mann helped to find as well involves SU (2), the group of two by two matrices with the determinant 1.

Finally, in one those wonderful ironies with which physics is replete, the development of QFT also resurrected the Klein–Gordon equation as one of its important tools, as the equation obeyed by the components of all free quantum fields. The equation describes spinless relativistic particles, usually composite ones, like the pion. However, because the Higgs boson is a spin-zero particle, it is the first elementary particle to be described by the Klein–Gordon equation. No less than Dirac's, or Schrödinger's equation, or the use of group theory (which accompanies all these equations in quantum theory), the Klein–Gordon equation, which—it is worth reiterating—helped Dirac's work on his equation, brought together calculations and concepts, calculations leading to new concepts and new concepts enabling new calculations, in QFT theory.

5.5 Conclusion

Schrödinger's time-dependent equation, I have argued here, is an extraordinary example of the interplay of calculations and the invention of concepts, the complex wave function among them. While he acknowledged and appreciated the power of his calculations, Schrödinger himself was not happy with these concepts. They were, however, a great gift of his thought to quantum theory, including eventually QED. Bohr said as much in his famous conversation with Schrödinger at Bohr's home in Copenhagen in August of 1926, as reported by Heisenberg. Schrödinger, in defending his "continuum theory" questioned the idea of quantum jumps:

> Schrödinger: Surely you realize that the whole idea of quantum jump is bound to end up in nonsense. Is this jump supposed to be gradual or sudden? If it is gradual, the orbital frequency and energy of the electron must change gradually as well. But in this case, how do you explain the persistence of fine spectrum lines? On the other hand, if the jump is sudden … then we must ask ourselves precisely how the electron behaves during the jump. Why does it not emit a continuous spectrum, as electromagnetic theory demands? And what laws govern its motion during the jump? In other words, the whole idea of quantum jump is sheer fantasy.
>
> Bohr: What you say is absolutely correct. But it does not prove that there are no quantum jumps. It only proves that that we cannot imagine them, that the representational concepts with which we describe events of daily life and experiments in classical physics are inadequate when it comes to describing quantum jumps. Nor should we be surprised to find it so, seeing that the processes involved are not the objects of direct experience. … We have known what Planck's formula means for twenty-five years. And, quite apart from that, we can see the inconstancies, the sudden jumps in atomic phenomena quite directly, for instance when we watch sudden flashes of light on a scintillation screen or the sudden rush of an electron through a cloud chamber. You cannot simply ignore these observations and behave as if they did not exist at all.
>
> Schrödinger [conceding as much, although not yet quite his hopes of his program of wave mechanics]: If all this damned quantum jumping were really here to stay, I should be very sorry that I ever got involved with quantum theory.

> Bohr: But the rest of us are extremely grateful that you did. Your wave mechanics has contributed so much to mathematical clarity and simplicity that it represents a gigantic advance over all previous forms of quantum mechanics. (Heisenberg 1971, pp. 73–76)

Bohr's argument concerning quantum jumps is in accord with the argument of this study: they represent a *physical* concept (as defined in Chap. 2) that reflects the ultimate nature of the reality responsible for quantum phenomena as a reality without realism, and hence cannot be captured by a daily concept, any more than anything that happened between experiments. As discussed in Chaps. 3 and 4, from where Bohr stood by that time, all quantum transitions were in effect quantum jumps, "transitions without connections," between quantum states. As Bohr argued later, "experiments in quantum physics" are equally defined by classical physical concepts, as a refinement of our daily concepts, a refinement no longer possible, in the RWR view, for representing the ultimate constitution of reality responsible for quantum phenomena. Bohr's comments here contain the kernel of his future RWR-type interpretation, even though such key ingredients as the uncertainty relations, complementarity, or even Born's probabilistic interpretation of the wave function were not yet in place. On the other hand, as noted, Bohr's brief and ambivalent return to a partial realism, still different from Schrödinger's program, for one thing, because Bohr retained quantum jumps, in the Como lecture was influenced by wave mechanics, vis-à-vis his preceding position, essentially shaped by Heisenberg's approach.

Bohr was just as much grateful for the concepts (no longer daily concepts!) arising from Schrödinger's formalism, the complex wave function in particular, as for its extraordinary calculational power. While Schrodinger had perhaps fewer things to regret than he thought, no qualifications are necessary concerning Bohr's feelings. Schrödinger's formalism was a gift to Bohr's work on his interpretation of quantum phenomena and QM. It would not be the same without Schrödinger's thinking, which led Schrödinger, even in part against his own grain, to "a gigantic advance over all previous forms of quantum mechanics." It was another triumph of mathematics in quantum theory.

References

Bitbol, M.: Schrödinger's Philosophy of Quantum Mechanics. Kluwer, Dordrecht, Netherland (1996)

Bohr, N.: The Philosophical Writings of Niels Bohr, vol. 3. Ox Bow Press, Woodbridge, CT, USA (1987)

Born, M.: Zur Quantenmechanik der Stoßvorgänge. Z. Phys. **37**, 863–867 (1926a)

Born, M.: Quantenmechanik der Stoßvorgänge. Z. Phys. **38**, 803–827 (1926b)

Dirac, P.A.M.: The fundamental equations of quantum mechanics. In: van der Warden, B.L. (ed.) Sources of Quantum Mechanics New York, 1968, pp. 307–320. Dover, NY, USA (1925)

Dirac, P.A.M.: The physical interpretation of the quantum dynamics. Proc. R. Soc. Lond. A **113**, 621–641 (1927)

Dirac, P.A.M.: The quantum theory of the electron. Proc. R. Soc. Lond. A **177**, 610–624 (1928)

Dirac, P.A.M.: Report KFKI-1997–62, Hungarian Academy of Science (1962)

Heisenberg, W.: Quantum-theoretical re-interpretation of kinematical and mechanical relations. In Van der Waerden, B.L. (ed.) Sources of Quantum Mechanics, pp. 261–277. Dover, New York, NY, USA. Reprint 1968 (1925)
Heisenberg, W.: The Physical Principles of the Quantum Theory (tr. Eckhart, K., Hoyt, F.C.). Dover, New York, NY, USA (rpt. 1949) (1930).
Heisenberg, W.: Physics and Beyond: Encounters and Conversations. G. Allen & Unwin, London, UK (1971)
Karam, R.: Schrödinger's original struggles with a complex wave function. Am. J. Phys. **88**, 433 (2020). https://doi.org/10.1119/10.0000852
Mehra, J., Rechenberg, H.: The Historical Development of Quantum Theory, vol. 6. Springer, Berlin, Germany (2001)
Newton, T.D., Wigner, E.: Localized states for elementary systems. Rev. Mod. Phys. **21**, 400–406 (1949)
Pauli, W.: Zur Quantenmechanik des magnetischen Elektrons. Z. Phys. **43**, 601–625 (1927)
Plotnitsky, A.: The Principles of Quantum Theory, from Planck's Quanta to the Higgs Boson: The Nature of Quantum Reality and the Spirit of Copenhagen. Springer/Nature, New York, NY, USA (2016)
Schrödinger, E.: Zur Einsteinschen Gastheorie. Physicalische Zeitschrift **27**, 95–101 (1926)
Schrödinger, E.: Four Lectures on Wave Mechanics. Blackie & Son, London, Glasgow, UK (1928)
Schrödinger, E.: The present situation in quantum mechanics. In: Wheeler, J.A., Zurek, W.H. (eds.) Quantum Theory and Measurement, 1983, pp. 152–167. Princeton University Press, Princeton, NJ, USA (1935)
Schrödinger, E.: Collected Papers on Wave Mechanics (tr. J. F. Shearer). Chelsea, New York, NY, USA (1982)
Schrödinger, E.: Interpretation of Quantum Mechanics: Dublin Seminars (1949–1955) and Other Unpublished Essay. Ox Bow Press, Woodbridge, CT, USA (1995)
Wigner, E.P.: On unitary representations of the inhomogeneous Lorentz group. Ann. Math. **40**, 149–204 (1939)

Chapter 6
Niels Bohr and the Character of Physical Law: "A Radical Revision of Our Attitude Toward the Problem of Physical Reality"

It is just this entirely new situation as regards the description of physical phenomena that the notion of complementarity aims at characterizing.
—Niels Bohr, "Can Quantum-Mechanical Description of Physical Reality Be Considered Complete?" (1935) (Bohr 1935, p. 700).

Abstract The chapter considers Bohr's thinking leading him to "a radical revision of our attitude toward the problem of physical reality" and to his ultimate, strong RWR-type, interpretation, and his key concepts, complementarity and phenomena, which shaped this interpretation. Complementarity was central to all versions of Bohr's interpretation, but the concept was revised by Bohr in his ultimate interpretation. The concept of phenomena was new. After a general introduction in Section 6.1, Section 6.2, by way of a prologue, considers Bohr's controversial appeal to the "irrationality" brought into quantum physics by Planck's discovery of the quantum of action, *h*. Section 6.3 outlines Bohr's ultimate, strong RWR-type, interpretation. Section 6.4 discusses the structure of quantum measurement as an entanglement between the quantum object considered and the measurement instrument used. Section 6.5 is devoted to an analysis of complementarity. Section 6.6 considers the concept of probabilistic causality (in the absence of classical causality) and relates this concept to complementarity, which Bohr saw as a "generalization of causality."

Keywords Discreteness · Causality · Complementarity · Entanglement · Individuality · Phenomenon · Quantum objects · Quantum causality · Quantum measurement · Reality · Reality without realism · Uncertainty relations

6.1 Introduction

I borrow the phrase "the character of physical law" in my title from Feynman's title of his book *On the Character of Physical Law* (Feynman 1965). There are several reasons for doing so. First of all, Bohr's concepts, such as and in particular that of complementarity, were designed to "provide room," a conceptual space, for laws of quantum physics as different from those of classical physics or relativity, but fully

A. Plotnitsky, *Reality Without Realism*,
https://doi.org/10.1007/978-3-030-84578-0_6

consistent with "the basic principles of science," a task that led Bohr to the strong RWR view by the late 1930s. This task, however, also required him to reassess the character of physical law in classical physics and relativity vis-à-vis those of quantum physics, for it was not only a question of replacing old laws with new ones, but also that of properly delimiting old laws in view of new ones. Accordingly, a reassessment of the character of physical law in general was at stake. As Bohr said:

> It is only the mutual exclusion of any two experimental procedures, permitting the unambiguous definition of complementary physical quantities, which provides room for new physical laws the coexistence of which might at first sight appear irreconcilable with the basic principles of science. It is just this entirely new situation as regards the description of physical phenomena that the notion of *complementarity* aims at characterizing. (Bohr 1935, p. 700)[1]

While the first sentence is not the easiest sentence to parse, its meaning is clear: the unambiguous definition of complementary physical quantities, such as those associated, respectively, with the position and momentum measurement, requires considering the measuring procedures by which they are determined, and thus two mutually exclusive phenomena, as defined by Bohr's concept of complementarity. Otherwise, the new physical laws, such as and in particular, the uncertainty relations, would appear to be in an irreconcilable conflict with the basic principles of science. The same argument applies to Bohr's other key concepts, especially that of phenomena, and for his interpretation, ultimately of the strong RWR-type, with which the concept of complementarity, used by Bohr earlier, was brought into accord. His concept of phenomena was introduced along with his ultimate interpretation.

Feynman's discussion of QM in his book is also famous for his pronouncement that "nobody understands quantum mechanics" (Feynman 1965, p. 129). Feynman clearly refers as much to quantum phenomena as to QM. It is not so difficult to understand, and Feynman certainly did, QM as a mathematical theory predicting quantum phenomena. QFT is more complex mathematically, but Feynman, who won his Nobel prize for the renormalization of QED, was certainly among those who understood its mathematics best. It is, however, a different question why either theory works, and works so well, and this question is indissociable from understanding quantum phenomena. While I do not think that Feynman's statement is literally true, it suggests the possibility of the RWR view of quantum phenomena and QM or QFT. If one adopts this view, it is, by definition, impossible to understand the ultimate nature of the reality responsible for quantum phenomena and, thus, why QM or QFT work. I do not attribute this view to Feynman, who never appeared to comment on this possibility. On the other hand, his dissertation director, Wheeler, for whom Bohr has often served as an inspiration, was open to the RWR view and might have even adopted a form of this view, at least at some points of his lifelong engagement with quantum theory, a form of this view. Just as those of Bohr,

[1] Sometimes, Bohr also used the term "complementarity" to designate his overall viewpoint on or his interpretation of quantum phenomena and QM. For clarity, however, I shall, in referring to this interpretation, speak of Bohr's interpretation (in one or another of its versions), and by complementarity only refer to the concept of complementarity.

Wheeler's views changed, sometimes significantly. Wheeler's own versions of the RWR view, in particular, as defined by his concept of "participatory universe," were different from those of Bohr, especially Bohr's ultimate version of it. The degree of this difference, or proximity, would depend on an interpretation of Wheeler's views and concepts, that of a participatory universe in particular (e.g., Wheeler 1981, 1983, 1990). Of course, it would also depend on an interpretation of Bohr's interpretation (in its various versions), which, too, may be different from that offered in this study.

There are different forms of the RWW view and the corresponding interpretations of quantum phenomena and quantum theory. Thus, as stated from the outset of this study, even Bohr's ultimate interpretation, with which I will be especially concerned in this chapter, is different from the one offered in this study, even both are of the strong RWR-type and are close in some of their key aspects, because the present interpretation builds on Bohr's ultimate interpretation. Arguably, the most significant difference is, again, that in the present interpretation the concept of a quantum object is an idealization applicable only at the time of measurement, which does not appear to be the case in any of Bohr's interpretations, except by implication. In addition, as discussed in Chap. 8, this concept of a quantum object is linked by this study to the concept of quantum *field*, which, while physical and quantum, is not a quantum *object*. No concept of quantum field of this type is considered by Bohr either, even though the genealogy of this concept still extends from Bohr's concept of a quantum jumps as transition without connections, discussed in Chap. 3, and tangentially on Bohr and Rosenfeld's analysis of measurement in quantum field theory (Bohr and Rosenfeld 1933, 1950).

Bohr's key concepts, such as complementarity and phenomena, will also be primarily considered here as they figure in his ultimate interpretation, in the case of his concept of phenomena by definition, because it was introduced by Bohr as part of this interpretation. This interpretation may be seen as Bohr's response to the demand, stated by him (twice) in his 1935 reply to EPR, for "a radical revision of our attitude toward the problem of physical reality," in view of the character of quantum phenomena (Bohr 1935, pp. 697, 702). As discussed in Chap. 7, Bohr's interpretation in his reply was still short of his ultimate interpretation.

I have considered different versions of Bohr's interpretation in detail in (Plotnitsky 2012), and shall only comment on some of them here. That Bohr has never commented on these changes is of some interest. Thus, in referring to his earlier publications in his reply to EPR, Bohr says: "I shall therefore be glad to use this opportunity to explain in somewhat greater detail a general viewpoint, conveniently termed 'complementarity,' which I have indicated on various previous occasions" (Bohr 1935, p. 696). This is not entirely accurate. His reply to EPR went beyond merely explaining his previous views in "somewhat greater detail," although the significance of complementarity in quantum theory was indeed underappreciated by EPR. In particular, Bohr's statement, accompanied by a reference to his 1931 *Atomic Theory and the Description of Nature* (now Bohr 1987, v. 1), cannot apply to the Como lecture. It can, on the other hand, nearly apply to "The Quantum of Action and the Description of Nature," contained in that volume, or to Bohr's "Introductory Survey" there and in and the last article in the volume (both dated 1929). Still, his reply to EPR introduces further changes that helped Bohr to reach his strong RWR view shortly thereafter. I am hesitant to assume that Bohr aimed to develop a synthesis of his various views, as Bitbol argues in considering different versions of complementarity or rather different instantiations of the concept, rightly emphasizing their difference (Bitbol 1996, pp. 213–219). There does not appear to me to be much evidence in Bohr's published work or his *Nachlass* that he ever aimed at such a synthesis, especially given the significance of some of the changes of his views. It appears to me more likely that he saw his different interpretations and different version of complementarity as manifestations of a broader philosophical way of thinking, which need not imply that this way of thinking is a synthesis of these manifestations. In any event, Bohr's thinking eventually led him to his ultimate, strong RWR-type, interpretation. This interpretation also establishes a different relationship between Bohr's thought and Kant's philosophy than those proposed by Bitbol more recently (Bitbol 2019). Bitbol, in my view, makes Bohr too much of a Kantian. By contrast, my reading of Bohr suggests that, while, as suggested earlier, there is a Kantian vein in Bohr's thinking, one might see their relationship as that of both proximity and difference, even as both a close proximity and yet a radical difference. The radical nature of this difference arises by virtue of placing the ultimate nature of the reality responsible for quantum phenomena not only beyond knowledge, as Kant would, but also beyond conception, at least as things stand now.

The chapter will proceed as follows. The next section, Sect. 6.2, offers, by way of a prologue, a discussion Bohr's appeal to the "irrationality" brought into quantum physics by Planck's discovery of the quantum of action, h, the concept that is part of the genealogy of the concept of the unthinkable adopted by this study, a genealogy extending, through the idea of the irrational or (the irrational is a Latin term) for the ancient Greeks the incomprehensible (*alogon*) to pre-Socratic and specifically Pythagorean thinking. Section 6.3 outlines Bohr's ultimate, RWR-type, interpretation. Section 6.4 discusses Bohr's understanding of quantum measurement as an entanglement between the quantum object considered and the measuring instrument used. Section 6.5 is devoted to complementarity. Section 6.6 introduces the concepts of probabilistic causality and quantum causality (in the absence of classical causality) in connection with complementarity.

6.2 From the Irrational to the Unthinkable, from the Pythagoreans to Bohr

I would like first, by way of a prologue, to discuss one, sometimes controversial, aspect of Bohr's argumentation (found in several on his works), his appeal to the irrationality found in quantum theory. Beginning with the Como lecture, which introduced both complementarity and his first full-fledged interpretation of quantum phenomena and QM, Bohr sees this irrationality as unavoidable, "inherent," in quantum theory given the role of Planck's quantum of action, h, in quantum phenomena (Bohr 1987, v. 1, p. 54). Bohr's appeal to the irrationality of quantum theory, especially as something inherent in it, has more often been criticized, even by Bohr's advocates, than carefully considered in relation to Bohr's argumentation. By contrast, I shall argue here that this appeal is a reflection of Bohr's thinking and gives it an additional, Pythagorean, angle, on this thinking. The genealogy of this concept in Bohr appears to be mathematical by virtue of its connection to the concept of an irrational number, which is one of several mathematical concepts (that of a Riemann surface is another) that shaped Bohr's thinking concerning quantum phenomena. There is, thus, also yet another type of connection between mathematics and theoretical physics, the type defined by translating abstract mathematical concepts, unconnected to physics, into physical concepts, which might have mathematical components unrelated to the mathematical concepts thus translated.

As noted earlier, Bohr's Como argument, retreated to a more realist view from the (weak) RWR-type view that Bohr adopted in the wake of Heisenberg's discovery of QM in 1925, represented by his statement at the time that "in contrast to ordinary mechanics, *the new quantum mechanics does not deal with a space–time description of the motion of atomic particles*" (Bohr 1987, v. 1, p. 48; emphasis added). While, however, assuming that the independent behavior of quantum objects could be represented (moreover, classically causally) by the mathematical formalism of QM, with the recourse to probabilistic considerations only introduced by measurement, the Como lecture was ambivalent concerning this assumption (Bohr 1987, v. 1, p. 52). Thus, Bohr maintained that "radiation in free space as well as isolated material particles are abstractions, their properties ... being definable and observable only through their interactions with other systems" (Bohr 1987, v. 1, pp. 57, 76). It is not easy to reconcile this claim with the view that the independent behavior of quantum objects is represented by QM. In any event, Bohr was never satisfied with the lecture and famously delayed and even tried to avoid its publication. By the time it was finally published Bohr had changed his views under the impact of the discussion with Einstein that took place at the Solvay Conference in Brussels just one month after the Como Conference (Bohr 1987, v. 2, pp. 41–47). Perhaps not coincidentally, the 1931 republication of the Como lecture in (Bohr 1987, v. 1, pp. 52–91) dates the lecture 1927, which is when it was given, and not 1928, when it was published. Bohr's next published article, "The Quantum of Action and the Description of Nature" (Bohr 1987, v. 1 pp. 92–101), manifested this change and was a key step on the trajectory leading Bohr to his ultimate interpretation, with the help of his subsequent exchanges with Einstein.[2]

Bohr's invocation of the "irrationality" introduced by h was another symptom of his ambivalence attitude toward his realist argument in the Como lecture. This invocation occurs immediately before Bohr introduces complementarity, which he

[2] My quotations from the Como lecture follow the published version. Although one cannot be certain how definitively it represents Bohr's views presented in Como, it appears to be as definitive as any available drafts of the lecture. It was so treated by Bohr in his "Introductory Survey" to *Atomic Theory and the Description of Nature*, published in 1931 (Bohr 1987, v. 1, pp. 9–15). For the history and drafts of the article, beginning with the draft of the lecture, and commentaries by J. Kalckar, see volume 5 of Bohr's collected works (Bohr 1972–1996).

connects to this irrationality. Bohr speaks of it as "inherent" in "the quantum postulate," symbolized by *h*. The quantum postulate is a new concept, different from (although related to) the quantum postulates of his 1913 theory. The quantum postulate defined *any* "atomic [quantum] process," or in terms of Chap. 3, any transition between quantum states, rather than only between stationary states as in his 1913 theory. As Bohr says:

> Notwithstanding the difficulties which, hence, are involved in the formulation of the quantum theory, it seems … that its essence may be expressed in the so-called quantum postulate, which attributes to any atomic process an essential discontinuity, or rather individuality, completely foreign to the classical theories and symbolized by Planck's quantum of action. (Bohr 1987, v. 1, p. 53)

This somewhat hesitant, and yet necessary, "or rather" introduces one of the most essential features of quantum phenomena, making each strictly individual—unique and unrepeatable—alongside their equally necessary essential discontinuity, discreteness, relative to each other. This study makes the first feature the quantum individuality, QI, postulate, and the second the quantum discreteness, QD, postulate. Both features will, by the late 1930s, be reinterpreted by Bohr in terms of his concept of phenomena, defined strictly by what is observed in measuring instruments (e.g., Bohr 1937, 1938, 1987, v. 2, p. 64). This reinterpretation makes this discontinuity or discreteness and this individuality refer to that between phenomena in Bohr's sense, rather than to the conventional, Democritean, atomicity of quantum objects, thus positioning the ancient Greek part of the genealogy of Bohr's thinking between the physics of the ancient Greek atomists and the mathematics of the Pythagoreans. Quantum objects (or in the present view, the ultimate, RWR-type, reality, which defines quantum objects, still as RWR-type idealizations applicable only at the time of measurement) are unobservable and unrepresentable, either as discrete or as continuous, and in the strong RWR view, are inconceivable. The kernel of the idea is, however, found already in the Como lecture, which, its realist dimensions notwithstanding, grounds Bohr's interpretation (in any version) in the irreducible role of measuring instruments in the constitution of quantum phenomena and, thus, in the impossibility of disregarding this role in the way it is possible to do, for all practical purposes, in classical physics or relativity.

Bohr's appeal to the symbolic nature of *h* reflects his view of all quantum–mechanical formalism as symbolic—a view developed following Heisenberg's introduction of his matrix formalism, but equally applied by Bohr to Schrödinger's formalism, including in the Como lecture. As discussed in Chap. 2, this formalism is symbolic because, while it is formally analogous to or even adopts the equations of classical mechanics, it does not describe the physical behavior of quantum objects in space and time, but only serves to predict the outcomes of quantum experiments, manifested in measuring instruments. Similarly, Planck's quantum of action (symbolizing the quantum postulate) is symbolic because it is not seen as implying the discrete and indivisible, "atomic," character of quantum objects themselves. The discreteness, again, pertains only to quantum phenomena, defined by effects of the interactions between quantum objects and measuring instruments, effects registered in measuring instruments. The effects are always individually discrete, and in this sense are particle-*like* (i.e., analogous to the interaction of particle-like objects with measuring instruments in classical physics). Under certain circumstances, they may also be collectively wave-*like*, by virtue of exhibiting certain wave-like patterns, such as the interference pattern in the double-slit experiment in the corresponding setup. These patterns, however, are composed of discrete individual events or effects. Bohr introduces "the inherent 'irrationality'" of the quantum postulate as follows:

> [The quantum postulate] implies a renunciation as regards the [classically] causal space–time co-ordination of atomic processes. Indeed, our usual description of physical phenomena is based entirely on the idea that the phenomena concerned may be observed without disturbing them appreciably. … Now, the quantum postulate implies that any observation of atomic phenomena will involve an interaction with [between a quantum object and?] the agency of observation not to be neglected. Accordingly, an independent reality in the ordinary physical sense can neither be ascribed to the phenomena nor to the agencies of observation. After all, the concept of observation is in so far arbitrary as it depends upon which objects are included in the system to be observed. Ultimately, every observation can, of course, be reduced to our sense perceptions. The circumstance, however, that in interpreting observations use has always to be made of theoretical notions entails that for every particular case it is a question of convenience at which point the concept of observation involving the quantum postulate with its inherent "irrationality" is brought in. (Bohr 1987, v. 1, pp. 53–54)

"A renunciation as regards the [classically] causal space–time co-ordination of atomic [quantum] processes" implies that spacetime coordination can only be applied to observed atomic (quantum) phenomena, which would, again, make it difficult to assume, as Bohr does in the Como lecture, the classically causal nature of these processes considered independently, with probability introduced only by acts of measurement. The RWR view is much more in accord with this renunciation. As explained in Chap. 2, the "cut" between the observed quantum phenomena and the ultimate constitution of the reality responsible for them is a matter of convenience only within a certain limit: it could only be made in the region where the quantum and classical predictions coincide (Bohr 1935, p. 701). This restriction, however, does not affect the fact that "the quantum postulate and its inherent 'irrationality'" do have to be brought into any possible observation.

Why does the quantum postulate contain this inherent irrationality and what is the meaning of this irrationality? Bohr's use of quotations marks indicates that one has to be careful in interpreting it. As noted, Bohr's appeals to "irrationality" have been used against him by his critics and have troubled even some of his advocates. I would argue, however, that this criticism is misplaced and is a result of misunderstanding Bohr's meaning. First of all, this "irrationality" is clearly not any "irrationality" of QM, which Bohr always saw as a rational theory, beginning with Heisenberg's discovery of it, characterized by Bohr "the *rational* formulation of the new quantum mechanics" (Bohr 1987, v. 1, p. 48; emphasis added). Bohr's phrase also, and I suspect, deliberately, echoes Newton's characterization of his mechanics as "rational" in the *Principia* (Newton 1999). This echo implies that, although QM does not represent the physical processes considered in space and time, it is, as a physical theory, as rational as is classical mechanics, or relativity, or as is mathematics that deals with irrational numbers. However, in contrast to classical mechanics or relativity, QM is a rational theory of something, specifically, the ultimate nature of the reality responsible for quantum phenomena, that is "irrational" in the sense of being inaccessible to a (logical) representation or even thought itself, in contrast to classical physics or relativity, where the ultimate constitution of the reality considered is rationally representable. This "irrationality" is that of the RWR-type reality ultimately responsible for quantum phenomena, which can be rationally handled (in terms of probabilistic or statistical prediction) by a rational physical theory, such as QM. If, however, one makes realist assumptions, "the quantum of action … appears as *an irrational element*" (Bohr 1931, p. 458). Hence, one encounters a boundary, "cut," up to a point (but, again, only up to a point) arbitrarily placed, between the world of observed phenomena and the RWR-type reality that, while beyond rational thought, is ultimately responsible for these phenomena, described by means of classical physics.

This argumentation helps us better to understand Bohr's often-cited (reported) statement, apparently in response to the question "whether the algorithm of quantum mechanics could be considered as somehow mirroring an underlying quantum world": "There is no quantum world. There is only an abstract quantum-physical description. It is wrong to think that the task of physics is to find out how nature is. Physics concerns what we can say about nature" (reported in Petersen 1985, p. 305). As I said, one must exercise caution in considering such extemporaneous comments, especially when they are reported. It is not known, for example, what else Bohr said then, as he might have done, or what he actually had in mind in making or supporting (as he would do in his writing) this statement. With these qualifications and Bohr's written works in mind, the statement may be read not as denying the existence of a reality ultimately responsible for quantum phenomena but, on RWR lines, as denying the applicability of any conceivable description or conception to this reality, such as that of "the quantum world." "An abstract quantum-physical description," essentially a mathematical description provided by QM, does not represent this reality or quantum phenomena, which belong to the world we observe classically, but only provides a set of procedures, algorithms, for predicting, probabilistically, the outcomes of quantum experiments. Accordingly, the last two sentences would apply to quantum *physics*. The statement could still apply more generally if read on Kantian lines or, more radically, by adopting the U-RWR view, insofar as we cannot assume that *phenomena* described by these theories or these theories themselves represent the ultimate constitution of nature, or "how nature is." This reading, too, is supported by statements found in Bohr's published writings. Thus, he said: "we meet [in quantum physics] in a new light the old truth that in our description of nature the purpose is not to disclose the real essence of the phenomena but only to track down, so far as it is possible, relations between the manifold aspects of our experience" (Bohr 1987, v. 1, p. 18).

According to Bohr, writing a few years after the Como lecture, by which time he is on his way to his RWR-type interpretation, "the quantum of action, which appears as *an irrational element* from the point of view of the classical mechanical physics, taken together with the existence of elementary particles, forms the foundation of atomic physics" (Bohr 1931, p. 458; emphasis added). Bohr's qualification "taken together with the existence of elementary particles" is important, for it is the existence of quantum objects that is ultimately responsible for quantum irrationality. As he explained in the "Introductory Survey," to his *Atomic Theory and the Description of Nature*, the situation had emerged already in his 1913 atomic theory. He said: "A conscious resignation in this respect [as concerns a classically causal description of quantum phenomena] is already implied in the form, *irrational* from the point of view of the classical theories, of those postulates, … upon which the author [Bohr] based his application of the quantum theory to the problem of atomic structure" (Bohr 1987, v. 1, p. 7; emphasis added). Later in the same article, he added, again invoking the irrationality of the quantum of action:

> Moreover, the purpose of such a technical term [complementarity] is to avoid, so far as possible, a repetition of the general argument as well as constantly to remind us of the difficulties which, as already mentioned, arise from the fact that all our ordinary verbal expressions bear the stamp of our customary forms of perception, from the point of view of which the existence of the quantum of action is an irrationality. Indeed, in consequence of this state of affairs, even words like "to be" and "to know" lose their unambiguous meaning. (Bohr 1987, v. 1, pp. 19–20)

Bohr's appeal to the irrational is, thus, defined by the difference between the rationality of a theory and the irrationality of that which this theory rationally deals with, by this time approaching the strong RWR view, suggested by his claim that "even words like 'to be' and 'to know' lose their unambiguous meaning" when applied to the ultimate nature of reality responsible for quantum phenomena.

Bohr's invocation of "irrationality" also suggests an analogy with and, I would argue, is derived from the concept of irrational numbers in mathematics. The Pythagoreans, who discovered the (real) irrationals, could not find an arithmetical, as opposed to geometrical, form of representing them, because their arithmetic was strictly that of rational numbers. The ancient Greek term "*alogon*," used in this connection, is better rendered as "incomprehensible" (by logical thought) rather than as irrational, which is a Latin word. The analogy is reinforced by the role of complex numbers in the mathematical formalism of QM. Complex numbers, such as $i\,(\sqrt{-1})$, are irrational in the strict sense, that is, insofar as they cannot be represented by fractions or any combination of real numbers, although they are solutions of algebraic equations with real coefficients. Although in mathematics the term "irrational numbers" is usually reserved for real irrationals rather than complex numbers, Bohr appears to have thought of complex numbers as "irrational" in both the mathematical and epistemological sense. In both cases (real irrational and complex numbers) we deal with the solutions of polynomial equations that do not belong to the field in which the equations themselves are defined, as in the case of $x^2 + 1 = 0$, the solutions of which are $\pm i$. (The field of complex numbers has no further algebraic extension: it is algebraically closed.) In mathematics itself the status and legitimacy of complex numbers was under debate for a long time, which is why they were called "imaginary numbers," the term now reserved for complex numbers that have no real component. At stake is the impossibility of representing or even conceiving of something that is nevertheless essentially related to a given, rationally formulated, theoretical framework, the situation also defining QM. The solution of its equations or, more accurately, what these solutions relate to (the ultimate, RWR-type, nature of reality responsible for quantum phenomena) is beyond the representational capacity of the theory and even beyond the reach of thought altogether. This relation is possible only via probabilistic predictions concerning what is observed in measuring instruments impacted by quantum objects. As Bohr said, amplifying the parallel between complex numbers and QM, and connecting them, as they are in QM, the "symbols (used in the quantum–mechanical formalism), as indicated already by the use of imaginary numbers, are not susceptible to pictorial representation," and ultimate to any representation or even conception (Bohr 1972–1996, v. 7, p. 314). Even if they were susceptible to it, it would not ultimately matter, because the formalism does not, in RWR-type interpretations, represent the ultimate nature of reality responsible for quantum phenomena.

The situation, thus, exhibits a parallel with the case of the diagonal of the square in Pythagorean mathematics, especially given that our relationships to the ultimate nature of reality responsible for quantum phenomena in QM are strictly mathematical. The parallel merits a brief commentary as an instructive case of the Pythagorean spirit of QM, which I invoked in connection with Heisenberg's invention of it in Chap. 4. The Pythagoreans' incommensurables must have been on Bohr's mind, not the least because in both cases one deals with the problem of measurement, with the mathematics and physics of *measurement*. As noted from the outset of this study, the latter concept acquires a radically new meaning in quantum physics, as against classical physics. One can no longer speak of measuring anything in the ultimate constitution of the reality responsible for quantum phenomena but only of constructing these phenomena by means of the interactions between this reality and measuring instruments and then measuring what is observed, as part of these phenomena, classically. Admittedly, this meaning emerged only with Bohr's ultimate interpretation and his concept of phenomena in the later 1930s. Earlier, even though the role of the measuring instruments was still seem as irreducible, Bohr's view contains the remnants of the classical understanding of measurement.

This meaning of measurement makes the connections between quantum measurement and the incommensurables of the ancient Greek mathematics even more pertinent. The deliberate nature of this analogy in Bohr is confirmed by the fact that Bohr expressly reflected on these connections between the mathematical and quantum-theoretical concepts of the irrational, via the idea of a Riemann surface (which was part of the genealogy of complementarity) in his final interview, on which I shall comment later in this chapter. It is true that in QM we deal with physical and not mathematical reality. At stake, however, is also a difference from classical physics or relativity, where mathematics functions representationally and related to measurement in its conventional sense of measuring the properties of the ultimate constitution of the reality responsible for classical or relativistic phenomena. For the Pythagoreans, too, numbers or geometrical entities would describe the ultimate nature of the reality, material, or spiritual, of the world, and measure this nature. QM no longer offers either algebraic or geometrical means of representing this irrationality. One can only, by using the geometry of Hilbert spaces (over $\mathbb{C}$) relate, probabilistically, to its effects on what is observed and classically measured in measuring instruments. The irrationality is that of the unrepresentable or inconceivable nature of the ultimate constitution of reality, while the geometry is that of the Hilbert space. Although this geometry no longer represents the incommensurable, thus, leaving it incomprehensible, it relates to it by means of the algebra of probability, enabled by the relationships between the geometry and algebra of Hilbert space, cum Born's rule, added to but, because it ultimately amounts to the complex conjugation, related to this algebra. This is the Pythagorean logos of QM, a logos elevated to the RWR thinking: it is a logos that contains the *alogon*, the unthinkable, within it.

In ancient Greek mathematics, the problem was resolved geometrically by a new theory of proportion, accommodating both commensurable and incommensurable magnitudes. The arithmetical problem, that of rigorously defining irrational numbers, was only resolved, by essentially modern mathematical means, in the nineteenth century, after more than 2000 years of effort, with Richard Dedekind and others. In view of the undecidability of Cantor's continuum hypothesis, however, this problem may still be ultimately unresolvable. It remains to be seen whether quantum–mechanical "irrationality" will ever be resolved by discovering a way to represent the ultimate constitution of the reality responsible for quantum phenomena. Besides, not everyone sees this situation as a problem that needs to be resolved. Bohr did not. On the other hand, one could not exclude that the quantum irrationality will be resolved by way of establishing a realist, rather than RWR-type, theory of quantum phenomena.

For Bohr, then, QM was a *rational* theory of something that is *irrational* in the sense of being inaccessible to a representation or knowledge, or even to thought. QM was a replacement, necessary in the case of quantum phenomena, of a *rational representational* theory, classical mechanics, with a *rational probabilistically or statistically predictive* theory (Bohr 1987, v. 1, p. 48). Thus, QM is both a reversal of what happened in the crisis of the incommensurable in ancient Greek mathematics, which compelled it to move from arithmetic to geometry, in their theory of measurement, and a new form of the relationship between geometry and measurement, which relates a mathematical continuity to a physical discontinuity in terms of probabilities.

"Thought … is the measure of the Universe," the greatest gift of Prometheus and the highest ambition of measurement, Percy Bisshe Shelley says in *Prometheus Unbound* (II, iv. 73). He actually says: "He [Prometheus] gave man speech, and speech created thought,/Which is the measure of the Universe" (II, iv. 73). Even assuming that such an ontological or anthropological primacy of language is possible (which is doubtful), it would not affect my point. If, to return to Heisenberg's argument, "we cannot speak about the atoms in ordinary language," we cannot think of atoms in terms of ordinary concepts either, concepts from which ordinary language is indissociable (Heisenberg 1962, pp. 178–179). In the RWR view, we cannot speak of them even in terms of physical concepts, assuming that they can be dissociated from ordinary concepts. Language and thought are indissociable from and shape each other even in mathematics, although the latter has the capacity, at least in principle, of representing physical (or mental) reality apart from ordinary or physical concepts. This capacity is, however, disabled by the RWR view. The phrase "Thought … is the measure of the Universe" echoes Protagoras's "man is the measure of all things," but gives it the Pythagorean meaning of a logos that must contain the *alogon* within it, because, as the poem also says, "the deep truth is imageless" (II, iv. 116). Human thought is only the measure of all things that it can actually think, even if, as in Kant, only as things-in-themselves, which it cannot know. In QM, or QFT, in strong RWR-type interpretations, beginning with that of Bohr, the ultimate nature of reality is beyond being only imageless. It is *unthinkable*, even in terms of abstract algebraic structures (which do not depend on images), at least as things stand now. While, however, the thought of the unthinkable is still only a thought, it is a testimony to thought's capacity to think beyond itself. Without this capacity thought could never be a true measure of the Universe, the Universe that we may ultimately not be able to *measure*.

6.3 Bohr's Ultimate Interpretation: Phenomena and Reality Without Realism

Bohr's ultimate interpretation was first presented in Bohr's (1937) article, "Causality and Complementarity" (Bohr 1937), impacted by the Bohr-EPR exchange and building Bohr's argument in his reply (Einstein et al. 1935; Bohr 1935).[3] Bohr does not use the language of reality without realism, but his understanding of quantum measurement, as defined by the irreducible role of the interactions between quantum objects and measuring instruments in the constitution of quantum phenomena, clearly amounts to the strong RWR view. According to Bohr:

> The renunciation of the ideal of causality in atomic physics which has been forced on us is founded logically only on our not being any longer in a position to speak of the autonomous behavior of a physical object, due to the unavoidable interaction between the object and the measuring instruments which in principle cannot be taken into account, if these instruments according to their purpose shall allow the unambiguous use of the concepts necessary for the description of experience. In the last resort an artificial word like "complementarity" which does not belong to our daily concepts serves only briefly to remind us of the epistemological situation here encountered. (Bohr 1937, p. 87)

I shall discuss complementarity, which does more than serve as such a reminder, including in allowing "the unambiguous use of the concepts necessary for the description of experience," in Sect. 6.5. The concept of causality that grounds this ideal of causality is clearly that designated here as "classical causality." It is defined by the claim that the state, X, of a physical system is determined, in accordance with a law, at all future moments of time once it is determined at a given moment of time, state A, and A is determined in accordance with the same law by any of the system's previous states. This assumption, thus, implies a concept of reality, which defines this law and makes this concept of causality ontological. By contrast, in accordance with Bohr's statement, RWR-type interpretations are not classically causal because of the absence of realism, necessary for such a law, in considering the behavior of quantum objects.

---

[3] The argument was then presented (without essential changes) in subsequent communications beginning with the so-called Warsaw lecture (Bohr 1938).

Along with several other Bohr's statements, considered earlier (e.g., Bohr 1987, v. 2, p. 62), Bohr's (1937) statement, cited above, concerning the impossibility of accessing the ultimate nature of physical reality responsible for quantum phenomena, may be read as making claim that is stronger than a claim merely pertaining to his *interpretation* of the situation. Instead, such as access would be "*in principle* excluded" by quantum phenomena for any theory or analysis (Bohr 1987, v. 2, p. 62). This stronger claim is also suggested by an earlier statement in the same article to the effect that the uncertainty relations would "preclude the possibility of a future theory taking both attributes [at stake in the uncertainty relation] into account on the lines of the classical physics" (Bohr 1937, p. 86). Could the uncertainty relations *preclude* this possibility for *any* future theory, even assuming that the uncertainty relations remain in place as pertaining to quantum phenomena themselves? In the present view, it is difficult to exclude such a possibility. In any event, both statements just cited and similar statements by Bohr are viewed in this study as pertaining to Bohr's interpretation, and all Bohr's philosophical positions are seen as manifestations of his interpretation, from 1937 on, in its ultimate RWR-type version. In the present view, all philosophical positions, the present one included, concerning quantum phenomena and quantum theory are interpretations, only practically justified, regardless of how those who hold these positions see them.

Bohr's ultimate interpretation was his response to the task announced in his 1935 reply to EPR (still alongside the same appeal to "a final renunciation of the classical ideal of causality") of "a radical revision of our attitude toward the problem of physical reality" in view of the nature of quantum phenomena (Bohr 1935, p. 697). A revision of an *attitude* toward the problem of physical reality is not the same as a revision of a given *concept* of reality, a point to which I shall return in closing this chapter. Bohr, however, clearly undertook such a revision, by adopting the strong RWR view, as confirmed by the passage under discussion and other key statements representing his views considered in this study, such as, again, that "in quantum mechanics, we are not dealing with an arbitrary renunciation of a more detailed analysis of atomic phenomena but with a recognition that such an analysis is *in principle* excluded" (Bohr 1987, v. 2, p. 62). Bohr's position, stated his 1937 "Causality and Complementarity," cited above, represents the strong RWR view, placing the ultimate nature of reality responsible for quantum phenomena beyond conception. For, if, as Bohr says there, we are no "longer in a position to speak of the autonomous behavior of a physical object, due to the unavoidable interaction between the object and the measuring instrument," this behavior must also be beyond conception. If we had such a conception, we would be able to say something about this behavior. It is true that there is a difference between forming some conception of this reality and forming a rigorous conception that would enable us to provide a proper representation of it. Bohr, however, clearly makes a strong claim, without such qualifications, that implies the impossibility of any such conception: we are no longer in a position to speak of the autonomous behavior of quantum objects *at all*.

As indicated in Chap. 2, given that there is no definitive statement to that effect on Bohr's part, it is a matter of an interpretation whether for Bohr our inability to do so only (A) characterizes the quantum–mechanical situation as things stand now, while allowing that quantum phenomena or whatever may replace them will no longer require this assumption and thus will no longer make RWR-type interpretations viable; or more strongly (B) reflects the *possibility* that this reality will never become available to thought. (B) might be suggested some of his stronger statements, such as those just considered, to the effects that a conception of the ultimate character of reality responsible for quantum phenomena "*in principle* excluded" for any theory now or in the future. In any event, it may be safely assumed that in his ultimate interpretation, Bohr at least adopts (A), and I shall assume this here, although virtually all of my argument would equally apply if he adopted (B). As explained earlier, the qualification "as things stand now" still applies to (B). It applies, however, not because the unthinkable nature of the ultimate constitution of the reality responsible for quantum phenomena as defined by (B) becomes available to thought, which it cannot, by definition, under (B). It applies because a return to realism in quantum theory or whatever might replace it is possible, either on experimental or theoretical grounds. This will make (B), or (A), obsolete even for those who hold it because there is no physical theory to justify it. The U-RWR view of all physics or even all human thinking about nature in its ultimate constitution may still be assumed. It will just not be necessary or desirable in physics because a realist view will be practically justified. No other justification than practical is, again, assumed as possible in this study for a physical theory or even physical phenomena, in which case our claims may have a broader scope, even, *for all practical purposes*, a universal scope. At the moment, however, the RWR view, either (A) or (B), may be practically justified in an interpretation of quantum phenomena and QM or QFT.

Bohr's ultimate interpretation is grounded, along with complementarity (adjusted to this interpretation), in a new concept, that of phenomena, defined strictly in terms of effects *observed* in measuring instruments as a result of their interaction with quantum objects (Bohr 1937, 1938). As I have argued here, in Bohr's ultimate interpretation or the one adopted here, a quantum measurement is not a measurement of some pre-existing property of the quantum object considered (or, in the present view, the ultimate constitution of the reality responsible for quantum phenomena, as well as quantum objects at the time of measurement). Instead, a quantum measurement, establishes, creates, a quantum phenomenon, whose *observed* physical properties can then be classically measured. As Bohr explained later (in 1949):

> I advocated the application of the word phenomenon exclusively to refer to *the observations obtained under specified circumstances*, including an account of the whole experimental arrangement. In such terminology, the observational problem is free of any special intricacy since, in actual experiments, all observations are expressed by unambiguous statements referring, for instance, to the registration of the point at which an electron arrives at a photographic plate. Moreover, speaking in such a way is just suited to emphasize that the appropriate physical interpretation of the symbolic quantum-mechanical formalism amounts only to predictions, of determinate or statistical character, pertaining to individual phenomena appearing under conditions defined by classical physical concepts [describing the observable parts of measuring instruments]. (Bohr 1987, v. 2, p. 64; emphasis added)

Referring to "observations" is precise, because only the classically observed properties of measuring instruments affected by these observations could be measured.

(Here, by "quantum measurement" I refer to this whole process.) As defined by "*the observations* [already] *obtained under specified circumstances,*" phenomena refer to events that have already occurred and not to possible future events predicted on the basis of previous events defined by already established phenomena. This is a crucial point, including, as discussed in Chap. 7, in considering EPR-type experiments. Bohr sometimes speaks of a phenomenon as defined by two measurements (the first of which is sometimes referred to as a preparation, although both instruments need to be *prepared* in accordance with what type of measurement we decide to perform):

> The essential lesson of the analysis of measurements in quantum theory is thus the emphasis on the necessity, in the account of the phenomena, of taking the whole experimental arrangement into consideration, in complete conformity with the fact that all unambiguous interpretation of the quantum mechanical formalism involves the fixation of the external conditions, defining the initial state of the atomic system concerned and the character of the possible predictions as regards subsequent observable properties of that system. Any measurement in quantum theory can in fact only refer either to a fixation of the initial state or to the test of such predictions, and it is first the combination of measurements of both kinds which constitutes a well-defined phenomenon. (Bohr 1938, p. 101)

This statement need not mean that Bohr's *concept* of phenomenon applies to two measurements, or, and in particular, that it can refer to a prediction, which is why Bohr's speaks of "the *test* of ... predictions," that is, already performed experiments. The point is that, if we want to establish and communicate our findings, we must specify two measurements and the instruments prepared accordingly: the first is the initial, *actual*, measurement or phenomenon and the second is a, *possible*, future measurement or phenomenon that would enable us to verify our probabilistic prediction, or our statistical predictions, after repeating the experiments many times. As Bohr said in the same article, in, essentially, defining his concept of a phenomenon: "It is certainly far more in accordance with the structure and interpretation of the quantum mechanical symbolism, as well as with elementary epistemological principles, to reserve the word phenomenon for the comprehension of the effects observed under given experimental conditions" (Bohr 1938, p. 105). This strongly suggests that a phenomenon is defined by an already performed measurement, as an effect of the interactions between quantum objects and the apparatus, but never by a prediction. Then, the first description above is contextualized as referring to "that all unambiguous interpretation of the quantum mechanical formalism involves the fixation of the external conditions, defining the initial state of the atomic system concerned and the character of the possible predictions as regards subsequent observable properties of that system." The second refers to the test of any such prediction. One needs both arrangements and both phenomena (defined when both measurements are performed) to test our predictions, in quantum physics, in repeated experiments, because our predictions are in general probabilistic or statistical.

Referring, phenomenologically, to observations explains Bohr's choice of the term "phenomenon." This "idealization of observation" (Bohr 1987, v. 1, p. 55) is the same as that of classical physics, which allows one to identify phenomena with the physical objects (here measuring instruments), because our observation does not interfere with their behavior, in contrast to the way an observation by means of a measuring instrument interferes with the (RWR-type) ultimate constitution of the reality responsible for a phenomenon thus observed. On the other hand, given that a quantum object is, in the present view, an idealization applicable only at the time of measurement, it is a product of this interference rather than something that is interfered with, as is the ultimate constitution of the reality responsible for quantum phenomena.

Quantum discreteness, too, is that of *phenomena*, rather than the Democritean atomic discreteness of the ultimate constitution of the RWR-type reality responsible for quantum phenomena or of quantum objects, which are beyond conception and cannot be seen either as continuous or as discrete (Bohr 1987, v. 2, pp. 32–34). Around the time when he introduced his concept of phenomena, Bohr also introduced a concept of "atomicity," pertaining to quantum phenomena rather than quantum objects (Bohr 1938, p. 94, 1987, v. 2, p. 34). This concept is essentially equivalent to that of phenomenon and only highlights some aspects of the latter concept, as discussed in detail in (Plotnitsky 2012, pp. 138–150). Briefly, it transfers to the level of observable phenomena, manifested in measuring instruments, the key "atomic" features of quantum physics—discreteness, discontinuity, individuality, and atomicity as indivisibility—previously associated with quantum *objects*. "Atomicity" is thus a feature of physically complex and divisible entities, and not of physically indivisible entities, such as elementary particles. It is possible that when Bohr refers to *atomic* (rather than quantum) phenomena in his later works, he also has this atomicity in mind, rather than only phenomena considered in quantum physics.[4]

In Bohr's ultimate strong RWR-type interpretation, no physical quantities are assumed to correspond to properties of quantum objects, even at the time of measurement. This is the case even in considering single properties, rather than only certain joint properties, not attributable simultaneously by virtue of the uncertainty relations. Bohr's earlier views (even after the Como lecture, for example, in his reply to EPR) allowed for this type of attribution of properties *at the time of measurement* and only then and under the constraints of the uncertainty relations, thus, still in accord with the assumption that such properties cannot be considered independently of measurements. Even this, less radical, view still implied that the physical state

[4] While the genealogy of Bohr's concept of phenomenon appears to have been more conventional, extending from Kant's philosophy, Bohr's concept of atomicity is unusual and its genealogy is less clear. An intriguing possibility is Alfred North Whitehead's concept of atomism in his 1929 *Process and Reality* (Whitehead 1929), although there is no evidence that this concept was familiar to Bohr. Whitehead's overall ontological position is different from that of Bohr. Whitehead's *discrete* atomicity of experience is underlain by the *continuous* ultimate reality, linked to potentiality. By contrast, as a strong RWR-type interpretation, Bohr's ultimate interpretation, which grounds his concept of atomicity, precludes any ontological conception, discrete or continuous, of this reality and does not appeal to potentiality, unlike Heisenberg in his later works, although Heisenberg's concept of potentiality is not the same as that of Whitehead (Heisenberg 1962). Accordingly, while Whitehead's ontological scheme has been applied to QM or QFT, it is difficult to transfer it to Bohr's interpretation, or even see them as philosophically proximate. See, for example, (Epperson 2012) on the connections between Whitehead's philosophy and quantum theory, although the book, I would argue, misrepresents Bohr's views, especially in their radical (RWR) aspects, where he differs from Whitehead most.

of an object cannot be defined on the model of classical physics. This is because classical physics requires an unambiguous determination of both conjugate quantities for a given object at any moment of time and independently of measurement, which is no longer possible in quantum physics because of the uncertainty relations. In Bohr's ultimate interpretation, however, an attribution *even of a single property* to any quantum object is *never possible—before, during, or after measurement*. Even when we do not want to know, say, the momentum of a quantum object and thus need not worry about the uncertainty relations, the *position* of this object itself is never determinable. The uncertainty relations remain valid, of course, as do other standard laws, such as conservation laws. But all these laws now apply to the corresponding (classical) variables of measuring instruments impacted by quantum objects. Thus, in the case of the uncertainty relations, one can either prepare our instruments to be able to measure a change of the momentum of certain parts of them or to locate the spot that registers an impact of a quantum object, but never both. The uncertainty relations are correlative to the mutually exclusive nature of these arrangements, in accord with Bohr's concept of complementarity, as discussed in Sect. 6.5.

It also follows that, while predicting, probabilistically or statistically, the data physically observed as part of phenomena, the formalism of QM is, in Bohr's interpretation, entirely dissociated in physical terms from phenomena themselves, described by classical physics. In other words, the relevant mathematical elements of the formalism of QM are merely our means, our *symbolic* means, of predicting the probabilities or statistics of the data to be observed, as part of possible future phenomena and experiment or experience associated with it. They have no *physical* correspondence with anything observed as part of such possible future phenomena, given that these data and phenomena are described by classical physics, thus using real functions and quantities. By contrast, the formalism of QM only uses complex entities (Hilbert-space vectors, operators, and so forth), supplemented by Born's or analogous rules, to connect them to the probabilities or statistics, which are real numbers, of quantum events.

This aspect of Bohr's concept of phenomenon and Bohr's interpretation—a dissociation of the abstract symbolic formalism of QM, its *abstraction*, from the classical physical description of phenomena and the data observed there, which the formalism predicts in probabilistic or statistical terms—as well as, in part correlatively, the role of human agents in Bohr's interpretation are worth considering further, because they are easy to miss or misunderstand. Difficulties of reading Bohr on both points (or in general) are not uncommon and in part have to do, as Bohr's noted, "the inefficiency of expression which must have made it very difficult to bring out the essential ambiguity involved in a reference to physical attributes of objects when dealing with phenomena where no sharp distinction can be made between the behavior of the objects themselves and their interaction with the measuring instruments" (Bohr 1987, v. 2, p. 61). The aspects of Bohr's concept of phenomena and his interpretation under discussion at the moment are clearly related to this RWR-type epistemology, in the way that I shall now clarify. Admittedly, this remains a matter of interpretation of Bohr's *interpretation.* That does not mean, however, that one cannot advocate one's interpretation, such as the one proposed here, on the basis of Bohr's writings, which are not always adequately considered in making claims concerning Bohr's views or his interpretation, including as concerns different versions of this interpretation, the existence of which, often disregarded, I have emphasized from the outset of this study.

Obviously, all of Bohr's interpretations, including his ultimate, strong RWR-type, interpretation (on which I focus here), establish an essential connection between the formalism of QM and the data observed in quantum experiments and thus phenomena in Bohr's sense, a concept that is part of his ultimate interpretation. The question is the nature of this connection, as defined by the nature of QM as a symbolic theory in Bohr's understanding (found in all of his interpretations), as discussed here and in Chap. 2 (Sect. 2.5). It is symbolic because, as emphasized here, unlike those of classical mechanics or relativity, the mathematical symbols, comprising the formalism of QM, do not *represent* or have any physical contact with "what we have done and what we have learned" in quantum experiments themselves, explaining which implies that "the account of the experimental arrangement and the results of the observation must be expressed in unambiguous language with suitable application of the terminology of classical physics" (Bohr 1987, v. 2, p. 39). This account represents unambiguously communicable experimental "facts," such as spots on a photographic plate. While the formulation just cited ostensible refers to the Como argument, which indeed introduced this point, Bohr's careful expression is clearly inflected by his subsequent thinking, leading to his concept of phenomenon and strong RWR-type interpretation in place at the time of this article, "Discussion with Einstein," written in 1949. As the preceding discussion in this chapter explains, in Bohr, the nature of the contact between the (symbolic) formalism of QM and phenomena and hence these, classically explainable, "facts" is as follows. The classical explanations thus used only refer to these "facts," as phenomena manifested in the observable parts measuring instruments. These instruments, again, also have quantum strata through which they interact with quantum objects or the ultimate constitution of the reality responsible for quantum phenomena. Both phenomena and the observable parts of measuring instruments are treated classically, but the quantum–mechanical formalism and its symbols do not represent, and in fact have no connections whatsoever to, this classical treatment. They are only used for predicting the data or information thus observed, which information is classical in nature, although its structure cannot be predicted by classical physics. In Bohr's or the present interpretation, the symbols of QM, which belong, along with QM itself, to our thinking, only *relate* to the probabilities of the outcomes of quantum experiments, via Born's rule, added, as a separate postulate, to the formalism. These symbols do not in any way represent "what we have done and what we have learned" in quantum experiments in considering which "the account of the experimental arrangement and the results of the observation must be expressed in unambiguous language with suitable application of the terminology of classical physics." Thus, while, of course, these symbols, too, can be unambiguously communicated, as can any mathematical symbols, they have nothing to do with the results of the observation, which are communicated between different human agents "in terms of in unambiguous language with suitable application of the terminology of classical physics." For one thing, QM only predicts, by using its symbols, the probabilities or statistics of the outcomes of possible future experiments, and does not at all relate to what has already happened.

This situation is clearly illustrated by Bohr's statement, cited in Chap. 2, in his reply to EPR: "In accordance with this situation [that at stake in QM] there can be no question of any unambiguous interpretation of the *symbols* of quantum mechanics other than that embodied in the well-known rules which allow us to *predict* the results to be obtained by a given experimental arrangement *described in a totally classical way*" (Bohr 1935, p. 701; emphasis added). Obviously, given their strictly predictive roles, these symbols—"embodied *in the well-known rules which allow us to predict [these] results*" but *not in these classically-described results themselves*—in no way represent the classical and in part daily-language or narrative accounts of "what we have done and what we have learned" (Bohr 1987, v. 2, p. 39). These accounts, again, only pertain to the outcome of quantum experiments, and they can be communicated unambiguously, as can, again, be the formalism of QM, as an abstract mathematical scheme enabling these predictions. This unambiguous communication is the meaning of objectivity for Bohr (1987, v. 2, pp. 68–69, v. 3, p. 7).

It might be added that, in this view, the symbols of QM for Bohr are not that far from what they are for QBism, insofar as they enable probability assignments by human agents, albeit for Bohr more likely (in the present interpretation expressly) reflecting statistical predictions of QM, given that, as noted in Chap. 2, Bohr appears to have been inclined to the statistical view of QM and in any event had never invoked a Bayesian view. These symbols could, however, also be interpreted on QBist lines, the relations of these symbols to classically observed phenomena in Bohr's sense will be the same. This is manifested, for example, in the QBist view of quantum states, as, *mathematically,* state vectors in the Hilbert space formalism of QM, as part of such probability assignments by human agents and thus belonging to a mental, rather than physical, reality. These symbols, in the QBist view, enable one to predict the probabilities of what is expected to be observed in future quantum experiments on the basis of what has been observed in previously performed quantum experiments. Nobody, however, in the Göttingen-Copenhagen circle ever believed that these symbols have any other role, even before Born's papers on the probabilistic interpretation of the wave function (Born 1926a, b). As explained in Chap. 4, Heisenberg, in his paper introducing QM, already used in this way the corresponding elements of the formalism (there was no language of quantum states then) and, in effect as another postulate, a version of Born's rule for a limited case of transitions between stationary states of electrons in atoms. It is only from in the late 1950s, following Bohmian mechanics, spontaneous collapse theories, many worlds interpretations, and so forth, and, along different lines, Bell's theorem and related developments, that realist views of the formalism of QM or other quantum theories gained more currency in foundational discussions, even though realism as such was preferred by most, beginning with Einstein, all along. QBism is one of the nonrealist views that emerged following quantum information theory, where, too, realist views, for example, via many worlds interpretation, remain wide spread. I mean "nonrealist" *in the present definition*, given that, as noted earlier, QBists themselves sometimes speak of QBism as a form of realism, Fuchs, in particular, in terms of "participatory realism," via Wheeler (Fuchs 2016). According to the present definition, however, their view is a form of the RWR view, insofar as QBism does not aim to provide a representation of the ultimate character of the reality responsible for quantum phenomena but (while assuming this reality) only to estimate (subjective) probabilities of quantum events manifested in observed quantum phenomena. It is true that Bohr would emphasize that, just the outcomes of experiments, the use of the symbols of the formalism *should* be communicable unambiguously (now, again, as parts of our mental reality) in order to maintain the status of QM as a mathematical-experimental science, rather than emphasize, as QBists do, the subjective aspects of probability assignments by human agents. He did not, however, disregard the latter, on which I shall comment presently. But then, QBists, at least some of them, might agree with the pertinence of this communicational aspect of scientific practice, notwithstanding the subjectivity of our probabilistic assignments concerning outcomes of possible experiments (e.g., Pienaar 2020).[5]

The subjectivity or human nature of our probability assignments is not in conflict with the unambiguous communication of them or the outcomes of the experiments leading to or confirming these assignments. While we always can, by decision, control the setups of our tests, we can never control their outcomes, especially in quantum physics, where, in general, we can only predict them probabilistically or statistically, even in dealing with the most elementary individual quantum processes. As noted earlier, however, for Bohr this fact also helps to ensure the objectively verifiable nature of our experiments, the setups of which but not their outcome we can control. It also helps to maintain the disciplinary character of QM (or QFT) as a

[5] Intriguingly, in his lucid reflections on the differences between QBism and the Copenhagen interpretation, Mermin, while extracting a few potentially relevant statements from Bohr's writings, refrains from offering a sustained reading of these statements as grounded in Bohr's argumentation. Mermin focuses on other versions of the Copenhagen interpretation (manifestly different from that of Bohr), such as that of Lev Landau and Alexander Lifshitz, and on the ideas of Bell and Rudolf Peierls, whom Mermin proclaims to be his heroes (Mermin 2016, pp. 232–248). In the absence of a real reading of Bohr or, for that matter, Schrödinger, neither of which is offered by Mermin, it is not clear why, like Peielrs and Bell (who, as discussed below, is hardly sympathetic to Bohr's view), "Bohr and Schrödinger ... might have found a common ground in QBism," as Mermin "like[s] to think" (Mermin 2016, p. 246). While acknowledging Bohr's significance for QBism, Mermin and most other advocates of QBism, such as Fuchs, dissociate it from Bohr's interpretation (it appears, in any of its versions, which, are, however, not always properly distinguished in these arguments, often having difficulties with reading Bohr as well), in part, one might surmise, in response to claims that QBism is not essentially different from Bohr's interpretation or the Copenhagen interpretation. One can readily concede that most of such claims, often arising from negative views of QBism and Bohr alike, are not based on careful assessments of either QBism or Bohr, or adequately discriminating between different versions of "the Copenhagen interpretation." While it may well be true that Bohr's views, at any stage of his work, are different from those of QBism (which has evolved as well), I shall, apart from indicating several relevant points, put aside the subject of QBism as such, given that my main concern is Bohr's thinking and not QBism. Accordingly, my comments should not be seen as a criticism of QBism, with which I have more sympathy than with most other currently advanced views of quantum phenomena and quantum theory. On the other hand, as their works addressing Bohr cited here manifest, QBists, as many others, have difficulties (acknowledged by them) in reading Bohr and, I would argue, sometimes miss subtler aspects of his argumentation, such as those under consideration at the moment.

mathematical-experimental science. The world does what it does and tests us in our experiments, even as we test it through these experiments, and the world may do so in defiance of our subjective expectations. While the outcomes of quantum measurements, as phenomena in Bohr's sense, are personal experiences by human agents, they are not always only that, because they can be and commonly are unambiguously communicated as part of the scientific practice, although they need not necessarily be. They can also become part of permanent records, existing independently, although, again, only a personal experience can assign meaning to these records, including as records, or use language in the first place. It is true that, because QM is not only indeterministic but is also not a classically causal theory, probability plays a more fundamental role in it. But, as QBists contend as well, as does, again, this study (in adopting the U-RWR view), deterministic predictions, predictions with probability one, including in classical physics and relativity, are still only predictions and not statement concerning reality. In addition, as discussed in Sect. 6.6, quantum phenomena bring new complexities to such predictions, making it even more difficult to assume that they correspond to any physical reality at the time of prediction, as opposed to the reality of our thought.

Science is a human enterprise. But sharing and communicating our estimates of possible events and experiences is also human, and doing so is helpful and even unavoidable in human life. Science capitalizes on this fact and on the possibility that the communication involved may be made unambiguous, helped by the use of mathematical symbols, central to modern physics, from Galileo on. As Bohr says: "Just by avoiding the reference to the conscious subject which infiltrate daily language, the use of mathematical symbols secures the unambiguity of definition required for objective [unambiguously communicable] description" (Bohr 1987, v. 2, p. 68). This statement confirms that for Bohr mathematical symbols *do not* represent a 'classical physical explanation... of what we have done and what we have learned," which would entail a reference to the conscious subject and additional explanations, by using common language, even in the mathematical aspects of physics. This is the case even in classical physics and relativity, but in quantum physics, in Bohr's view, the use of mathematical symbols has no representational connections to observations possible in classical physics and relativity.

With these considerations in mind, one can also better understand, Bohr's famous drawing of quantum experiments, depicting heavy, unmovable measuring instruments with traces of their interactions with quantum objects, without any depiction of what happens between experiments, thus representing the RWR view (e.g., Bohr 1987, v. 2, pp. 48–49, 54). The absence of any human agent in this drawing need not mean that human observers are not involved or, in the first place, do not stage these experiments, as, in Bohr's apt locution, "actors as well as spectators" (Bohr 1987, v. 1, pp. 119, v. 2, p. 11). This absence only means that what is so staged and observed does not depend on any particular observer or a group of observers, *as far as what is relevant for physics is concerned*. Beyond physics, an individual human subject could and in fact always does experience such observable phenomena quite uniquely. Bohr's readers could be seen as representing such observers, as observers of *physical phenomena*, which was likely intended by Bohr. Or more accurately, what is so observed is unambiguously communicable, which enables different observers to repeat the same experiments with the same outcomes, confirming the same statistics of predictions. Bohr does say that "the appropriate widening of our conceptual framework [in either relativity or QM] [does not] imply any appeal to the observing subject, which would hinder unambiguous communication of experience" (Bohr 1987, v. 3, p. 7). That might appear to justify claims that Bohr's interpretation disregards human agents. But this is not the case. To say that a *conceptual* framework of a theory does not imply any appeal to the observing subject does not mean that there are no observing subjects, whose presence is clearly suggested by the reference to "communication of experience," or that these subjects do not play their role in using this framework or in experiences associated with it. Whose experience or a communication between whom, would it be otherwise? It is an experience (either of a given experiment and its outcome or a given theoretical argument) of an observing subject communicated to other observing subjects. Bohr often refers to "experience" rather than only to an experiment, including in several passages cited by this study (e.g., Bohr 1937, p. 87, 1987, v.1, pp. 12–18, v. 2, p. 57, v. 3, p. 7). These are not casual uses of the word "experience." This is suggested in particular by the passage, cited earlier, to the effect that "in our description of nature the purpose is not to disclose the real essence of the phenomena but only to track down, so far as it is possible, relations between the manifold aspects of our experience" (Bohr 1987, v. 1, p. 18). Bohr's uses of words are rarely casual. This communication must be unambiguous for the conceptual framework used to conform to the requirement of modern physics as a mathematical-experimental science of natural phenomena, which are, as phenomena, experienced by human agents. While we, as humans, use physical theories and assign probabilities, we must share their verification in order for a physical theory, such as QM, to be a mathematical-experimental science.

At the same time, there could be plenty of individual factors, some of which are unconscious, that each of us bring in observing quantum phenomena registered (as the same material, physical features) in measuring instruments, which instruments and registrations are independent of us, once the devices used are set up by us or others. Our interpretation of them and our predictions concerning future experiments do of course depend on us. Such considerations or Bayesian priors that they may determine may affect our estimates of future events in physics as well. We are, however, better or even best off, as things stand now, by betting on the probabilistic or statistical predictions by using QM or QFT. If we bet otherwise, we are certain to lose or at least not to win at least in a long run, just as we are if we bet on outcomes different from those defined by classical physics or relativity. These theories, however, allow us to bet, ideally, with certainty, while QM gives us better probabilistic or statistical bets, even ideally, science always aims at betting on the best and most consistent predictions, exact if possible (which still remain predictions and not reality) or probabilistic or statistical. Quantum physics, as all physics, also aims, experimentally, at betting on the best measurements on which these predictions are based, measurements that are made or used by human agents or possibly technological agents. The latter can make measurements or even make predictions, but, at least thus far, both need to be set up by human agents. Science is about giving us better bets or, if possible, best bets, if one wants to speak in these terms. Bohr does not use them, but his argument is in accord with this view.

In one of his passages, Bohr confirms the essential role of human decisions, invoking a parallel with art. He says: "even in written papers, where we have the possibility of reconsidering every word, the question whether to let it stand or change it demands for its answer a final decision essentially equivalent to improvisation" (Bohr 1987, v. 2, p. 79). Bohr, famous for his tendency of endlessly reconsidering every word of his papers, including, one surmises in writing this very sentence, undoubtedly thought here of his own practice. But the point is general and implies an essential individuality, singularity of experience and decisions, akin to that of art. And yet, a scientific paper, experimental or theoretical, must contain an unambiguously communicable stratum, even though any reading, or writing, of it also has personal experiential elements, some of which may not be communicable. This double character of scientific practice is also found in the case of mathematical papers, where, as Bohr noted in the statement cited above, from the same essay, "avoiding the reference to the conscious subject which infiltrate daily language, the use of mathematical symbols secures the unambiguity of definition required for objective [unambiguously communicable] description" (Bohr 1987, v. p. 68). Avoiding the reference to the conscious subject does not mean that there are no conscious (or unconscious) subjects between whom mathematics is communicated unambiguously, and whose experience of mathematics also has individual aspects, some of which may not be communicable.

Pauli argued, in part in juxtaposing his view to Bohr, that a measuring instrument can or even must be seen as an extension of the agent using it, rather than something independent, a view adopted by QBism as well, again, as against that of Bohr (e.g., Fuchs 2016, 2017).[6] At one level this is true: it is true as concerns any given case of measurement as an actual experience, and it is doubtful that Bohr would deny this. On the other hand, as Bohr's drawings show as well, a measurement device has to be made by somebody, usually by somebody else, as a material object in order for others (or of course its makers) to have this actual experience. In this respect they are separate, at least insofar as there are other human beings around, some of whom could repeat the experiment and have the corresponding experience, different overall. If there are no longer any humans or their equivalents, such as computers or robots using these instruments, then there are no instruments either, even things previously built as such instruments. In fact, there are no "things" either, which is true for robots and computers in any event, unless they are assumed to have or to be eventually able to acquire consciousness or experiences of the type we do. If one adopts the U-RWR view, as Bohr appears to have done, or even if one adopts Kant's view, our phenomenal experience of measuring instruments as heavy solid physical bodies may not correspond to what they or anything solid (or liquid or gaseous, as parts of our measuring

[6] I have considered Pauli's views, including in juxtaposition to those of Bohr, elsewhere (Plotnitsky 2013, 2016a, pp. 174–176). On the other hand, Pauli is not a Bayesian and his view of probability and statistics in quantum theory is actually closer to that of Bohr or the one adopted in this study (Plotnitsky 2016a, pp. 174–176).

devices may be) actually are. On the other hand, the U-RWR view still allows us to treat them as such, as we, it appears, must do in order for the experiments and our experiences defined by and defining them to be possible at all. These are manmade devices, as Bohr's drawings, again, depict them, including often noted heavy bolts, by which they are attached to the ground. Such bolts must be made by somebody. They can become part of the experience and thus of our phenomenal worlds, and in this sense become subjective, but still they are different from any subjective probability assignments, insofar as there are human beings around, because we make probability assignments by our thought, while these devices are made by "our hands." This is what Bohr means. There must, again, be somebody in order to experience measuring instruments as heavy solid entities pictured in Bohr's drawings, although evolutionarily all animals on this planet would experience them as something solid (although not as measuring instruments). Experientially, the devices and the agents—technically, the phenomenal representation of devices and the experience of the agents, defined by their bodies and brains, but phenomenal in nature—are essentially and irreducible connected. These devices are, however, not merely one with the agent, as Pauli and his followers, including QBists, thought, because they could be used by others. Bohr's concept of phenomenon reflects this double nature or structure of the quantum–mechanical situation.[7]

One could speak, in accord with Pauli's view, of the indivisibility of the agent's experience in the case of Bohr's phenomena, the experience of phenomena by human agents, in each given case. In Bohr's or the present view, however, this indivisibility of the (human) agent's experience is still underlain by the indivisibility of each phenomena in Bohr's sense, which is a different type of indivisibility of wholeness.

[7] Curiously, in contrast to Pauli or QBism, Rovelli faults Bohr for giving too much space to human subjects, which is equally problematic, but from the other side of the issue (e.g., Rovelli 2021, pp. 139–141). Rovelli's, essentially realist, relationist approach, which he developed in numerous articles and presented in several popular books, such as (Rovelli 2021), is very different from QBism, as he stresses, correctly, although his analysis of QBism is not entirely accurate (Rovelli 2021, pp. 65–7, 87). It is not the place to discuss Rovelli's overall relationist approach. It is clear, however, that his view is in a radical conflict with the present view or set of assumptions, or those of Bohr, whose views Rovelli's is trying to adapt to his own, an adaptation in which the key aspects of Bohr's views, often problematically glossed by Rovelli, are lost (Rovelli 2021, pp. 139–141). "Phenomena are action by one part of the natural world upon another part of the natural world," Rovelli asserts against Bohr, for whom quantum "phenomena" are observed in measuring instruments by human agents (Rovelli 2021, p. 141). Actions are only phenomena when we observed them, and they may also be ascribed to the natural world by us, as conception applicable apart from observation, but just as "reality," "action" is only a human word, that, just as "something happened," could only be applicable to what is observed in an experiment. Bohr knew all this, which is why he said what he said. At the very least, Rovelli should have distinguished between phenomena and objects, which he could have done with the help of Bohr, on whom he comments, if not Kant, whom he does not mention. Rovelli never defines "phenomena," even though he speaks of "phenomenal reality" and "phenomenal realism" (Rovelli 2021, pp. 144–149). Admittedly, this is a popular book, aimed at a broad readership, which make a philosophical rigor difficult, although it could have been more careful even as a popular book. The problems just indicated, however, tangibly reflect Rovelli's philosophical views, also found in his technical works and often equally problematic there.

This indivisibility is that of the behavior of quantum objects and measuring instruments, defined by "our not being any longer in a position to speak of the autonomous behavior of a physical object, due to the unavoidable interaction between the object and the measuring instruments" (Bohr 1937, p. 87). Bohr manifestly differentiates them from classically defined or experienced "objects" observed, as phenomena in Bohr's sense, in measuring instruments, or measuring instruments themselves. In Bohr's or the present view, as the RWR view, quantum objects are never part of any experience qua human phenomenal experience. They only have effects on measuring instruments and, through them, on our experience. In the present view, although, again, not that of Bohr, quantum objects are, moreover, idealizations applicable at the time (in fact after) measurements and, hence, in this experience. But they are never experienced. Only phenomena are, but phenomena are defined by classical objects emerging in measuring instruments under the impact of quantum objects or the ultimate constitution of the reality responsible for quantum objects and quantum phenomena. Furthermore, as discussed in Sect. 6.4, by the time of our observation of a given phenomenon, as the outcome of the measurement performed, and thus by the time of our experience, the quantum object under investigation is no longer there, being either destroyed or on its way elsewhere, possibly to some future measurement and thus a possible future experience.

If one extends the RWR view to the U-RWR scope, a scope that Bohr appears at least to allow for and the present study assumes, the ultimate constitution of the reality responsible for any physical phenomena is always a reality without realism and as a result cannot be assumed to be classically causal. Some physical phenomena, however, such as those considered in classical physics and relativity, allow the reality responsible for them to be treated in realist and classically causal ways, in some cases also allowing for ideally exact or deterministic predictions. Accordingly, on this point in affinity with the QBist view (without, again, equating Bohr's or the present view, as the RWR view, with QBism), one, in doing physics, never ultimately deals with anything other than measurements of or predictions concerning observed phenomena (which corresponds to one's experience), rather than with the ultimate constitution of the reality responsible for them. This is, again, the case even when these predictions are with probability one, which are, under certain circumstances, also possible in quantum physics, because, in either view, such predictions do not mean that one ever establishes the reality thus predicted, even that of phenomena, to which one can have an access. Predictions only belong to our mental reality and are never part of a physical reality, only measurements are. Quantum physics, if interpreted in accord with the Q-RWR view, helps to bring about the U-RWR view of all physics.

6.4 What Does a Measurement Measure and What Does Quantum Theory Predict?

As Bohr came to realize in the wake of the EPR experiment, a quantum measurement has a subtler nature, which is parallel to that of EPR-type measurements even in the standard case of quantum measurement. In any quantum experiment, the *object* under investigation and the measuring instrument become entangled as a result of their interaction with each other.[8] More accurately, entanglement is a feature of the formalism of QM which reflects the feature of quantum phenomena that is defined by this interaction and that distinguishes quantum phenomena from classical ones (although, as discussed earlier, it is not clear whether it does so uniquely). For simplicity, however, I shall refer by the term entanglement to this situation as a whole and speak of the entanglement between the objects and the instrument.

Yet further qualifications are necessary given the concept of quantum objects in the interpretation adopted here, as an idealization only applicable at the time of measurement. As a result, when one makes a prediction based on a given measurement, it can only concern a new possible quantum object, defined by the interaction between the ultimate constitution of the reality responsible for quantum phenomena and a measuring instrument used, and not an object that we measure in order to make the prediction. As explained in Chap. 2, however, in QM or low-energy QFT, although not in high-energy QFT, assuming that our predictions concern the same quantum object as registered by the initial measurement, on the basis of which these predictions are made, is a permissible idealization.

In any event, in the RWR view (in either version), any rigorous statements can only refer to observable events, with which the idealization of a quantum object is associated, while one cannot ultimately speak of the interaction between quantum objects between experiments. One can only speak of two quantum objects associated with the measurement or measurements performed initially and then two quantum objects associated with two measurements performed subsequently. Quantum entanglement can, however, be defined in terms of such measurements, the outcomes of which QM properly predicts. The concept is explained in detail in the next chapter. It suffices to say here, with the qualifications just given in mind, that an entanglement between two quantum objects, S_1 and S_2, forming an EPR pair, allows one by means of a measurement performed on S_1 to make predictions, with probability one, concerning S_2. In the present view, S_2 is only defined once the corresponding measurement is performed, but not when the prediction concerning it is made, which make it even more difficult and indeed rigorously impossible to speak of any independent properties of S_2, however, predicted, because there is no S_2, in the first place, until it is measured. Importantly, however, there is the independent, RWR-type, reality, ultimately responsible for the existence of S_2, when it is measured, and secondly, even then S_2 is still an RWR-type entity, which means that no physical properties can be attributed to it as such. These properties could only be attributed

[8] See (Haroche 2001) and (Haroche and Raimond 2006), for experimental illustrations of this point, although these works do not consider the implications of this entanglement discussed here.

to the measuring instrument used. As predictions at a distance, these predictions are "quantum-nonlocal," a concept explained in Chap. 7. As I argue here, following Bohr, all quantum predictions are predictions at a distance and, in this sense, are quantum-nonlocal, without, however, entailing any instantaneous transmission of physical influences between such events, "a spooky action at a distance" [*spukhafte Fernwirkung*], famously invoked by Einstein, defining in terms of this study, Einstein-nonlocality (Born 2005, p. 155). Einstein-locality would prohibit such an action, as would relativity, although Einstein-locality, sometimes referred to as the locality principle, is independent of relativity.

The interaction between the object and the measuring instrument, or the quantum stratum of the instrument, leading to their entanglement is not a measurement in the sense of *giving rise* to an observable quantity: this interaction occurs before the measurement takes place, or rather before the outcome of this interaction is registered as a quantum phenomenon. This interaction is part of a quantum measurement, as defined by Bohr and here, which is not a measurement of a *pre-existing* property of the quantum object or, in the present view, the ultimate constitution of the reality responsible for both quantum objects and quantum phenomena. A quantum measurement establishes a quantum phenomenon manifested in a measuring instrument, to which, or the data found in it, a measurement in the sense of measuring a physical property can then apply. In Bohr's earlier views (including in his reply to EPR) such properties could still be attributed to quantum objects *at time of measurement*, which, as explained, became no longer possible in his ultimate strong RWR-type interpretation. Once performed, the measurement, say, that of momentum (manifested only as a property of the instrument), disentangles the object and the instrument, with the observed outcome "irreversibly amplified" to the level of *the classically observed stratum of the apparatus* (Bohr 1987, v. 2, p. 73). This outcome is defined by the quantum stratum of the apparatus *after* this interaction, rather than by the object. It is, as Schrödinger explains in his cat-paradox paper, this disentangling that enables one to predict the probability that the momentum measurement at a given future moment in time will be within a certain range (Schrödinger 1935, pp. 162–163). Alternatively, if the initial measurement was that of the position, one could predict the probability that the position measurement at a given future moment in time will locate the trace of the interaction between the object and the instrument within a certain area. Quantum *phenomena* are never entangled. In the present view, not even quantum objects are entangled because they are idealizations only applicable at the time of measurement and thus always irreducibly associated with quantum phenomena. Rigorously speaking, two initial measurements associated with S_1 and S_2 lead to the situation in which possible future measurements can be handled by the mathematics of entangled states in the formalism of QM and the expectation catalogs they enable. Accordingly, only *quantum states*, ψ-functions, can be entangled. If one assumes an independent existence of quantum objects between measurements, which is possible even in RWR-type interpretations (as in that of Bohr, for example), then one could say that they become entangled, although, if one adopts an RWR-type interpretation, the nature of the reality defining this entanglement is beyond representation or knowledge or even conception, even conception expressed by a phrase "something happened." ψ-functions never represent either the ultimate, RWR-type, reality responsible for quantum phenomena or quantum phenomena and thus the outcomes of measurements. They do not represent these outcomes even if one adopts a realist view of ψ-functions as representing what happens between measurements, because one needs Born's rule added to the formalism to predict the probabilities (and only them) of these outcomes, described by classical physics.

Now, according to Bohr's remarkable observation, in effect describing quantum measurement as entanglement:

> After a preliminary measurement of the momentum of the diaphragm, we are in principle offered the choice, when an electron or photon has passed through the slit, either to repeat the momentum measurement or to control the position of the diaphragm and, thus, to make predictions pertaining to alternative subsequent observations. It may also be added that it obviously makes no difference, as regards observable effects obtainable by a definitive experimental arrangement, whether our plans of considering or handling the instruments are fixed beforehand or whether we prefer to postpone the completion of our planning until a later moment when the particle is already on its way from one instrument to another. (Bohr 1987, v. 2, p. 57)[9]

If, then, a measurement is always made after the object has left the location of the measurement, what does this measurement measure? How does it create the corresponding phenomenon? It "measures" the quantum state of the quantum stratum of the instrument, which interacted with the object in the past (however recently, but always in the past!), by amplifying this state to the classical level of observation, by registering the corresponding state of the measuring instrument.[10] This amplification leads to the phenomena in which the outcome of a measurement is registered. What will be registered is either the change in the momentum of certain observed parts of the apparatus or the position of one or another trace of this interaction, say, a spot on a silver screen, given that both can never be registered in the same arrangement, as reflected by the uncertainty relations. Such concepts as momentum or position can only rigorously apply at this classical level in the RWR view, adopted by Bohr by the time of this statement (in 1949).

This situation is consistent with the present interpretation, according to which a quantum object is an idealization only applicable at the time of measurement, insofar

---

[9] This observation, as Wheeler notes, anticipates the delayed choice experiment (Wheeler 1983, pp. 182–192).

[10] This point appears to have been missed either in commentaries on Bohr or by treatments of quantum measurement elsewhere. Subtle as it is, Schrödinger's analysis of quantum measurement in his cat-paradox paper does not consider it (Schrödinger 1935, pp. 158–159). Neither does Ozawa (1997), discussed in Chap. 2 (note 25), an analysis that is expressly realist and implies that the measured quantity is attributed to the object at the time of measurement. Von Neumann's analysis comes close, but, while it is possible that von Neumann realized this point, he did not comment on it, and some of his statements appear to attribute the measured quantity to the object at the time of measurement (Von Neumann 1932, pp. 355–356). On the other hand, this aspect of quantum measurement supports the point, made by von Neumann and others, that, for idealized measurements, an instantly repeated measurement will give one the same result as the initial measurement (Von Neumann 1932, pp. 214–215; Schrödinger 1935, pp. 158–160).

as one refers by measurement, as Bohr does, to the overall process in question, as leading to the emergence of an observed phenomenon: the interaction between the quantum object and the quantum stratum of the instrument, and the amplification of the resulting quantum state of this stratum (after this interaction and hence no longer in the presence of the quantum object) to the classical level of the observed stratum of the instrument, in which stratum the outcome is registered. Even though the quantum object is no longer there (or no longer exists) when the corresponding phenomenon is established and an observation and possibility a measurement of some observed quantity takes place, one might still see this quantum object as, by its interaction with the quantum stratum of this instrument, responsible for the effect observed. It follows that, a quantum object or what is so idealized, while still assumed to exist at the time of a measurement, only exists in the past of an observed quantum event. An observed effect could also be that of the measurement of the charge, mass, or spin of an electron. These quantities will be the same for all electrons and will define them as electrons, in the present view at the time of measurement. Such quantities as the position, velocity, momentum, or energy registered in a measurement will of course be different.

One might assume that, say, because of the exchange of momenta between the object and the instrument, the momentum of the object will correspond to the difference between two momentum measurements of the instrument before and after the interaction with the object. Physically, however, one never measures that momentum, given that the object has already left the location of the instrument and that one could have performed instead of the position measurement after it did. In any event, one can ascertain, *regardless of an interpretation*: (a) that one can perform either of the two complementary measurements concerning the state of the quantum stratum of the instrument, with the outcome amplified to the classical level of the observable part of this instrument; and correlatively (b) the quantum-nonlocal nature of quantum predictions, because by changing one's decision which measurement to perform after the interaction in question, one can make two alternative predictions concerning distant future events, to which one is not physically connected at the time of either measurement.

Thus, using the measurement of the state of the apparatus, one can predict, at a distance, by means of a ψ-function (cum Born's rule), a possible outcome of a future measurement of either variable, without "in any way disturbing the system," just as in any EPR-type experiment (Einstein et al. 1935, p. 138). It is true that there was an interaction between the object and the instrument before that measurement. But this is also the case for the two objects of the EPR pair, which have been in an interaction, entangling them. In a standard measurement, the probability of such a prediction will not be equal to one, as it would be in the case of the EPR-type experiments, which possibility, however, requires qualifications, discussed below. Besides, as Bohr realized, with some simple additional arrangements one can, at least in principle, reproduce the EPR case in considering the standard quantum measurement (Bohr 1938, pp. 101–103, 1987, v. 2, p. 60). It *might seem*, then, that, in either the standard or the EPR case, because either of the two complementary quantities could be predicted at a distance for one quantum object by an alternative measurement on another quantum object, the first object at a distance can be assigned both quantities, as, in EPR's language, "elements or reality" "without in any way disturbing" it (Einstein et al. 1935, p. 138). This was, essentially, EPR's argument, although the possibility of predicting either quantity with probability one in the EPR experiment strengthened their case that QM is incomplete. EPR also argued that an alternative, even the only alternative to this incompleteness, in the case of the EPR experiments, would be that QM is Einstein-nonlocal, because a measurement on one object would alternatively define the state of reality at a distance (Einstein et al. 1935, p. 141). As, however, should already be apparent and as will be discussed in detail in Chap. 7, this is not necessarily the case, because, in any *actual* experiment, only one of these quantities could be so predicted. There is no experiment that would allow one to physically realize the prediction of both quantities for the same object. At the same time, there is no need to assume that our predictions are Einstein-nonlocal by virtue of determining the quantity in question at a distance, because, even if one can predict this quantity with probability one, one could still measure the complementary quantity and thus establish a different element of reality from the one predicted (Einstein et al. 1935, p. 138). A measurement performed on one quantum object cannot be claimed to define an element of reality pertaining to another, spatially separated, quantum object, by means of a prediction with the probability one. A prediction even with probability one is, again, still only a prediction. It does not define the element of reality thus predicted, because an alternative, complementary, measurement on the object for which the original variable was predicted, would establish a different state of reality, which would disable the possibility of a verification of the original prediction. Accordingly, there is no rigorous basis for assuming that this prediction established the reality it predicts. In classical physics these limitations do not apply because we can always measure and predict, and simultaneously verify our predictions for conjugate variables (such that of position and that of momentum). The conjugate variables of classical physics are not the same as complementary variables of quantum physics, a fact not always sufficiently appreciated (Schwinger 1994).

The outline just given is only a sketch that requires a proper argument, to be offered in Chap. 7. The main point at the moment is that, even in a standard quantum measurement, one must always consider not only the object under investigation as in classical physics (where one can disregard the role of measuring instruments), but a composite entangled quantum system, consisting of the object and the quantum stratum of the instrument.[11] In each EPR-type experiment at the last stage of the experiment, when one of the two possible EPR predictions is made, one deals with two combined systems, each consisting of an object and an instrument, the first associated with an actual (already performed) measurement and the other with a possible future measurement, concerning which one makes a prediction, and thus with four systems in total. The analysis of quantum measurement just given confirms,

[11] The significance of considering composite systems in quantum theory was emphasized in D'Ariano's recent works (some of which are cited in this study), in contrast to other quantum-informational approaches to quantum foundations. Schrödinger, actually, suggested that this might be necessary in the end of the cat-paradox paper (Schrödinger 1935, p. 167).

however, that Bohr's ultimate interpretation has a complex conceptual structure. This architecture is not always properly considered by commentators, which often leads to a misunderstanding of Bohr's statements and concepts, because they lose their proper meaning if considered apart from this structure. I discuss next Bohr's most famous concept, complementarity, introduced in the Como lecture, but redefined by Bohr's subsequent interpretations, finally, by his ultimate, strong RWR-type, interpretation.

6.5 Complementarity

As stated in the preface, defined arguably most generally, complementarity is characterized by

(a) a mutual exclusivity of certain phenomena, entities, or conceptions; and yet
(b) the possibility of considering each one of them separately at any given point; and
(c) the necessity of considering all of them at different moments of time for a comprehensive account of the totality of phenomena that one must consider in quantum physics.

The concept was never given by Bohr a single definition of this type. However, this definition may be surmised from several of Bohr's statements, such as: "Evidence obtained under different experimental conditions cannot be comprehended within a single picture, but must be regarded as complementary in the sense that only the totality of the phenomena [some of which are mutually exclusive] exhaust the *possible* information about the objects" (Bohr 1987, v. 2, p. 40; emphasis added). In classical mechanics, we can comprehend all the information about each object within a single picture because the interference of measurement can be neglected, which is a permissible idealization even in the U-RWR view: this assumption allows one to (ideally) identify the phenomenon considered with the object under investigation and to establish the quantities defining this information, such as the position and the momentum of each object, in the same experiment. In quantum physics, this interference cannot be neglected and indeed defines any quantum phenomenon, which leads to different experimental conditions for each measurement on a quantum object (assumed to be prepared in some way, by a previous event, such as an emission of this object by some device) and their complementarity, in correspondence with the uncertainty relations. This is, again, keeping in mind that, while in the present view, a quantum object is an idealization that only applies at the time of measurement, speaking of the same quantum object in the case of these two measurements is a permissible idealization in QM (although no longer in QFT). The situation implies two incompatible pictures of what is observed, as phenomena, in measuring instruments. Hence, the *possible* information about a quantum object, the information *to be found* in measuring instruments, could only be exhausted by the mutually incompatible evidence obtainable under different experimental conditions.

On the other hand, once made, either measurement, say, that of the position, will provide the complete *actual* information about the system's state, as complete as possible, at this moment in time. One could never obtain the complementary information, provided by the momentum measurement, at this moment in time, because to do so one would need simultaneously to perform a complementarity experiment on it, which is never possible. In fact, as noted, if one repeats the first, position, measurement in a new experiment with the same initial preparation, the outcome, observed after the same time interval after the preparation, will be different.

It follows that one cannot assume that two complementary measurements represent *parts* of the same *whole*, of the same single reality. Each measurement establishes the only reality there is, and the alternative decision would establish a different reality, at all three levels of idealization assumed here—the ultimate nature of the reality responsible for quantum phenomena, quantum objects, and quantum phenomena themselves, even though in the first two cases this reality is each time unknowable and even unthinkable. It may still be assumed to be each time different because each of its effects, observed as a phenomenon, is different and indeed unique, by the QI postulate. Rather than arbitrarily selecting one or other part of a pre-existing physical reality, as is ideally possible in classical physics, our decisions concerning which experiment to perform establish the *single* reality which defines what *type* of quantity (although not its value, which can only be predicted probabilistically or statistically) can be observed or predicted and precludes the complementary alternative.

Accordingly, parts (b) and (c) of the above definition of complementarity are as important as part (a), and disregarding them can lead to a misunderstanding of Bohr's concept. That we have a free (or at least sufficiently free) choice as concerns what kind of experiment we want to perform is in accordance with the very idea of experimentation in science, including in classical physics (Bohr 1935, p. 699). However, contrary to the case of classical physics or relativity, implementing our decision concerning what we want to do will allow us to make only certain types of possible predictions and will irrevocably exclude certain other, *complementary*, types of possible predictions. We actively shape what will happen, define the course of reality.

Complementarity is, thus, a reflection of the fact that, in a radical departure from classical physics or relativity, the behavior of quantum objects of the same type, say, electrons, or, again, the ultimate nature of reality responsible for quantum phenomena handled by such concepts as electrons, is not governed by the same physical law, especially a representational physical law, in all possible contexts, specifically in complementary contexts. This leads to incompatible observable physical effects in complementary contexts. On the other hand, the mathematical formalism of QM offers correct probabilistic or statistical predictions of quantum phenomena *in all contexts*, in RWR-type interpretations, under the assumption that the ultimate nature of reality responsible for quantum phenomena is beyond representation or even conception.[12]

[12] This situation is also responsible for what is known as "contextuality," which I considered from the RWR perspective in (Plotnitsky 2019). I give a brief summary. The concept of contextuality

Speaking of a physical law here requires caution, because, in RWR-type interpretations, there is no physical law representing this behavior, or, again, in the present interpretation, representing the ultimate nature of physical reality that will, in a measurement, give rise to the quantum object or what is so idealized. One might speak, with Wheeler, of "law without law" (Wheeler 1983), along with, and as an effect of, "reality without realism," a connection to which I shall return in closing this chapter.

The analysis of complementarity just given is shaped by Bohr's ultimate, strong RWR-type interpretation, as confirmed by the statement cited above, which dates from his 1949 "Discussion with Einstein on Epistemological Problems in Atomic Physics" (Bohr 1987, v. 2, pp. 33–66), by which point this interpretation was in place for over a decade. It took Bohr a previous decade (1927–1937) to arrive at both this interpretation and this definition of the concept, from that point on instantiated in terms of complementary phenomena in Bohr's sense, defined by what is observed in measuring instruments. I would like to emphasize, following the argument of Chap. 2, that Bohr's complementarity is a new physical *concept*, although, as all concepts, it has its history, in physics and beyond, and had undergone development and changes in Bohr's own work.[13] Unlike the adjective "complementary," the noun complementarity has not been used and was introduced by Bohr, which

was introduced by Simon Kochen and Ernst Specker, and developed, following it, Bell's theorem, and related findings (discussed in Chap. 7) during the last half century, primarily dealing with the Bohm version of the EPR experiment for discrete variables and finite-dimensional Hilbert spaces. The primary reason for this is that the (idealized) thought-experiment proposed by EPR, which deals with continuous variables, cannot be physically realized because the EPR-entangled quantum state is not normalizable, although this fact does not affect the fundamentals of the case. Bohm's version of the EPR experiment, which deals with discrete variables, could and has been performed, confirming the existence of quantum correlations. These correlations can be ascertained experimentally, apart from QM. I summarize the Kochen–Specker theorem, following Held (2018), apart from considerations involving Gleason's theorem. The Kochen–Specker theorem considers the following three assumptions of the hidden variable theories: (1) value definitiveness: all observables defined for a QM system have definite values at all times; (2) noncontextuality: if a QM system possesses a property (corresponding to a value of an observable), then it does so independently of a measurement context, i.e., independently of how that value is measured; and (3) projector operator representation: assuming the standard association of properties of a quantum system with and projection operators on the corresponding Hilbert space, there is a one-to-one correspondence between properties of a quantum system and projection operators on this Hilbert space. Noncontextuality, as formulated in (2) is in accord with Einstein's concept of reality used in his EPR-type arguments, while complementarity implies the dependence of all QM-quantities on their measuring contexts, mutually exclusive in the case of complementarity variables. The Kochen–Specker theorem says that holding all three assumptions is incompatible with QM (assuming the dimension of the corresponding Hilbert space is three or higher). If one assumes an RWR-type interpretation, contextuality is automatic, given complementarity, with (3) retained. Locality (the sense of precluding an action at a distance) is preserved as well.

[13] Among the works addressing a more general philosophical background of the concept of complementarity and Bohr's thinking in general, are Faye (1991), Folse (1985), Honner (1987), and Murdoch (1987). See also a comprehensive volume of essays (Faye and Folse 2019). These studies offer important reflections on Bohr's thinking, and engaging with them in more detail would enrich the present discussion. On the other hand, while some of them offer readings of Bohr's interpretation of quantum phenomena and QM *similar* to the one presented by this study, including in its

is a reflection of the fact that it is a concept in the sense defined in Chap. 2. Using it as a noun also distinguishes it from the use of complementary as an adjective, as in James, to which use Bohr's concept is sometimes compared and even traced, questionably in my view (Plotnitsky 2016a, pp. 116–117). Most crucial, however, is the specificity of the structure of Bohr's concept, beginning, again, with the fact that parts (*b*) and (*c*) arise from physical considerations. Equally important, are the probabilistic or statistical, or correlatively, mathematical, aspects of it, which give the concept its quantum-theoretical specificity. When complementarity applies to *predictions* concerning (complementary) experimental arrangements, in which the corresponding measurements occur, this application is probabilistic or statistical. It was quantum physics that made the concepts of complementarity and phenomenon, in the specificity of their architecture, necessary for Bohr, although his thinking concerning quantum physics also helped him to define more general philosophical aspects of complementarity, thus, establishing a potential for defining it as a philosophical concept.[14] It is, accordingly, not surprising that Bohr speaks of complementarity as "an artificial word" that "does not belong to our daily concepts." It is a physical concept, eventually linked to Bohr's ultimate, strong RWR-type, interpretation of quantum phenomena and QM.

This connection to Bohr's interpretation defines the specificity of complementarity as *quantum-theoretical concept*, at any stage of Bohr's understanding of it. It is difficult to grasp complementarity as a concept apart from these connections. This may explain (although one cannot be certain) why Einstein, who, as discussed in Chap. 7 misread Bohr on some of the key points of his interpretation, in particular, as concerns Bohr's view of the Einstein-locality of QM in his reply to EPR, had difficulties with complementarity, admitting that he was "unable to attain... the sharp formulation... [of] Bohr's principle of complementarity" (Einstein 1949b, p. 674). As discussed in Chap. 7, Bohr's argument in his reply, misread by Einstein, is essentially grounded in complementarity and Bohr's contention that EPR disregarded this role. It would, accordingly, be difficult to understand Bohr's reply without properly considering complementarity as a concept or the principle that gives rise to this concept. Bohr's concept, however, has posed difficulties for many others, even for

---

nonrealist aspects, these readings do not really amount to treating Bohr's interpretation as an RWR-type interpretation, especially as a strong RWR-type interpretation. In addition, while indebted to Bohr, this study departs from Bohr's view in several key respects. It also pursues a broader set of agendas, such as those concerning QFT or quantum information theory, and the role of mathematics in quantum theory.

[14] The definition of complementarity given above is more general, rather than only defining it as a *quantum-theoretical* concept. This definition allows for applications of the concept elsewhere in physics and beyond physics. Bohr and others inspired by him, such as Pauli, Karl Gustav Jung, and Max Delbrück, proposed using the concept in psychology, biology, and philosophy, where the concept has become a subject of renewed attention recently, as part of the general surge of interest in the use of quantum–mechanical-like modeling (Haven and Khrennikov 2016). These extensions are beyond my scope here. See Pauli (1994, pp. 149–164), on connections between quantum theory and psychology. See also (Plotnitsky 2012, pp. 157–166)), (Wang and Busemeyer 2015), and (Faye and Folse 2019). On applying complementarity in biology in Bohr's works, see (Bohr 1987, v. 2, pp. 3–12; Bohr 1962, Bohr 1987, v. 3, pp. 23–29).

those sympathetic to his views, especially when it comes to the relationships between the whole and its parts. As noted, complementarity prevents us from ascertaining the "whole" composed from the complementary "parts," in conflict with the conventional understanding of parts (always) *complementing* each other within a whole. At any moment of time only one of these parts and not the other could be ascertained. This part is the only "whole" at this moment of time. Consider Bell's comments on complementarity, reflecting his general discontent with Bohr's views:

> [Bohr] seemed to revel in contradiction, for example, between 'wave' and 'particle.' ... Not to resolves these contradictions and ambiguities, but rather to reconcile us to them, he put forward a philosophy which he called 'complementarity.' ... There is very little I can say about 'complementarity.' But I wish to say one thing. It seems that Bohr used this word with the reverse of its usual meaning. Consider for example the elephant. From the front she is head, trunk, and two legs. From the sides she is bottom, tail, and two legs. They supplement one another, they are consistent with one another and they are all entailed by the unifying concept 'elephant.' It is my impression that to suppose Bohr used the word 'complementarity' in this ordinary way would have been regarded by him as missing his point and trivializing his thought. He seems to insist rather that we must use in our analysis elements which *contradict* one another, which do not add up or derive from a whole. By 'complementarity' he meant, it seems to me, the reverse: contradictoriness. ... Perhaps he had a subtle satisfaction in the use of a familiar word with the reserve of its familiar meaning. (Bell 2004, p. 190)

My concern is not Bell's general *discontent* with Bohr's views. Bell does not favor not only them but also QM itself. This attitude is, however, legitimate. Bell even admits, grudgingly, on an earlier occasion (Bell is less charitable to Bohr in his later articles), that Bohr's view could correspond to the ultimate character of physical reality, still misreading Bohr by attributing to him the claim that "there is no reality below some 'classical' 'macroscopic' level," (Bell 2004, p. 154). Bohr never made such a claim and, as the preceding discussion here makes clear, his argument does not imply it either. Bell, as he acknowledged, had difficulties in understanding Bohr's argument and, on one occasion, in considering a passage from Bohr's reply to EPR (Bohr 1935, p. 700), says that "he has very little idea what this means" (Bell 2004, p. 155). Understanding this passage, explained in Chap. 7, would be difficult apart from considering Bohr's argument preceding it (from which the meaning of this passage derives). Bell does not do so, thus making it difficult to assess his claims concerning Bohr's statements in this passage because the meaning of these statements is defined by Bohr's preceding argument. Indeed, Bell truncates the passage in quoting it by omitting, among other things, Bohr's statement on complementarity, crucial to the argument of his reply. As discussed in Chap. 7, Bohr's argument is based in his view that EPR disregard to the role of complementarity, a role indispensable for the understanding of the EPR experiment.

In any event, Bell's comments on complementarity in the passage cited here are beside the point, even apart from his attempt to psychologize Bohr's thinking. Bell is not incorrect in saying that "to suppose Bohr used the word 'complementarity' in this ordinary way would have been regarded by him as missing his point and trivializing his thought." It would have been, because it would indeed both miss his meaning and trivialize his thought. Besides, as noted, while "complementary" as an adjective is a familiar word, "complementarity" was never used as a noun before Bohr. Also, as explained below, Bohr's treatment of the question of "particles and "waves" in quantum theory is a more complex matter than Bell makes it appear, which compels Bohr to avoid speaking of wave–particle complementarity. Complementarity is, I argue, a new physical concept, which must be understood in the specific sense Bohr gives it. There is no point in attempting to relate it to a meaning it may be given in our daily life, as Bell does, by defining complementary parts as adding up to a whole. This is what Bohr wants to avoid. The reason is that he needed complementarity as a new concept, which "does not belong to our daily concepts," introduced in order to account for the epistemological situation defined by quantum phenomena (Bohr 1935, p. 700, 1937, p. 87). It is not, as Bell suggests, some "subtle satisfaction in the use of a familiar word with the reverse of its familiar meaning." Bell, to his credit, does qualify by "perhaps," but even this qualified surmise is hardly pertinent. There is no evidence that Bohr ever had such a satisfaction. Besides, even if he had, it is irrelevant in assessing his concept in its meaning and significance. There is plenty of evidence for his physical reasons for defining complementarity in the way he did, such as the uncertainty relations, the double-slit and other iconic quantum experiments, or the EPR experiment. Complementarity was introduced in the spirit of resolving contradictions, and not reveling in them, as it "seemed" to Bell. Bohr's concept of complementary "parts" that do not add up to a "whole" aims to do just that. One always deals with individual, each time unique, quantum phenomena, some of which are complementary. These phenomena are never parts of any whole beyond themselves.

As indicated earlier, part of the genealogy of the concept of complementarity is Bohr's early interest (before he began to study physics) in a complementarity-like way in which a Riemann surface allows one to remove ambiguity and properly define functions of complex variables, such as $f(z) = \sqrt{z}$. It is ambiguous, and hence not properly definable, when considered on the complex plane, as $\sqrt{z}$ has two meanings, but is well defined on the corresponding Riemann surface, because it has a single meaning on each of the two separate sheets of it. Bohr reflected on this genealogy of complementarity, also bringing together the mathematical and quantum-theoretical concepts of the irrational he used in accord with the analysis given in Sect. 2.1 of this Chap. 2, in his final interview and in the picture that he drew on the blackboard in the course of this interview. First, Bohr commented on the time before his interest in physics took over:

> At that time I really thought to write something about philosophy, and that was about this analogy with multivalued functions. I felt that the various problems in psychology—which were called big philosophical problems, of the free will and such things—that one could really reduce them when one considered how one really went about them, and that was done on the analogy to multivalued functions. If you have square root of x, then you have two values. If you have a logarithm, you have even more. And the point is that if you try to say you have now two values, let us say of square root, then you can walk around in the plane, because, if you are in one point, you take one value, and there will be at the next point a value which is very far from it and one which is very close to it. If you, therefore, work in a continuous way, then you—I am saying this a little badly, but it does not matter—then you can connect the value of such a function in a continuous way. But then it depends what you

> do. If in these functions, as the logarithm or the square root, they have a singular value at the origin, then if you go round from one point and go in a closed orbit and it does not go round the origin, you come back to the same (value). That is, of course, the discovery of Cauchy. But when you go round the origin, then you come over to the other (value of the) function, and that is then a very nice way to do it, as Dirichlet (Riemann), of having a surface in several sheets and connect them in such a way that you just have the different values of the function on the different sheets. And the nice thing about it is that you use one word for the function, $f(z)$. Now, the point is, what is the analogy? The analogy is that you say that the idea of yourself is singular in our consciousness—do you think it works; am I doing it sufficiently loud? Then you find—now it is really a formal way—that if you bring this idea in, then you leave a definite level of objectivity or subjectivity. For instance, when you have to do with the logarithm, then you can go around; you can change the function as much as you like; you can change it by 2π; when you go one time round a singular point. But then you surely, in order to have it properly and be able to draw conclusions from it, will have to go all the way back again in order to be sure that the point is what you started on.—Now I am saying it a little badly, but I will go on.—That is then the general scheme, and I felt so strongly that it was illuminating for the question of the free will, because if you go round, you speak about something else, unless you go really back again (the way you came). That was the general scheme, you see. (Bohr 1962, Session 5)

There are of course important differences between the two concepts. In the case of Riemann's surface, we have two mutually exclusive mathematical representations. In quantum theory, complementarity means that each complementary situation or context exhibits a different behavior of quantum objects of the same type, while the formalism of quantum mechanics predicts this behavior in both and indeed all contexts, admittedly in two mathematically different ways as well. In addition, these predictions are probabilistic or statistical, which is obviously not a feature of Riemann's concept. Nevertheless, this is a remarkable example of a *translation* of a mathematical concept into a physical one.

Bohr's first specific instance of complementarity was introduced, along with the concept itself, in the Como lecture, as the complementarity of spacetime co-ordination and the claim of causality:

> On one hand, the definition of the state of a physical system, as ordinarily understood [i.e., in classical mechanics], claims the elimination of all external disturbances. But in that case, according to the quantum postulate, any observation will be impossible, and, above all, the concepts of space and time lose their immediate sense. On the other hand, if in order to make observation possible we permit certain interactions with suitable agencies of measurement, not belonging to the system, an unambiguous definition of the state of the system is naturally no longer possible, and there could be no question of causality in the ordinary [classical] sense of the word. *The very nature of the quantum theory thus forces us to regard the space-time co-ordination and the claim of causality, the union of which characterizes the classical theories, as complementary but exclusive features of the description, symbolizing the idealization of observation and definition respectively. ... Indeed, in the description of atomic phenomena, the quantum postulate presented us with the task of developing a "complementarity" theory the consistency of which can be judged only by weighing the possibility of definition and observation.* (Bohr 1987, v. 1, pp. 54–55; emphasis added)

These formulations are short of rigorously defining complementarity as a concept and conveying its proper architecture, as it was eventually worked out by Bohr. In particular, it is not clear what complementary means here beyond "exclusive," corresponding, in terms of the definition of Bohr's concept in his ultimate form given above, to: *(a) (a mutual exclusivity of certain phenomena, entities, or conceptions).* Parts *(b)* (*the possibility of considering each one of them separately at any given moment of time*) and *(c)* (*the necessity of considering all of them at different moments of time for a comprehensive account of the totality of phenomena that one must consider in quantum theory*) of the concept that came to define it later are not stated in the lecture. While *(c)* appears to be implied and to be the reason why Bohr sees spacetime coordination and the claim of causality not only as "mutually exclusive" but also as "complementary," *(b)*, which became central for defining the concept later on, does not figure in the lecture at all. Indeed, the complementarity of spacetime co-ordination and the claim of causality makes it difficult to rigorously define this feature. Bohr is correct in saying that "[the quantum postulate] implies a renunciation as regards the causal space–time co-ordination of atomic processes" (Bohr 1987, v. 1, p. 53). But then, if the spacetime coordination and [classical] causality are mutually exclusive, what does causality mean, including as "symbolizing the definition"? In what sense is this causality physical, as Bohr appears to imply. None of this is, I would argue, worked out in the lecture.

The use of "disturbance," although properly handled by Bohr here, requires careful attention to avoid confusion as well. In his later writings, Bohr warned against using the language of "disturbing phenomena by observation" in considering quantum phenomena, along with "creating physical attributes of atomic objects by measurement" (Bohr 1938, p. 104, 1987, v. 2, p. 64, 73). It is worth registering both careful distinctions between phenomena and objects, implied but not made as sharply in the Como lecture, and between observation and measurement, within the concept of "quantum measurement" as defined in this study, following Bohr. To reiterate the main point, a "measurement," is defined by the establishment of a new quantum phenomena that we *observe* in the instruments used and that allows us to *measure* (classically) physical properties of the observable parts of said measuring instruments. These instruments also have quantum strata through which they interact with quantum objects or the physical reality ultimately responsible for quantum phenomena. No creation of physical attributes of atomic objects by measurement is possible in the strong RWR view adopted by Bohr at this point by 1937, because in this view no physical attributes could be assigned to quantum objects even at the time of measurement. All physical properties considered are only those observed, as parts of phenomena, in measuring instruments. Bohr's preferred term becomes "interference": measuring instruments interfere with quantum objects, or in the present view, with the ultimate reality responsible for quantum phenomena, a reality that at the time of a measurement manifests itself as a quantum object. Bohr still (rightly) retains "interaction."

In any event, the Como argument, which never satisfied Bohr, even though he republished the lecture without changes later in *Atomic Theory and the Description of Nature,* now Volume 1 of (Bohr 1987), was revamped by Bohr after his discussion with Einstein in Brussels in 1927, mentioned above. The main casualty of this revamping was the complementarity of the spacetime co-ordination and the claim of causality. It disappeared from Bohr's writings after the Como lecture, along with and correlatively to the key realist part of Bohr's argument, classical causality. Beginning with his next article "The Quantum of Action and the Description of Nature," published in 1929 (Bohr 1987, v. 1, pp. 92–101), Bohr abandoned the idea that the independent behavior of quantum objects is classically causal, although, as noted, the idea itself has appealed to others, Dirac and von Neumann (Dirac 1958; von Neumann 1932) in particular, and continues to do so, and is often associated with the Copenhagen interpretation, as discussed in (Plotnitsky 2009, pp. 202–218). By the time of his reply to EPR, Bohr, speaks of "a final renunciation of the classical ideal of causality" (Bohr 1935, p. 697). Eventually Bohr came to see complementarity as "a rational generalization of the... ideal of causality," which is very different from maintaining classical causality, even if only as complementary to something else, such as spacetime co-ordination (Bohr 1937, pp. 84–85, 1987, v. 2, p. 41). I discuss this generalization in the next section.

Wave–particle complementarity, with which the concept of complementarity is often associated, had not played a significant, if any, role in Bohr's thinking, especially after the Como lecture. Bohr thought deeply, even before QM, about the problem of wave–particle *duality*, as it was known then. De Broglie's formulas for matter waves (in which wave-properties could be understood, as they were by Bohr, symbolically rather than physically even in the Como lecture) and Schrödinger's wave mechanics, rather than only his equation (which, too, could be and were by Bohr, in the Como lecture, interpreted symbolically) might have affected Bohr's initial thinking concerning complementarity. Bohr used de Broglie's formulas in his elegant elementary derivation of the uncertainty relations in the Como lecture (Bohr 1987, v. 2, pp. 57–60). Bohr was, however, always aware of the difficulties of applying the concept of physical waves to quantum objects or of thinking in terms of the wave–particle duality as the assumption that both types of nature and behavior pertain to the same individual entities, such as each photon or electron itself, considered independently.

The wave–particle duality was thought of as representing the same thing in two different ways. By contrast, complementarity refers to two different, incompatible, things, like two different effects of an electron on a measuring instrument, but not two features of the electron itself, a major difference. The "*both*" (both types of properties) of the wave–particle duality is the opposite of complementarity, based on "*either or*" (either one or the other type of effects). Bohr's ultimate solution to the dilemma of whether quantum objects are particles or waves was that they were neither. Instead, either "picture" refers to one of the two mutually exclusive sets of discrete individual effects of the interactions between quantum objects and measuring instruments, particle-like, which may be individual or collective, or wave-like, which are always collective, but still composed of discrete individual effects.

An example of the latter is "interference" effects, composed of a large number of discrete traces of the collisions between the quantum objects and the screen in the double-slit experiment in the corresponding setup (when both slits are open and there are no means to know through which slit each object has passed). These two sets of effects may be seen as complementary, also when it comes to calculating the probabilities or statistics for each set of events, or, if one takes a Bayesian view, for each event of each set. Bohr's thinking took advantage of the fact these two types of effects are always mutually exclusive and require mutually exclusive experimental setups to be observed. Bohr, again, did use the idea of symbolic waves as related to probability or statistics of quantum predictions, in accordance with Born's interpretation (e.g., Bohr 1935, p. 697). As explained in Chap. 4, however, speaking of waves in this context requires caution. The concept is best understood as, in Schrodinger's terms, defining the character of "expectation-catalogs" concerning possible discrete events.

In Bohr's post-Como argumentation, the concept of complementarity becomes primarily exemplified by complementarities of spacetime co-ordination and the application of momentum or energy conservation laws, correlative to the uncertainty relations, in effect as complementarities of phenomena observed in measuring instruments and thus in effect in accord with Bohr's concept of phenomena, introduced a decade later. Technically, the uncertainty relations, $\Delta q \Delta p \cong h$ (where q is the coordinate, p is the momentum in the corresponding direction), only prohibit the simultaneous exact measurement of both variables, which is always possible, at least ideally and in principle, in classical physics, and allows one to maintain classical causality there. Even this statement needs further explication, however, and the physical meaning of the uncertainty relations is much deeper in Bohr's interpretation and a complex subject in its own right with a long history.[15] First of all, the uncertainty relations are not a manifestation of the limited accuracy of measuring instruments, because they would be valid even if we had perfect measuring instruments. In classical and quantum physics alike, one can only measure or predict each variable *within* the capacity of our measuring instruments. In classical physics, however, one can, again, in principle measure both variables simultaneously within the same experimental arrangement and improve the accuracy of this measurement by improving capacity of our measuring instruments, in principle indefinitely. The uncertainty relations preclude us from doing so for *both* variables in quantum physics regardless of this capacity. The uncertainty relations make each type of measurement complementary to the other, in conformity with the definition of complementarity given above, which definition, however, also implies that each by itself could be measured ideally exactly, just it can be in classical physics. According to Bohr: "we are of course not concerned with a restriction as to the accuracy of measurement, but with a limitation [in accord with complementarity] of the well-defined application of space–time concepts and dynamical conservation laws, entailed by the necessary distinction between measuring instruments and atomic objects" (Bohr 1937, p. 86, 1987, v. 3, p. 5 and v. 2, p. 73).

In Bohr's view, moreover, one not only cannot measure both variables simultaneously but also cannot define them simultaneously. It follows, that one cannot ever define such elements alternatively for the same quantum object. One always needs

---

[15] See Hilgevoord and Uffink (2014) for a useful, although one-sided survey. The uncertainty relations remain a subject of ongoing foundational discussions. Among standard technical treatments (both, ultimately, on realist lines) are Ozawa (2003) and Shilladay and Bush (2006). See also Peres 1993, p. 93), which is close to Bohr's view (Plotnitsky 2016a, p. 133).

two objects to define both variables. If, after determining, say, the position of the electron, emitted from a source, at time t_m after the emission, we want to determine the momentum, we need to repeat the same, identically prepared, emission (from the same source) of another electron, and then measure its momentum after the same time t_m after the emission. The uncertainty relations apply to these two measurements on two different quantum objects, still with further qualifications explained below, actually, to two sets of multiple experiments of both kinds (e.g., Peres 1993, p. 93). In the present view, at stake are two different interactions with the different, RWR, type reality responsible for each set of phenomena and measurements, each of which defines two different quantum objects, electrons, in each of these two experiments. Within each, it is, again, permissible, as a statistical idealization, to speak of the emitted and the measured electron as the same electron, or assume that the measured electron was emitted from the source. This assumption, however, is only statistically justified, rather than guaranteed, and it no longer holds even statistically in QFT.

Probabilistic or statistical considerations are unavoidable in considering both the uncertainty relations, which is commonly recognized, and complementarity, which is usually overlooked. According to Bohr, however: "the statistical character of the uncertainty relations in no way originates from any failure of measurement to discriminate within a certain latitude between classically describable states of the objects, but rather expresses an essential limitation of applicability of classical ideas to the analysis of quantum phenomena," a limitation defined by complementarity (Bohr 1938, p. 100). There is no contradiction between this statement and the fact that quantum phenomena are described classically, because the statement only says that their ultimate nature and emergence cannot be *analyzed* classically. The predictions involved will still be probabilistic or statistical, reflecting the probabilistic or statistical nature of the uncertainty relations and complementarity, which may be seen, as Bohr argued, as a "generalization of the very ideal of causality" (Bohr 1937, pp. 84–85, 1987, v. 2, p. 41). In RWR-type interpretations, this nature is in conflict with *classical causality*, rather than only determinism, which defines all quantum predictions on experimental grounds. Bohr never explained what he had in mind in speaking of this generalization. It can, however, be understood by means of the concept of *quantum causality*, sketched in Chap. 2. I shall now discuss this concept in detail.

6.6 Causality and Complementarity

The title of this section repeats that of Bohr's arguably final and the shortest exposition of his interpretation of quantum phenomena and QM in "Causality and Complementarity" (Bohr 1987, v. 3, pp. 1–7). It was published in 1958, thus, after Einstein's death (in 1955), but is still written in Einstein's shadow, which never left Bohr's thought. The article contains, it appears, for the first time in Bohr, a view of causality applicable in quantum physics, as opposed to classical causality to which, as discussed in this study, Bohr had previously referred by causality. Notably, this is the case in his

"Causality and Complementarity" (Bohr 1937), where causality referred to classical causality, renounced in view of "our not being any longer in a position to speak of the autonomous behavior of a physical object, due to the unavoidable interaction between the object and the measuring instrument," which by the same token, entails a "renunciation of the ideal of causality in atomic physics" (Bohr 1937, p. 87). A view of causality suggested by Bohr, even if without defining the corresponding concept of causality, is akin to and is, arguably, the first intimation of the concept probabilistic and quantum causality I shall define in this section, and it is also connected, as this concept is, to both the arrow of time ("the irreversibility of the recordings") and complementarity. Bohr says:

> Although, of course, the classical description of the experimental arrangement and the irreversibility of the recording concerning the atomic objects ensure a sequence of cause and effect conforming with elementary demands of causality, the irrevocable abandonment of the ideal of determinism finds striking expression in the complementary relationships governing the unambiguous use of the fundamental concepts on whose unrestricted combination the classical physical description rests [a combination no longer possible in quantum physics]. Indeed, the ascertaining of the presence of an atomic particle in a limited space-time domain demands an experimental arrangement involving a transfer of momentum and energy to bodies such as fixed scales and synchronized clocks, which cannot be included in the description of their functioning, if these bodies are to fulfill the role defining the reference frame. Conversely, any strict application of the law of conservation or momentum and energy to atomic processes implies, in principle, a renunciation of detailed space-time coordination of the particles. (Bohr 1987, v. 3, pp. 4–5)

One must keep in mind that "the classical description of the experimental arrangement" assumes that the instruments themselves used also have a quantum stratum, through which they interact with quantum objects. (This passage confirms that, unlike the present view in which quantum objects are only assumed to exist at the time of measurement, Bohr assumes the independent existence of quantum objects.) There is no conflict here with Bohr's previous appeal to the renunciation of "the classical ideal of causality," as classical causality, because the view of causality thus suggested is not classical and, thus, does not conform to this ideal. This renunciation is now replaced with "the irrevocable abandonment of the ideal of determinism," which is in effect the same or correlative ideal, because in considering individual or simple classical systems both ideals coincide. On the other hand, while, given an earlier comment in the article (Bohr 1987, v. 3, p. 4), Bohr must have had the connections between causality, thus understood, and probability in mind, he nevertheless only speaks of "ensur[ing] a sequence of cause and effects conforming with the elementary demands of causality," rather than defines a *concept* of causality that conforms to these demands. Quantum causality, as a form of probabilistic causality, is such a concept.

The concept of quantum causality was introduced by this author previously (Plotnitsky 2011, 2016a, pp. 203–207). The present definition, however, refines and explores new dimensions of this concept, and grounds it in a more general concept of probabilistic causality. As sketched in Chap. 2, quantum causality is defined as follows. An *actual* event that has happened determines which events *may or (in view of complementarity) may not* happen and be predicted with *one probability or another*, which is not the same that any of them *will* happen. This event, A, at time t_0 defines certain *possible*, but *only possible* future events, say, X, at time t_1. The concept can be automatically transferred to or, more accurately, is a version of a more general concept of probabilistic causality in the U-RWR view of any physical phenomena, in part in accord (there are differences discussed below) with the concept of "causality" proposed in D'Ariano (2018), D'Ariano et al. (2014). In the case of individual or simple (nonchaotic) classical or relativistic systems, probabilistic causality reduces to classical causality and, correlatively, determinism. The temporal precedence of A and the corresponding (local) arrow of time is crucial to quantum causality. All such predictions are quantum-nonlocal, are *predictions* at a distance, but they respect Einstein-locality, which prohibits any *action* at a distance. It follows, however, that quantum causality does so without, as in the case of classical causality, definitively establishing, or rather (since it would already be definitively established in advance) being definitively connected to any future state of the system considered. As indicated earlier, one can at t_1 perform an alternative, in particular, complementary measurement and thus, by an alternative decision, establish an event, Y, and a reality different from the one predicted, X, even if this prediction was made with probability one. In classical mechanics all quantities that could pertain to a given system can, at least in principle, be established simultaneously, and, correlatively, the role of measuring instruments in this determination can be disregarded, allowing one to treat these quantities as properties of the objects considered, as always determinable simultaneously. Only the temporal precedence of A vis-à-vis either X or Y is definitive. Quantum causality allows one to relate actual events in terms of statistical correlations, such as those of EPR-type between them, which events are, however, are specifically prepared by repeated initial measurements. By the quantum in definitiveness postulate, no definitive relationships between any two actual events, events that have already occurred, could be established. As noted in Chap. 2, quantum causality may be seen as a *symbolic* version of causality, just as QM itself may be seen, as it was by Bohr, as symbolic because it deploys the symbols of classical mechanics for a very different, fundamentally probabilistic, theory.

Thus, while classical causality is ontological or realist, quantum causality is probabilistic and as such is epistemological, as defined by our knowledge of past events and probabilities we assign to future events. It enables one to relate to physical reality at the level of observable phenomena rather than to represent the ultimate nature of reality responsible for these phenomena. It pertains strictly to our interactions with the world by means of experimental technology, in which each event or phenomenon is uniquely defined as such, rather than by a measurement of some pre-existing properties of the object or of the ultimate constitution of reality considered to be responsible for it, especially as defined by a classically causal process found in classical physics or relativity. Quantum causality, thus, essentially depends on this concept of event or phenomenon, which is strictly in accord with Bohr's concept of phenomenon, and thus the concept of quantum measurement as establishing such phenomena, rather than a measurement of a pre-existing property of the object, as in classical physics or relativity. Quantum causality has to do with our probabilistic knowledge concerning possible future event on the basis of our determinate knowledge, obtained in measurement, concerning past events, but the recourse to probability defined by it is not due to insufficient knowledge concerning how quantum events come about. Any definitive knowledge we have is no longer knowledge of how these events come about, as no knowledge or even conception concerning it is possible at all. The knowledge defining quantum causality only concerns the data obtained in already performed experiments (or quantum events that may be seen as experiments "performed" by nature) and predictions concerning future events. Hence, the recourse to probability is not due to the lack of knowledge concerning this reality. It is due to nature itself, or the *nature* of our interactions with nature.

In situations governed by classical causality what has happened determines, by means of a given law, what *will happen*, thus, in principle, connecting all events involved in a single causal chain or network, even though, due to a lack of knowledge of how this happens, our predictions could still be probabilistic or statistical.In dealing with individual or small system such predictions could be ideally exact or deterministic, determined by our measurement in the classical sense of measuring the properties of the objects considered. As explained earlier, classical causality does not imply that any such event is a cause of other events, but only that the law in question determines with certainty what happens in all of the events considered. The cause or (as they may be multiple) causes that define this law and thus governs this process as an effect or a set of effects may be something else. While actually establishing such (always assumed) causes, as discussed in Chap. 2, may pose problems, these difficulties do not undermine the determination of events with certainty in advance. In classical mechanics, when dealing with an individual or small systems (apart from chaotic ones), such laws also allow one to make (ideally) exact, deterministic predictions concerning future events or trace past events on the basis of the complete information concerning a given event, which information is always ideally possible to obtain. In other classically causal cases, our predictions can only be probabilistic or statistical, but only because the complete relevant information is not available, while the classically causal architecture defining the phenomena considered is assumed. By contrast, quantum causality determines, by our decision, what *may (or may not) happen*, as an "effect," an effect without a classical cause or causes, although *not necessarily what will happen*, possibly, and in the RWR view definitively, in the absence of the underlying classical causal connections between events. Correlatively, there is no longer measurement in the classical sense, but only in the sense of creating phenomena, containing data or information, a measurement as a number or bit generation, a quantum computer created by our interaction with nature.

It is instructive to consider with the concept of quantum causality in mind, the case of predictions with "probability equal to unity." As discussed in the next chapter, such predictions define EPR's argument, based on their criterion of reality: "*If, without in any way disturbing a system, we can predict with certainty (i.e., with probability equal to unity) the value of a physical quantity, then there exists an element of physical reality corresponding to this physical quantity*" (Einstein et al. 1935, p. 138). One might ask first: What does it mean to predict with probability equal to unity in physics?

It means that one assumes that, if one measures what is so predicted, the measurement will confirm the prediction. But does the predicted quantity correspond to an element of physical reality, unless the measurement is performed? Technically, this is not so even in classical physics or relativity. An outside interference could change the reality thus predicted. It is also difficult and even impossible to obtain the predicted value by a measurement exactly. These qualifications, however, do not pose difficulties for using EPR's criterion, given that by virtue of classical causality what one predicts, even as a distant event, is bound to happen (short of outside interference), regardless of what initial measurement we decide to perform, because all possible measurable quantities are determined in advance within the same experimental arrangement.

The situation is fundamentally different in the case of quantum phenomena because of the irreducible role of measuring instruments and complementarity, leading to the circumstances, considered here, defining all quantum predictions, including those with probability equal to unity. By quantum causality, it is our decision concerning what measurement to perform that defines what might happen with one probability or the other, in general not equal to one, while an alternative, complementary, measurement would define an incompatible expectation-catalog. One might add, along Bayesian lines, that predictions of anything with any probability are only meaningful insofar as those who made them or know of them are still alive. Suppose that one had predicted, on the basis of the position measurement at time t_1, a future value of the position of an object at time t_2 with probability equal to unity, by means of a measurement performed on a different quantum object and thus without in any way disturbing the first one, which is ideally possible in EPR-type experiments. This prediction can then be confirmed, ideally, by the corresponding position measurement at t_2. However, measuring at t_2 the value of the complementary variable, that of momentum, will make it impossible to assign the position variable to the object at t_2, even though if we had measured the position, the outcome would correspond to our prediction. Once such alternative measurement is performed, there is no experiment that could allow us to make this assignment, as opposed to classical physics, where one can, in principle, measure both variables simultaneously, within the same experimental arrangement, and assign both properties to the object independently of measurements. A prediction with probability equal to unity is meaningful only if it is in principle verifiable, which cannot be assured in considering quantum phenomena, as could be in classical physics. In quantum physics, establishing any measurable quantity unavoidably interferes with the object or the ultimate constitution of the reality responsible for quantum phenomena, and one can always interfere differently from the way necessary for verifying a given prediction and thus preclude establishing the predicted quantity as representing an element of reality.[16] Thus, in the case of quantum phenomena, a prediction with probability equal to unity, even ideally, only

---

[16] As noted earlier, that a prediction with probability one is not the same as establishing the reality of what is so predicted has been stressed by QBists, but primarily on the grounds of the subjective nature of Bayesian probability, rather than the reasoning used here (e.g., Fuchs et al. 2014; Mermin 2016, pp. 243–244; Fuchs 2016).

accords with quantum causality and not, as in classical physics or relativity, with classical causality (which also insures determinism in the case of individual or simple systems). Such a prediction also remains fully compatible with the requirements of relativity or more generally Einstein-locality. It is, however, a reflection of quantum nonlocality (as defined in Chap. 7), insofar as the corresponding *predictions* can be made, as in EPR-type experiments, concerning spatially separated objects, as predictions at a distance, without an action at a distance.

With the concept of quantum causality in hand, one can give meaning to Bohr's view of complementarity as a generalization of causality, of "the very ideal of causality" (Bohr 1987, v. 2, p. 41). On the one hand, "our freedom of handling the measuring instruments, characteristic of the very idea of experiment" in all physics, our "free choice" concerning what kind of experiment we want to perform is essential (Bohr 1935, p. 699). On the other hand, as against classical physics or relativity, implementing our decision concerning what we want to do will allow us to make only certain types of predictions and will exclude the possibility of *complementary* types of predictions. Complementarity generalizes causality, because it defines which events can or cannot be probabilistically or statistically predicted by our decision concerning which experiment to perform, without, however, being ever definitively established as future events. Hence, this generalization of causality excludes classical causality.

In addition to Bohr's remarkable anticipation, in 1958, to which the present view of quantum causality is indebted, several concepts of probabilistic and quantum causality were proposed in quantum information theory (e.g., Hardy 2010; Brukner 2014; D'Ariano 2018). Some philosophical treatments of causality considered the role of probability in causality as well, as noted by D'Ariano (e.g., Salmon 1998; Pearl 2000; D'Ariano 2018). D'Ariano defines causality in general (at least in physics) by means of this type of concept (D'Ariano 2018; D'Ariano et al. 2014). Classical causality or determinism (the term used, equivalently, by D'Ariano and coworkers) is a special case of causality in his probabilistic sense, which may allow, as in classical mechanics, for ideally exact predictions. This limit is equally in place the case in of quantum causality or, more generally, probabilistic causality, as defined here.[17]

This is an important concept. Just as the present concept of quantum or probabilistic causality, this concept tells us that our understanding of physics (classical, relativistic, or quantum) changes once we see the world in terms of probability, as, in Jaynes's title phrase, "the logic of science" (Jaynes 2003). This is also something that both D'Ariano and coworkers' and the present view shares with the Bayesian view, such as that of QBism (Jaynes was a Bayesian). Of course, in quantum physics, we cannot do otherwise, at least thus far. This perspective, however, changes our view of all physics, even if it is classically causal or deterministic, in part because probability is indissociable from temporality and the arrow of time. The work of D'Ariano and coworkers is an important contribution to this philosophy, in their

[17] An example of a toy theory (very difficult mathematically) that is deterministic but not causal in D'Ariano's sense is offered in D'Ariano et al. (2014). This theory is manifestly not classical. Indeed, such a theory cannot in principle be classical, because in the case of individual classical systems determinism and classical causality (as the classical limit of the concept of causality as defined by D'Ariano et al. or quantum causality, as defined here) coincide.

case (or that of QBism) as part of the quantum-informational philosophy. The latter also plays an important role in the present view, which arises from the history of quantum theory and the RWR view, beginning with Bohr 1913 theory and Heisenberg's thinking leading him to his discovery of QM. On the other hand, as noted in Chap. 4, Heisenberg's thinking could be seen as quantum-informational in spirit, and quantum information theory as Heisenbergian in spirit.

In my view, however, there is a question that needs to be considered—that of the idea of cause, to which the present concept of probabilistic or quantum causality, does not appeal, unlike, expressly, that of D'Ariano and coworkers. This question arises in considering the relationships between the preparation, the event designated above as A, and the observation, the event designated above as X, as those of cause and effect. This question is also pertinent to Bohr's view of causality, just discussed, as defined by a "sequence of cause and effect conforming with the elementary demands of causality" (Bohr 1987, v. 3, p. 5). According to D'Ariano and coworkers: "The preparation [A] plays the role of the cause and the observation that of the effect [X]" (D'Ariano et al 2014, p. 5). The question is: In what sense is this a cause-effect relationship and, if so, whether it can be guaranteed?

Ironically, this question arises because of the irreducible role of measurement in quantum phenomena, as defining each phenomena qua phenomena, rather than measuring any pre-existing property of a quantum system considered, a view consistent with their argumentation and enabling them to speak of the preparation as the cause of the observation. As discussed in Chap. 2, it is not always (and, as Hume already argued, perhaps ultimately never) possible to ascertain that any given event, A, is actually the cause of any other event, X, even if one assumes classical causality or a (deterministic) law that allows us to predict X, on the basis of A, with probability one. The existence of a classical cause or set of causes of X is assumed (even by Hume), but it may not be possible for us to claim this status for a given entity preceding X. In quantum physics, on the other hand, our intervention, as a preparation, defines quantum phenomena and as a result affects the course of reality, and prevents determinism, because in general our predictions, even concerning the most elementary quantum systems, do not have probability one. But, can one still speak of the preparation as the cause of the observation?

First of all, one cannot be certain that X will happen because of A in any individual case. One could only ascertain statistically that if one repeats A that X will in general happen, which would define causality as a statistical concept at least in the case of quantum phenomena, as D'Ariano and coworkers might accept, although they do not qualify the situation in this way. More importantly, as explained above, even when our prediction concerning X can be made, at time t_0, on the basis of the measurement (preparation) A, with probability one, as in EPR-type experiments, one can always make an alternative, complementary, measurement at time t_1 which will disable the original prediction and define a different event or reality Y at t_1. Unlike in classical physics, one cannot assume, at least in the RWR view (and D'Ariano and coworkers do not assume this either), that the system in question can simultaneously possess both quantities in the way it would in classical physics, or Bohmian mechanics. One cannot see A as the cause of Y, which, while an observation in its own right in its local environment, would no longer be an observation with the preparation A, as X would be. But, once Y is defined, X is no longer possible. All these events or phenomena, A, X, and Y, are effects of the interactions, at the time of measurement, between measuring instruments and quantum objects, or the ultimate constitution of the reality responsible for quantum phenomena. One cannot, however, say either that our preparation of measuring instruments is a cause of "effects" observed because the outcome is defined by this *interaction*, rather than by this preparation, and is never guaranteed.

If one retains here the term cause, there is only a possible cause, or better, a cause of the possible. Perhaps, one could also speak of "causality without causes." This statement could also in principle apply in classically causal or deterministic situations. There, however, a cause or a set of causes is always assumed, even if it is unknown or even unknowable, but not unthinkable, as it is in the present view of probabilistic or quantum causality as a strong RWR view. Hence, the expression of causality without causes is more fitting in considering quantum phenomena, because this causality without causes is a consequence of the ultimate constitution of the reality responsible for quantum phenomena as a reality without realism. For the moment, it does not appear that the cause-effect relationships between the preparation and observation are guaranteed (even, it follows, when one deals with EPR-type predictions with probability one), which make it difficult to see the preparation as the cause of the observation, because the outcome of an observation could be determined otherwise than by the preparation. The preparation is a "cause" of probable events of observation, but observations may not be its effects. It is of some interest that these complexities were also not addressed by Bohr in his view, considered above, of quantum measurement as "conforming with elementary demands of causality," given that these complexities reflect the considerations he used his earlier arguments. These arguments, however, did not refer to probabilistic forms of causality, but were considered as implying a renunciation of classical causality. In fairness, Bohr's 1958 article is a summation of Bohr's views rather than a full-fledged argument. D'Ariano's "causality principle: The probability of preparations is independent of the choice of observations" (D'Ariano 2018, p. 6) would fully still apply, also as correlative to (Einstein) locality. I am not sure either that "causality principle" is the best term given the aspects of the situation just considered and the history of the phrase from Kant on (apart from the use of causality in relativity, which is, however, not probabilistic). The present view avoids an appeal to causes, even though it cannot quite avoid an appeal to causality, as a causality without causes.

What remains in place is the temporal precedence of A relative to either X or Y. Either measurement or event, X or Y, would respect this precedence, as it cannot precede A, and thus the arrow of time, inherent in and crucial to D'Ariano's concept of causality. This precedence is equally inherent and crucial to quantum causality, as defined here, or assuming the corresponding concept of probabilistic causality, in the U-RWR view. If one adopts this concept, then, while probabilistically relating A and X, one need not speak of A or the preparation as a cause. One could still speak of both A and X, or any quantum event, as an effect of the interaction between measuring instruments and quantum objects, or the ultimate constitution of the reality responsible for quantum phenomena. For the reasons just explained, however, this effect is without an ascertainable cause either. So is X. While A enables us, and *only us*, probabilistically to predict X, it is not the cause of X in the present view.

In this view, as an RWR-type view, the arrow of time, along with quantum causality itself, is only manifested classically in observable phenomena. D'Ariano, on the other hand, appears to see the arrow of time as found in the ultimate workings of reality responsible for quantum (or classical) phenomena (D'Ariano 2018). What can, in the RWR view, be objectively ascertained is that the ultimate nature of reality responsible for quantum phenomena is such that all our interactions with it, on all scales, by means of experimental technology entail the arrow of time, which quantum causality reflects in the case of quantum phenomena. As indicated earlier, however, the present view does not imply that at the ultimate level of reality there is no change or multiplicity but only permanence and oneness, sometimes suggested in literature, as in (Barbour 1999), although Julian Barbour revised his position in (Barbour 2020). In the RWR view, this concept would not apply to the ultimate constitution of reality any more than those of change or motion, or any other concepts, such as space or time, or the arrow of time. By the same token, as stated, the equations of QM or QFT, such as Schrödinger's or Dirac's equation, are not equations of the motion of quantum objects, which are, in the present view, defined only at the time of measurement. They only provide (with the help of Born's or rule) expectation-catalogs concerning the outcomes of future experiments, which implies the arrow of time. Accordingly, their formal time reversibility, sometimes used to argue against the arrow of time or even temporality, has no physical significance.[18] Rigorously, what is reversible is not *time* but the value of the parameter *t*, which refers to measurements, which respect the arrow of time as "the irreversibility of recordings" (Bohr 1987, v. 3, p. 5).

In the U-RWR view, this argumentation is extendable to all physics. In the U-RWR view, the ultimate constitution of nature, at least as considered in physics, is *always* a reality without realism, while allowing that, in considering some physical phenomena, such as those of classical physics and relativity, the reality responsible for them could be *treated in realist terms*, for all practical purposes within the scope of these theories. By the same token, the relationships between all phenomena or events are probabilistic, defined by *probabilistic causality*, underlain by the ultimate constitution of nature, as a reality without realism, while the connections between certain physical phenomena, such as those of classical physics and relativity, can be *treated in a classically casual way*. In some cases, as in considering individual or small systems in classical mechanics or relativity, this treatment also allows for ideally exact or deterministic predictions. Accordingly, in the U-RWR view, one, in doing any physics (which in this respect becomes no different from QM or QFT in the Q-RWR view), never ultimately deals with anything other than measurements of and predictions concerning certain observed phenomena, rather than representing

[18] While I briefly mention a few works, it is not my aim to assess these views or the question of time in general. See Emery et al. (2020) for a basic review and references, which, however, bypasses entirely anything akin to the RWR view and the impact it has on this question, which is, I argue, significant. There are arguments for the arrow of time, based on realist and classically causal views, such as offered in (Ungar and Smolin 2014) based on Smolin's earlier work.

the ultimate reality responsible for them, even when these predictions are predictions with probability one. The possibility of such predictions, even in classical physics or relativity, still does not mean that we determine the ultimate constitution of nature, and as discussed above, in the case of quantum phenomena, such predictions, while possible, bring with them new complexities. A prediction does not define a physical reality, only a measurement does.

From this perspective, just as in the case of Schrödinger's and Dirac's equations, the formal time reversibility of the equations of classical mechanics and relativity, again, often used to argue against the arrow of time or even temporality in general, has no physical significance, either. The difference, which is of course important, is the possibility of, in general, assuming both classical causality and, for all practical purposes, ideally exact predictions in classical mechanics (of individual or simple systems) and relativity, while the predictions of QM and QFT are, in general, probabilistic even in considering the most elementary quantum objects. There is, however, no time reversibility or backward-in-time causality, sometimes claimed on the basis of the formal structure of the equations of classical physics and relativity, or again, those of QM and QFT. First of all, there is, thus far, no experimental evidence to support such claims. There are hypothetical arguments that are legitimate if one assumes a realist view vs. the RWR view of these theories. Among them is the existence of closed time loops in Kurt Gödel's solutions of the equations of general relativity ("Gödel's metric"), Kip Thorne's wormhole "time-machines," the hypothetical existence of tachyons (particles that travel only faster than light in a vacuum, which is not technically forbidden by relativity), and a few others. It is, however, a different question *how* compelling these arguments are and to whom, even under the assumption of realism. While most realist arguments against retroaction in time do not rule it out, even in these arguments, the problems of the assumption remain serious on well-known logical and physical grounds. Its main appeal appears to be that it may solve certain (sometimes presumed rather than real) problems of QM or QFT, or some versions of quantum gravity, all of which are hypothetical in any event.

By contrast, the RWR view, of either Q or U type, again, rules out retroaction in time and backward-in-time causality automatically, given the status of all physical theories in this view and the (probabilistic) nature of causality it assumes. The arrow of time itself is, however, only manifested classically, as an objective experimental fact, in observable phenomena, as an effect of our interactions with the ultimate nature of reality by means of experimental technology, beginning with our bodies and brains.

QM brought with it both the role of consciousness and what may be called "futural" temporality, associated with the arrow of time, in defining our predictions concerning quantum events in accord with quantum causality. With each new experiment, we define anew what *may or may not happen*, and with a quantum theory in hand, the probabilities of such possible events, even if *not what is bound to happen*, which would amount to classical causality and, again, the corresponding concept of measurement. That does not mean that the concept of past is lost in dealing with quantum phenomena: the past is defined by measurements performed in the past, just as the present is defined by the measurement performed at a given present ("now") moment and possible future moments are defined by possible future measurements. All of these events or expectations pertain to the corresponding quantum phenomena, at which level our phenomenal and physical concepts apply. Records of the past are always classical, while predictions concerning the future are always quantum-theoretical. These records are part of the spatial–temporal continuum of our experience, where we can also use clocks (or rods) in the way we do in classical physics or relativity, keeping in mind that relativity changes the behavior of rods and clocks depending on their local frame of reference, thus precluding the possibility of the universal clock. But our classical concepts, including time, cannot, in RWR-type interpretations, apply to the ultimate constitution of the reality responsible for quantum phenomena.

The irreducible discreteness of quantum phenomena, and, in the strong RWR view, the impossibility of assuming a continuous and especially classically causal process of linking them, place time "out of joint," technically, vis-à-vis any concept of time, but in any event time that can be measured by clocks in the way it is in classical physics or relativity and registered by our consciousness as such. In quantum physics, "time is out of joint," as Hamlet famously said ("Time is out of joint. O cursed spite/That ever I was born to set it right" [*Hamlet*, I.v. 190–192]). Unlike, however, in Shakespeare's play, where this statement refers to a single juncture of the narrative, in quantum physics, in *narratives* of quantum physics, this statement applies every time the concept of time applies. To "set it right" means something very different from restoring the spacetime continuum to the ultimate constitution of reality, because, in RWR-type interpretations, quantum events cannot be considered as connected by any continuous spatiotemporal process. In the strong RWR view, no concept of time can ultimately apply to this constitution, any more than any other concept, such as "constitution." "Reality" itself is, again, a name without a concept associated to it. What allows us to "set it right" is quantum causality.

At the same time, the fact that QM, or QFT, only predicts future phenomena or events rather than traces any past events, determined only by irreversible measurements, suggests local "time arrows," perhaps also indicating a global "time arrow" for the available world-manifold, say, the observable universe, assuming the latter is quantum in its ultimate constitution, or if one adopts the U-RWR view, even otherwise. I shall mention this as a possibility because this constitution is currently unknown and such a concept (sporadically entertained in modern cosmology) may not be applicable: we do not have a microscopic gravity theory, which may prove not to be quantum. I shall, accordingly, limit myself to the arrows of time in quantum physics.

If one adopts the RWR view, one can, again, only speak of time and time arrows as effects manifested in quantum phenomena, and not as pertaining to the ultimate nature of reality responsible for these effects. The concept of time cannot apply to this reality anymore than any other concept, such as space or motion. This reality is that of matter (which is ultimately beyond conception), while time belongs to thought, possibly only to conscious thought, from which, moreover, the time used in physics, the time defined by clocks, is constructed by idealization, as Einstein recognized. From this, RWR-type, perspective, claims, made by, famously, Einstein and others, to the effect that such concepts as "now," "before," and "after" only belong to consciousness and not to physical reality, may be correct, but not because the equations of classical physics, relativity, or quantum theory, as mathematically not sensitive to such concepts, represent physical reality. It is, in the RWR view, because the ultimate nature of physical reality cannot be represented either by these concepts or by these equations, while any actual application of these equations in classical physics, relativity, and especially quantum physics, to physical phenomena observed always involves these concepts and with them the arrow of time, along with the necessary use of consciousness.

The ultimate "temporality," if the term applies, of the unconscious may, however, be unthinkable in this sense. Sigmund Freud's German term for the unconscious is *das Unbewusste*, literally the unknowable. The unthinkable? Quite possibly, even if not for Freud himself! The time of consciousness, conceptualized by Husserl's phenomenology as a linear sequence of presences (*Now* B would be constituted by the retention of *Now* A and the protention of *Now* C) may be affected, from the unconscious, by what may be called time-unthinkable, which may even be as responsible for these conscious sequences as it is for their disruption by random events, akin to what happens in the case of quantum phenomena. As Derrida observed: "the temporality to which [Freud] refers [in the case of the unconscious] cannot be that which leads itself to a phenomenology of consciousness or of [now] and one may indeed wonder by what right all that is in question here should still be called time, now, anterior present, delay, etc.," in short any name we associate with Husserlian and related forms of temporality (Derrida 1974, p. 67). If so, there is a parallel between two situations, that of phenomena of consciousness and that of quantum phenomena, or if one adopts the U-RWR view, all physical phenomena, in the nature of their emergence.[19] In addition, Husserl's model of temporal succession, of the present or "now" in the first place, is at most an idealization of experience and is not experience itself. The time used in physics, measured and, as Einstein argued, defined by clocks, is still idealized from this model, essentially by mathematizing Husserl's

[19] This parallel should not confused with von Neumann's "psycho-physical parallelism" (Von Neumann 1932, pp. 428–421) and related concepts, which are, in the first place, related to consciousness. Apart from this brief comment, still dealing only with time-unthinkable, I put aside the application of quantum or quantum-like (physically, mathematically, or philosophically) theories and concepts, including those of Bohr, such as complementarity or phenomena, to human thinking or social phenomena in such fields as consciousness studies, cognitive psychology, or economics. This is a different subject that I have considered elsewhere (e.g., Plotnitsky 2016b, 2021). See Khrennikov (2015) for an argument for a quantum-like view of conscious–unconscious dynamics, and (D'Ariano and Faggin 2021) for an application to this problematic of D'Ariano and coworkers' quantum-informational framework of operational quantum theory, discussed in Chap. 9. The article does not address the question of temporality or presence. It is of some interest that, apart from those by the present author, most works on the subject, with (Khrennikov 2015) being a rare exception, deal primarily and even exclusively with consciousness, which, while essential, especially as concerns our experience of the present or "now," is a very limited part of thinking. However critically one might see Freud's psychoanalytic theory, he deserves credit for his realization that our thinking is primarily unconscious and, as such, it may not be governed by the same temporality as our conscious thinking. As our dreams, memories, and other manifestations of the unconscious tell us, the unconscious may adopt a variety of temporal or atemporal modes.

concept, as Weyl argued (Weyl 1952, pp. 7–10). This idealization has served physics well from Galileo or even Aristotle to Einstein and beyond, including in quantum physics. While, however, the latter uses clocks and the corresponding concept of time in dealing with quantum phenomena, in the RWR view, this or any conception of time cannot apply to the ultimate constitution reality responsible for them.

The U-RWR view extends this view to the ultimate constitution of nature as responsible for all physical phenomena. Consider the claim that the (known) Universe is about 13.8 billion years old, a time arrow extending from the Big Bang. Is this claim, which is classical, objective? Yes, but only in the following sense. There is something existing or real in the constitution of the Universe that in the interaction with our measuring instruments, beginning with our bodies and brains, in this case especially clocks, allows us to make this claim, as objectively communicable and verifiable. But the physical reality that is ultimately responsible for the possibility of doing so is, in the RWR view, beyond our capacity to know or even to conceive of. This assumption does not require the U-RWR view of this ultimate constitution, but only the Q-RWR view, if the origin of the Universe is quantum, as it appears to be, at least at this point, even if not without some dissenting views. Such concepts as space and time, as all concepts that we can form, even those that place certain entities, such as a reality without realism, beyond conception, are human, due to our evolutionary biological and neurological nature. They need not and, in the RWR view, do not belong to the independent physical constitution of the Universe to the degree one can, again, use any of these terms, "reality," "entity," "independent," "physical," "constitution," or "universe," and one ultimately cannot, given that they are human terms, too, as is the RWR view.

6.7 Conclusion: Law Without Law, Reality Without Realism, and It Without Bit

If combined with the RWR view, complementarity may, I have argued here, be related to Wheeler's idea of "law without law" (Wheeler 1983). On the one hand, complementarity as a concept may be seen as a reflection of the fact that the behavior of quantum objects of the same type, say, electrons, is not governed by the same physical law, especially a representational one, in complementary contexts. In the present view, one would speak of the ultimate character of reality responsible for quantum phenomena and, along with them and thus at the time of measurement, quantum objects, which would allow one to keep the relational structure of Bohr's concept as pertaining to quantum phenomena. On the other hand, in RWR-type interpretations, there is no physical law representing this character, given that, in these interpretations, it is beyond representation or even conception. Hence, it is fitting to speak of law without law, thus connecting it to reality without realism, and, I shall now suggest, to "it without bit."

Wheeler eventually linked this "law without law" to quantum information theory, which he helped to usher in, along with Feynman, his student. It was Feynman, too, who gave his lectures on "the character of physical law" with which I started this chapter. The lectures were given in 1965, and thus decades before quantum information theory or the idea of quantum computing was suggested by several figures, such as, in addition to Feynman, Paul Benioff, and Yuri Manin. As noted in Chap. 4, it became apparent already with Heisenberg's thinking, leading him to his discovery of QM, that quantum objects, in their interactions with measuring instruments, create specifically organized, including, in certain circumstances, complementary, structures of information, composed of classical bits, and allow one to use theories, such as QM, to predict this information. The very nature of quantum measurement, as understood in this study, is defined by this situation, making the data observed in quantum phenomena generated by a quantum computer created by our interaction with nature. The ultimate constitution of matter is, to adopt Wheeler's famous maxim (even if not quite his meaning), "it from bit," in the present, even if not Wheeler's view, as "it" *inferred* from "bit" (Wheeler 1990, p. 3). The present view, it follows, also makes this it "it without bit," while, at the same time, no bit is possible without this "it." According to Wheeler:

> [E]very it—every particle, every field of force, even the spacetime continuum itself—derives its function, its meaning, its very existence entirely—even if in some contexts indirectly—from the apparatus elicited answers to yes or no questions, binary choices …, bits. It from bit symbolizes the idea that every item of the physical world has at bottom—at a very deep bottom, in most instances—an immaterial source and explanation; that what we call reality arises in the last analysis from the posing of yes-no questions and the registering of equipment-evoked responses; in short, that all things physical are information-theoretic in origin and this is a participatory universe. (Wheeler 1990, p. 5)

In Bohr's ultimate view or the present view, as the strong RWR view, this is only true insofar as all things in physics are human, and, in the first place, insofar as any specific concept designated "reality" is inferred from such responses to the questions posed by thought. There are no responses, any more than questions, other than those belonging to thought. The term "reality," defined as a reality without realism, has no concept associated with it, and the strong RWR view implies that no such a concept can, in principle, be formed. Still, however, this concept without a concept is human and is inferred by thought from phenomena observed in measuring instruments: it is a product of the interaction among matter (as a reality without realism), thought, and technology. Bohr is reported to have said: "We must never forget that 'reality' too is a human word" (Kalckar 1967, p. 234). As I qualified earlier, one should always be cautious concerning such reported statements, especially in Bohr's case. This statement is, however, consistent with Bohr's argumentation in his writings, just as it is with the RWR view. This would still be true even if there is no concept associated to reality or any other word, such as "it", difficult as it may be to use a word, even "it," unlike, a mathematical symbol without a concept. Even to deprive a word of a concept is still human, however.

Wheeler's "it-from-bit" manifesto was inspired by Bohr, whom Wheeler cited at the outset. Wheeler's very definition of "it from bit" may be seen as a variation of Bohr's view, with which I began this study, of the irreducible role of measuring technology (the "apparatus" invoked by Wheeler) and the constitution of quantum phenomena and information it contains, and a derivation of the existence of quantum objects from the character of this information. It is a separate question whether Wheeler's view converges with that of Bohr or diverges from it, as in my view, it ultimately does. The reason for this assessment is that one could think of "it from bit" in more than one way, in particular, in the following two ways. The first is that of "it" as defined by or even created by "bit," which appears to be closer to Wheeler's own meaning, correlative to his concept of "a participatory universe," to which he refers next in his manifesto as well. He says: "It from bit symbolizes the idea that every item of the physical world has at bottom—at a very deep bottom, in most instances—an immaterial source and explanation; that what we call reality arises in the last analysis from the posing of yes–no questions and the registering of equipment-evoked responses; in short, that all things physical are information-theoretic in origin and this is a participatory universe" (Wheeler 1990, p. 5). This may still be somewhat ambiguous, because Wheeler might mean that "all things physical are information-theoretic in origin" only insofar as they are constructed by us, even if constructed as unconstructible, which would be true in Bohr's and the present view. I am willing to grant that this may have been Wheeler's meaning, even though he does not elaborate on this point. This would bring him closer to Bohr. In this case, "it" as inferred from "bit," which is also the reading of "it" as that which, while giving rise to "bit," is beyond our capacity to define or even think *it* (a pronoun that is difficult to avoid). This will make the word "it" inapplicable either. In the RWR view, this "it," while real, is beyond thought, and as such, cannot be called "it," any more than anything else, including reality, unless one defines it, as in the phrase reality without realism, as a word without a concept associated to it. This will make this "it" "it without bit."

Wheeler invokes Bohr at the outset of his "it-from-bit" manifesto: "The overarching principle of 20th-century physics, the quantum—and the principle of complementarity that is the central idea of the quantum—leaves us no escape, Niels Bohr tells us, from 'a radical revision of our attitude as regards physical reality'" (Wheeler 1990, p. 4). The phrase is used twice, with a minor variation, by Bohr in his reply to EPR, on the second occasion, cited by Wheeler, by way or establishing a parallel revision of our concept of physical reality in general relativity (Bohr 1935, p. 697, 702). This revision of attitude led Bohr to his ultimate, RWR-type, interpretation, in part in responding to Einstein's questioning of QM, in particular in EPR's paper, Bohr's reply to which is cited by Wheeler. In RWR-type interpretations, QM is incompatible with the concept of a proper, "complete" in a realist sense, physical theory advocated by Einstein and defining his discontent with QM. This is because QM offered no representation or even conception of the behavior of the ultimate individual constituents of nature and, correlatively, no deterministic predictions concerning this behavior in the way classical mechanics or relativity did. As such it was not in accord with Einstein's concepts of the completeness of a fundamental theory, which I call "Einstein-completeness." For Einstein, QM was at most a correct statistical theory of ensembles. QM may, however, be seen, as it was by Bohr, as complete in a different sense, which I call Bohr-complete. It is as complete as nature allows our theory of (nonrelativistic) quantum phenomena to be, as things stand now. This is a broader concept, with Einstein-completeness being a special case, which would apply in classical physics or relativity.

As discussed in the Introduction, Einstein admitted that "the belief that [QM] should offer an exhaustive [complete] description of individual phenomena" by only providing statistical predictions concerning the outcome of repeated experiments was "logically possible without contradiction," but found this belief "so very contrary to [his] scientific instinct that [he could not] forego the search for a more complete conception" (Einstein 1936, p. 375). He did not believe either that QM or QFT could provide a proper starting point for finding such as theory (Einstein 1949a, p. 83). And yet, QM and QFT remain our standard theory of quantum phenomena to this day.

Perhaps, then, the question is not what we require from a fundamental theory, unless experimental evidence leads to such requirements (which was not the reason for Einstein's view), but what a fundamental theory, either one already in place or one we need to develop, requires from us. One of the things it may require is a change of our *attitude* toward problems, such as that of physical reality, that we confront. I would argue, given Bohr's customarily careful way of expressing his points, that his invocation of "attitude" in speaking of "a radical revision of our *attitude* toward the problem of physical reality" (Bohr 1935, p. 697; emphasis added) need not mean that one should necessarily adopt any particular concept of reality, even though Bohr did adopt an RWR-type one, as against a realist one. More important is our attitude itself toward this problem or other problems: we should not be bound by previously established views, no matter how ingrained or cherished, and be ready to change our ways of thinking, including the RWR view, if physics requires it.

References

Barbour, J.B.: The End of Time: The Next Revolution in Physics. Oxford University Press, Oxford, UK (1999)

Barbour, J. B. The Janus Point: A New Theory of Time, Basic Books, New York (2020)

Bell, J.S.: Speakable and Unspeakable in Quantum Mechanics. Cambridge University Press, Cambridge, UK (2004)

Bitbol, M.: Schrödinger's Philosophy of Quantum Mechanics. Kluwer, Dordrecht, Netherland (1996)

Bitbol, M.: On Bohr's transcendental research program. In: Faye, J., Folse, H. (eds) Niels Bohr and the Philosophy of Physics: Twenty-First-Century Perspective, pp. 47–66. Bloomsbury, London, UK (2019)

Bohr, N.: Space–time continuity and atomic physics (Bristol lecture). In: Bohr, N. (eds.) Niels Bohr: Collected Works, 10 vols, 1972–1996, vol. 6, pp. 361–370. Elsevier, Amsterdam, Netherland (1931)

Bohr, N.: Can quantum-mechanical description of physical reality be considered complete? Phys. Rev. **48**, 696–702 (1935)

Bohr, N.: Causality and complementarity. In: Faye, J., Folse, H.J. (eds.) The Philosophical Writings of Niels Bohr, Volume 4: Causality and Complementarity, Supplementary Papers, 1999, pp. 83–91. Ox Bow Press, Woodbridge, CT, USA (1937)
Bohr, N.: The causality problem in atomic physics. In Faye, J., Folse, H.J. (eds.) The Philosophical Writings of Niels Bohr, Volume 4: Causality and Complementarity, Supplementary Papers, 1999, pp. 94–121. Ox Bow Press, Woodbridge, CT, USA (1938)
Bohr, N.: Interview with Thomas Kuhn, Aage Petersen, and Eric Rüdinger, 17 November 1962. In: Niels Bohr Archive, Copenhagen and American Institute of Physics, College Park, MD, USA (1962)
Bohr, N.: The Philosophical Writings of Niels Bohr, 3 vols., Ox Bow Press, Woodbridge, CT, USA (1987)
Bohr, N.: Niels Bohr: Collected Works, 10 vols. Elsevier, Amsterdam, Netherland (1972–1996)
Bohr, N., Rosenfeld, L.: On the question of the measurability of electromagnetic field quantities. In: Wheeler, J.A., Zurek, W.H. (eds.) Quantum Theory and Measurement, 1983, pp. 479–522. Princeton University Press, Princeton, NJ, USA (1933)
Bohr, N., Rosenfeld, L.: Field and charge measurements in quantum electrodynamics. In: Wheeler, J.A., Zurek, W.H. (eds.) Quantum Theory and Measurement, 1983, pp. 523–534. Princeton University Press, Princeton, NJ, USA (1950)
Born, M.: Quantenmechanik der Stoßvorgänge. Z. Phys. **38**, 803–827 (1926a)
Born, M.: Zur Quantenmechanik der Stoßvorgänge. Z. Phys. **37**, 863–867 (1926b)
Born, M.: The Einstein-Born Letters (trans. Born, I.). Walker, New York, NY, USA (2005)
Brukner, C.: Quantum causality. Nat. Phys. **10**, 259–263 (2014)
D'Ariano, G.M.: Causality re-established. Philos. Trans. Royal Soc. A **376**, 20170313 (2018). https://doi.org/10.1098/rsta.2017.0313
D'Ariano, G.M., Faggin, F.: Hard problem and free will: an information-theoretical approach (2021). arXiv:2012.06580v2 [quant-ph], 28 Jan 2021
D'Ariano, G.M., Manessi, F., Perinotti, P.: Determinism without causality. Physica Scripta **T63,** 14013 (9pp) (2014). https://doi.org/10.1088/0031-8949/2014/T163/014013
Derrida, J.: Of Grammatology (trans. Spivak, G.C.). Johns Hopkins University Press, Baltimore, MD, USA (1974)
Dirac, P.A.M.: The Principles of Quantum Mechanics, rpt 1995, 4th edn. Clarendon, Oxford, UK (1958)
Einstein, A.: Physics and reality. J. Franklin Inst. **221**, 349–382 (1936)
Einstein, A.: Autobiographical Notes (trans. Schilpp, P.A.). Open Court, La Salle, IL, USA (1949a)
Einstein, A.: Remarks to the essays appearing in this collective volume. In: Schilpp, P.A. (ed.) Albert Einstein: Philosopher-Scientist, 1949, pp. 663–688. Tudor, New York, NY, USA (1949b)
Einstein, A., Podolsky, B., Rosen, N.: Can quantum-mechanical description of physical reality be considered complete? In: Wheeler, J.A., Zurek, W.H. (eds.) Quantum Theory and Measurement, 1983, pp. 138–141. Princeton University Press, Princeton, NJ, USA (1935)
Emery, N., Markosian, N., Sullivan, M.: "Time", Stanford Encyclopedia of Philosophy (Zalta, E.N., ed.), 2020 Edn. Winter (2020). https://plato.stanford.edu/archives/win2020/entries/time/
Epperson, M.: Quantum Mechanics and the Philosophy of Alfred North Whitehead. Fordham University Press, New York, NY, USA (2012)
Faye, J.: Niels Bohr: His Heritage and Legacy. An Anti-Realist View of Quantum Mechanics. Kluwer, Dordrecht, Netherland (1991)
Faye, J., Folse H. (eds.) Niels Bohr and the Philosophy of Physics: Twenty-First-Century Perspective. Bloomsbury, London, UK (2019)
Feynman, R.: The Character of Physical Law, Reprint 1994. MIT Press, Cambridge, MA, USA (1965)
Folse, H.J.: The Philosophy of Niels Bohr: The Framework of Complementarity. North Holland, Amsterdam, Netherland (1985)
Fuchs, C.A.: On participatory realism (2016). arXiv:1601.04360v3 [quant-ph], 28 June 2016

Fuchs, C. A., Nowithstanding Bohr, Reasons for QBism. Mind and Matter, Volume 15, Number 2, 245-300 (2017)
Fuchs, C.A., Mermin, N.D., Schack, R.: An introduction to QBism with an application to the locality of quantum mechanics. Am. J. Phys. 82:749–754 (2014). https://doi.org/10.1119/1.4874855
Hardy, L.: A formalism-local framework for general probabilistic theories, including quantum theory (2010). arXiv:1005.5164 [quant-ph]
Haroche, S.: Entanglement and decoherence in cavity quantum electrodynamics experiments. In: Gonis, T., Turchi, P.E.A. (eds.) Decoherence and Its Implications in Quantum Computation and Information Transfer, 2001, pp. 211–223. IOS Press, Amsterdam, Netherlands (2001)
Haroche, S., Raimond, J.-M.: Exploring the Quantum: Atoms, Cavities, and Photons. Oxford University Press, Oxford, UK (2006)
Haven, E., Khrennikov, A. (eds.) The Palgrave Book of Quantum Models in Social Science: Applications and Grand Challenges, pp. 335–357. Palgrave-MacMillan, London, UK (2016)
Heisenberg, W.: Physics and Philosophy: The Revolution in Modern Science. Harper & Row, New York, NY, USA (1962)
Held, C.: The Kochen-Specker theorem. In: Zalta, E.N. (ed.) Stanford Encyclopedia of Philosophy (2018). https://plato.stanford.edu/archives/spr2018/entries/kochenspecker/
Hilgevoord, J., Uffink, J.: The uncertainty principle. In: Zalta, E.N. (ed.) Stanford Encyclopedia of Philosophy, Spring, 2014 edn. http://plato.stanford.edu/archives/spr2014/entries/qt-uncertainty/
Honner, J.: The Description of Nature: Niels Bohr and the Philosophy of Quantum Physics. Clarendon, Oxford, UK (1987)
Jaynes, E.T.: Probability Theory: The Logic of Science. Cambridge University Press, Cambridge, UK (2003)
Kalckar, J.: Niels Bohr and his youngest disciples. In: Rozental, S. (ed.) Niels Bohr: His Life and Work as Seen by his Friends and Colleagues. North-Holland, Amsterdam, Netherland (1967)
Khrennikov, A.: Quantum-like model of unconscious-conscious dynamics. Front. Psychol. **6**, Art. N 997 (2015)
Mermin, N.D.: Why Quark Rhymes with Pork: And Other Scientific Diversions. Cambridge University Press, Cambridge, UK (2016)
Murdoch, D.: Niels Bohr's Philosophy of Physics. Cambridge University Press, Cambridge, UK (1987)
Newton, S.I.: The Principia: Mathematical Principles of Natural Philosophy (trans. Cohen, I.B., Whitman, A.). University of California Press, Berkeley, CA, USA (1999 [1687])
Ozawa, M.: An operational approach to quantum state reduction. Ann. Phys. **259**, 121–137 (1997). https://doi.org/10.1006/aphy.1997.5706
Ozawa, M.: Universally valid reformulation of the Heisenberg uncertainty principle on noise and disturbance in measurements. Phys. Rev. **A67**, 042105 (2003)
Pauli, W.: Writings on Physics and Philosophy. Springer, Berlin, Germany (1994)
Pearl, J.: Causality: Models, Reasoning and Inference. Cambridge University Press, Cambridge, UK (2000)
Peres, A.: Quantum Theory: Concepts and Methods. Kluwer, Dordrecht, Netherland (1993)
Petersen, A.: The philosophy of Niels Bohr. In: French, A.P., Kennedy, P.J. (eds.) Niels Bohr: A Centenary Volume, pp. 133–151. Harvard University Press, Cambridge, MA, USA (1985)
Pienaar, J.: Extending the agent in QBism. Found. Phys. **50**, 1894–1920 (2020). https://doi.org/10.1007/s10701-020-00375-z
Plotnitsky, A.: Epistemology and Probability: Bohr, Heisenberg, Schrödinger and the Nature of Quantum-Theoretical Thinking. Springer, New York, NY (2009)
Plotnitsky, A.: 'Dark materials to create more worlds': on causality in classical physics, quantum physics, and nanophysics. J. Comput. Theor. Nanosci. **8**(6), 983–997 (2011)
Plotnitsky, A.: Bohr and Complementarity: An Introduction. Springer, New York, NY, USA (2012)
Plotnitsky, A.: "It's best not to think about it at all-like the new taxes": reality, observer, and complementarity in Bohr and Pauli. In: Khrennikov, A., Atmanspacher, H., Migdal, A., Polyakov,

S. (eds.) Quantum Theory: Reconsideration of Foundations, vol. 6, pp. 22–47. American Institute of Physics, Melville, NY, USA (2013)

Plotnitsky, A.: The Principles of Quantum Theory, from Planck's Quanta to the Higgs Boson: The Nature of Quantum Reality and the Spirit of Copenhagen. Springer/Nature, New York, NY, USA (2016a)

Plotnitsky, A.: The visualizable, the representable, and the inconceivable: realist and non-realist mathematical models in physics and beyond. Philosophical Trans. Royal Soc. A **374**(1) (2016b). https://doi.org/10.1098/rsta.2015.0101

Plotnitsky, A.: Spooky predictions at a distance: reality, complementarity and contextuality in quantum theory. Philosophical Trans. Royal Soc. A **377**, 20190089 (2019). https://doi.org/10.1098/rsta.2019.0089

Plotnitsky, A. On "Decisions and Revisions Which a Minute Will Reverse": Consciousness, The Unconscious and Mathematical Modeling of Thinking. Entropy, 23, 1026 (2021). https://doi.org/10.3390/e23081026

Rovelli, C.: Helgoland: Making Sense of the Quantum Revolution. Riverhead, New York, NY, USA (2021)

Salmon, W.: Causality and Explanation. Oxford University Press, Oxford, UK (1998)

Schrödinger, E.: The present situation in quantum mechanics. In: Wheeler, J.A., Zurek, W.H. (eds.) Quantum Theory and Measurement, 1983, pp. 152–167. Princeton University Press, Princeton, NJ, USA (1935)

Schwinger, J.: Quantum uncertainty principle: no Loopholes. Science **264**, 1830 (1994)

Shilladay, C., Busch, P.: Complementarity and uncertainty in Mach-Zehnder interferometry and beyond. Phys. Rep. **435**, 1–31 (2006)

Ungar, R.M., Smolin, L.: The Singular Universe and the Reality of Time: A Proposal in Natural Philosophy. Cambridge University Press, Cambridge, UK (2014)

Von Neumann, J.: Mathematical Foundations of Quantum Mechanics (trans. Beyer, R.T.), rpt. 1983. Princeton University Press, Princeton, NJ (1932)

Wang, Z., Busemeyer, J.: Reintroducing the concept of complementarity into psychology. Front. Psychol. **6**, 1822 PMCID: PMC4661229 (2015). https://doi.org/10.3389/fpsyg.2015.01822

Weyl, H.: Space Time Matter (trans. Brose, H.L.). Dover, Mineola, NY, USA (1952)

Wheeler, J. A.: The participatory universe. Science **81** 2(5), 66–67 (1981)

Wheeler, J.A.: Law without law. In: Wheeler, J.A., Zurek, W.H. (eds.) Quantum Theory and Measurement, pp. 182–216. Princeton University Press, Princeton, NJ, USA (1983)

Wheeler, J.A.: Information, physics, quantum: the search for links. In: Zurek, W.H. (ed.) Complexity, Entropy, and the Physics of Information, pp. 3–28. Addison-Wesley, Redwood City, CA, USA (1990)

Whitehead, A.N.: Process and Reality: An Essay in Cosmology, rpt. 1978. Free Press, New York, NY, USA (1929)

Chapter 7
"Without in Any Way Disturbing the System": Reality, Probability, and Nonlocality, from Bohr to Bell and Beyond

> *If, without in any way disturbing a system, we can predict with certainty (i.e., with probability equal to unity) the value of a physical quantity, then there exists an element of physical reality corresponding to this physical quantity.*
> —Albert Einstein, Boris Podolsky, and Nathan Rosen, "Can Quantum-Mechanical Description of Physical Reality be Considered Complete?" (1935) (Einstein et al. 1935, p. 138)

> *From our point of view we now see that the wording of the above mentioned criterion of physical reality proposed by Einstein, Podolsky, and Rosen contains an ambiguity as regards the meaning of the expression "without in any way disturbing a system.*
> —Niels Bohr, "Can Quantum-Mechanical Description of Physical Reality be Considered Complete?" (1935) (Bohr 1935, p. 700)

Abstract While continuing the exploration of Bohr's thinking, this chapter aims to contribute to the ongoing task of clarifying the relationships among reality, probability, and nonlocality in quantum theory from the RWR viewpoint. The history of these relationships was inaugurated by EPR's 1935 paper, "Can Quantum–Mechanical Description of Physical Reality be Considered Complete?" By considering a thought experiment, the EPR experiment, EPR argued that QM is either incomplete or nonlocal, in the sense of allowing an instantaneous action at a distance, "a spooky action at a distance," as Einstein called it. Their argument was challenged by Bohr in his reply under the same title. The exchange decisively shaped the subsequent debate concerning the subject, rekindled by Bell's theorem and related findings. After a brief introduction (Sect. 7.1), Sect. 7.2 offers an outline of the problematic considered in this chapter. Section 7.3 focuses on completeness, and Sect. 7.4 on (non)locality, via the difference between "Einstein-nonlocality" and "quantum nonlocality." Section 7.5 discusses quantum correlations in the context of Bell's theorem via Fine's and Mermin's works. Section 7.6 considers the relationships between complementarity and entanglement via Schrödinger's discussion of entanglement, a concept that he introduced.

A. Plotnitsky, *Reality Without Realism*,
https://doi.org/10.1007/978-3-030-84578-0_7

Keywords Bohr-completeness · Einstein-completeness · Einstein-nonlocality · Entanglement · EPR experiment · Quantum correlations · Quantum ignorance · Quantum nonlocality

7.1 Introduction

While continuing this book's exploration of Bohr's thinking, this chapter also aims to contribute to our understanding of the relationships among reality, probability, and nonlocality in quantum theory, by considering them from the RWR viewpoint. Although there are earlier indications of the significance of the question of locality in quantum theory, including at the earlier stages of the Bohr–Einstein debate (e.g., Bohr 1987, v. 2, pp. 52–58), the history of this question may be seen as inaugurated in 1935 by EPR's paper, "Can Quantum–Mechanical Description of Physical Reality be Considered Complete?," via the question of the completeness of QM (Einstein et al. 1935). It took a few decades, until the 1960s, for the question of nonlocality to take the center stage in the debate concerning quantum foundations, in the wake of Bell's and the Kochen–Specker theorems and related findings. The question of probability entered into this problematic in an indirect way because the EPR experiment deals with, and EPR's argument depends on, predictions with probability equal to unity. As discussed in Chap. 6, however, such predictions pose additional complexities in the case of quantum phenomena. These complexities were, Bohr argued, underappreciated by EPR but assisted Bohr in his counterargument in his reply, published under the same title soon thereafter (Bohr 1935).

EPR proposed a concept of reality based on the following "reasonable" or "sufficient" criterion, equally applicable, they believed, in both classical and quantum theory: "*If, without in any way disturbing a system, we can predict with certainty (i.e., with probability equal to unity) the value of a physical quantity, then there exists an element of physical reality corresponding to this physical quantity*" (Einstein et al. 1935, p. 138). In order for a theory to be complete, according to EPR's, now "necessary" criterion, "every element of the physical reality must have a counterpart in the physical theory" (Einstein et al. 1935, p. 138). The theory would then represent every such element in a realist way and, as a result, predict every future element with a probability equal to unity and thus establish it as real, on the model of classical mechanics (for individual or small, nonchaotic, systems). It would predict, with certainty, what would happen to the system considered as an *individual* system, barring possible outside interferences, from which the system could, however, be ideally isolated. By using the famous thought experiment, the EPR experiment, EPR argued that QM is either incomplete by their criterion, because it could not predict all that was possible to ascertain as real, or else is nonlocal in the sense of allowing an instantaneous action at a distance, or, in the present definition, Einstein-nonlocal. EPR's argument was challenged by Bohr, on both counts, completeness and nonlocality, Bohr-completeness and Einstein-nonlocality. Bohr argued that this "criterion of reality ... contains—however cautious its formulation may appear—an essential ambiguity when it is applied the actual problem with which we are here concerned" (Bohr 1935, p. 697). It followed that QM was in fact able to predict all that one could ascertain as real in the case of quantum phenomena, or at least that EPR did not demonstrate otherwise. EPR did, however, change the nature of the debate concerning quantum theory, although took a while and a few other events for this change to become fully manifested, sometimes in the 1960s. Einstein and Bohr were no longer part of this debate by then, but the shadow of their debate, including their exchange concerning the EPR experiment, has continued to spread over this debate during the last half-century and still does.

Section 7.2 offers a general outline of the problematic considered in this chapter. Section 7.3 focuses on the question of completeness and Sect. 7.4 on the question of (non)nonlocality, via the difference between Einstein-nonlocality (an action at a distance) and quantum nonlocality (predictions at a distance). Section 7.5 discusses quantum correlations in the context of Bell's theorem via Fine's and Mermin's works. Section 7.6 considers the relationships between complementarity and entanglement via Schrödinger's discussion of entanglement in his 1935 papers, including the cat-paradox paper (Schrödinger 1935a), written in response to EPR's paper.

7.2 Confronting EPR: Completeness, Complementarity, and Quantum Nonlocality

Neither EPR's article nor Bohr's reply spoke of "correlations" and "entanglement." The concept of entanglement was introduced, in German (*Verschränkung*) and English, by Schrödinger in three papers, including his cat-paradox paper, written in response to EPR's paper (Schrödinger 1935a, b, 1936), discussed in the final section of this chapter. The language of correlations came into prominence, in the wake of Bell's and the Kochen–Specker theorems, in the 1960s.

In view of these theorems and related findings, the main focus of the debate concerning quantum foundations during the last half-century shifted to the question of locality of QM or quantum phenomena themselves, rather than that of the completeness of QM, although this assessment may depend on one's concept of completeness. This concept may, for example, be realist, as was that of Einstein, representing, ideally (classically) causally, the behavior of the ultimate constitution of the physical reality considered, which I term Einstein-completeness. QM, in RWR-type interpretations, such as that of Bohr, is manifestly Einstein-incomplete because it places the ultimate constitution of physical reality responsible for quantum phenomena beyond representation or even conception. On the other hand, QM may be argued to be, in terms on this study, Bohr-complete, a broader conception, with Einstein-completeness being a special case, found in classical physics and relativity: QM is as complete as nature allows our theories of quantum phenomena to be, as things stand now, insofar as it predicts all that is possible to predict in accordance with quantum experiments. Even this Bohr-completeness was, however, questioned by EPR. They argued that QM could not predict all elements of reality that could be established as such in considering any individual quantum systems, no matter how elementary, unless one allows for Einstein-nonlocality, an instantaneous transmission of physical influences between spatially separated physical systems, famously called by Einstein "a spooky action at a distance" [*spukhafte Fernwirkung*] (Born 2005, p. 155). Bohr contested this argument on both counts, by arguing that QM could in fact predict all that was possible to ascertain as real, that is, all elements of reality that could be unambiguously defined, without violating Einstein-locality.

Einstein's sentence, where the phrase "a spooky action at a distance" occurs, merits a full citation: "I cannot seriously believe in [QM] because the theory cannot be reconciled with the idea that physics should represent a reality in time and space, free from spooky actions at a distance" (Born 2005, p. 155). This combination is also known as "local realism." Relaxing either requirement was unacceptable to Einstein. For Bohr, by contrast, only Einstein-locality was required, but not a spacetime representation of the ultimate constitution of the reality responsible for quantum phenomena, even in considering the individual behavior of quantum objects, no matter how elementary. Such a representation is required for Einstein-completeness, but is not necessary for Bohr-completeness, although EPR and Einstein in his other arguments concerning EPR-type experiments, again, claimed that QM is Bohr-incomplete as well.

What is known, more generally, as the principle of locality states that no instantaneous transmission of physical influences between spatially separated physical systems ("action at a distance") is allowed or that physical systems can only be physically influenced by their immediate environment. While relativity conforms to locality and gave rise to the concept, locality is independent of relativity theory. First of all, it is independent of the concepts with which it is associated in relativity, in particular the Lorentz invariance of special relativity. For one thing, the Lorentz invariance is violated in general relativity, where it is only infinitesimally valid, while locality is strictly maintained there. Secondly, relativity prohibits a propagation of physical influences not only instantaneously but faster than the (finite) speed of light in a vacuum, a requirement that could, in principle, be violated (as a different speed limit on physical action may be discovered), while still allowing for Einstein-locality. Einstein's "spooky action at a distance," too, refers to an instantaneous action, which is in conflict with relativity, but is not defined by it. Accordingly, while the Einstein-locality of QM would imply that it is in compliance with the requirements of relativity, it may be a deeper fact, as Bell's and the Kochen–Specker theorems indicate. The compatibility with relativity does, however, imply locality. In his reply to EPR, Bohr argued for the compatibility of quantum phenomena and QM, at least in his (RWR-type) interpretation and hence abandoning realism, for "all exigencies of relativity theory" and thus for its Einstein-locality (Bohr 1935, p. 701n). Hence, the inference that QM or nature itself is Einstein-nonlocal may be avoided, although this inference and the corresponding interpretations of the EPR-type experiments or quantum correlations have been adopted.

In all of his arguments concerning the subject, Einstein only considered locality or nonlocality in this sense, as against "quantum nonlocality," the term never used by Einstein. It was introduced, along with several definitions of it, in the wake of

Bell's theorem. I shall adopt the following definition, which takes advantage of the fact that one might in the case of quantum phenomena speak of "spooky *predictions* at a distance," without assuming a spooky action at a distance (e.g., Plotnitsky 2009, pp. 269–271, 315, 2012, pp. 128–130, 2016, pp. 33–34, 136–137, 148, 2019a, 2020). There are, experimentally confirmed, statistical correlations between certain, specifically prepared, distant quantum events, correlations properly predicted by QM (cum Born's or a similar rule). In some idealized cases of this type, such as that of the EPR experiment, involving a pair of spatially separated quantum systems (S_1, S_2), one can also predict with certainty certain measurable quantities for one of them, S_2, on the basis of measurements performed on the other, S_1. These predictions are "spooky" insofar as, against classical physics or relativity, there is, at least in RWR-type interpretations, no story to be told and no concept to be formed of how these correlations or, in the first place, quantum phenomena come about or why these predictions are possible. At the same time, these predictions need not entail a spooky *action* at a distance or Einstein-nonlocality, including in the EPR-type cases, where they are possible with certainty. As discussed in Chap. 6, all quantum predictions, including in standard cases of measurement, are predictions at a distance, without implying an action at a distance. I define "*quantum* nonlocality" as the existence of such related or correlated *distant* quantum events and the possibility of predicting, it follows, at a distance, these relations or correlations.

At the same time, as discussed later in this chapter, it is impossible to ascertain that the relationships between *any single pair* of events, associated with the EPR-type pair (S_1, S_2), involved in the EPR experiment or quantum correlations of the type considered in Bell's and the Kochen–Specker theorems, are Einstein-local. This fact is in accord with the quantum indefinitiveness postulate, which precludes making definitive statements of any kind, concerning the relationship between any two individual quantum phenomena or events, and even to ascertain the existence of any such relationship. The postulate, thus, also precludes one from ascertaining that the relationships between any two quantum events, each defined only as something, an event, that has already happened, is either Einstein-local or Einstein-nonlocal, in the EPR-type situation, or otherwise. The postulate allows statements concerning the relationships between multiple events, in this case statements statistical in nature, such as those that correspond to correlations between distant events in the case of quantum entanglement, and as such, are in accord with quantum nonlocality, as defined here, which only concerns predictions at a distance, and not relationships between individual events. The postulate only applies to statements concerning events that have already happened, rather than possible future events, in which case one can make probabilistic statements, on Bayesian lines, concerning individual future events. The concept of locality or nonlocality, Einstein or quantum, only pertains to actual events, as that which has already happened, and not possible events and predictions concerning possible events.

Quantum nonlocality is sometimes defined differently, for example, in terms of violations of Bell's or related inequalities, or still by other mathematical features, dealing with the data obtained in corresponding experiments.[1] These features need not be those of QM, given that these inequalities pertain to these data as such. Such definitions, however, still leave space for their physical interpretation, and quantum nonlocality, as just defined, provides one such interpretation, among other possible interpretations, some of which interpret quantum nonlocality in terms of Einstein-nonlocality. On the other hand, interpretations of quantum phenomena and QM, such as those of the RWR-type, which only entail quantum nonlocality in the present or similar definitions and avoid Einstein-nonlocality, show that the latter is not necessarily a feature of quantum phenomena or QM, while quantum nonlocality may be.

It is true that predictions at a distance and correlations between distant events are also possible, without violating Einstein-locality, in classical physics. This circumstance may compel one to argue that quantum nonlocality may not be significant for defining the difference between classical and quantum correlations. This type of argument was suggested by Khrennikov under the headings "eliminating the issue of quantum nonlocality," while arguing that the difference between classical and quantum correlations is essentially defined by the discreteness of quantum phenomena and complementarity (Khrennikov 2019a, 2020a, b, c). These two features are indeed crucial in considering quantum correlations or quantum phenomena, in the first place, and they are not found in classical physics. It can easily be shown, however, that both features are correlative to quantum nonlocality, as defined here. The reasons are as follows.

First of all, as explained in Chap. 6 (Sect. 6.4), discreteness, complementarity, and quantum nonlocality are essentially connected already in the standard quantum measurement. This is because either of the two complementary measurements (say, that of position and that of momentum) can be set up, so as to enable the corresponding types of predictions, *after* the object has left the location of measurement and, hence, is at a distance at the time of the prediction. At the same time, each such measurement in principle precludes the other (and the corresponding prediction) in the same experimental arrangement and even for the same quantum object. As also explained there, this situation is defined by the entanglement between the object and the instrument and thus is, in effect, equivalent to an ERP-type situation, apart from the fact that in the standard case such predictions are, in general, not with the probability one. This, however, does not affect their essential character as defined by discreteness, complementarity, and quantum nonlocality. By contrast, in classical physics, one can always measure and predict both variables in the same experimental arrangement, and always with probability one in the case of individual or simple (nonchaotic) classical systems. (In quantum theory, this is only possible in special situations, such as those of the EPR type, and then with qualifications, discussed in Chap. 6 and further considered below.) This is because both measurements are connected by a continuous and classically causal process, which can be assumed to be observable without disturbing it appreciably, thus also allowing one to assume

[1] The literature on the subject is extensive, and my limits here only allow me to mention a very small portion of it.

that all physical properties considered could be assigned to the object itself. There is neither discreteness akin to that of quantum phenomena nor complementarity, which is due to the irreducible role of measurement instruments in the constitution of quantum phenomena, thus, always, in principle, different from quantum objects. (In the present view, moreover, the concept of a quantum object only applies at the time of measurement.) Quantum predictions concern discrete events, which cannot, at least in RWR-type interpretations, be connected by a continuous and classically causal process. This makes them, as *always* predictions at a distance, essentially different from classical predictions, including those that are ostensibly at a distance, ostensibly, because of the continuous and classical causal connections between classical events. It could be added that, while, under the assumption of such continuous or classical causal connections between quantum phenomena, there is no longer real discreteness, complementarity is still retained, as it is, along with the uncertainty relations, in Bohmian mechanics. The latter is, however, Einstein-nonlocal.

Complementarity remains in place insofar as quantum measurements and predictions are defined, as they are in the case of quantum phenomena, by the irreducible role of measuring instruments. This is also true in EPR-type experiments, when our predictions concerning one system, are made, at a distance, by performing measurements on another system, with which the first system has previously interacted. It is, accordingly, not surprising that its role, along with and correlatively to that of measuring instruments, which is subtler, but still essential, in the case such predictions, grounds Bohr's reply to EPR, who, Bohr argued, did not sufficiently appreciate this role. As discussed below, EPR's criterion of reality in effect presumes and their argument is based in the possibility of assigning both complementary properties to the same quantum object in the same experimental arrangement, which possibility is precluded by complementarity. This assumption, Bohr argues, makes this criterion, which they assume to be equally applicable in classical *and* quantum physics, ambiguous when applied to quantum phenomena.

For a classical parallel with the EPR experiment, consider the prediction, with probability one, enabled by the momentum conservation law, of the momentum of a classical object, S_2, on the basis of the measurement performed on another classical object, S_1, after both objects collided and exchanged their momentum. Nothing, however, prevents one, in principle, from measuring, along with its predicted momentum, the position of S_2, or of S_1, and thus simultaneously predict and then measure both the position and the momentum of S_2, in a single measuring arrangement. This is possible because, the physical state of either object may be assumed to be determined by a continuous trajectory in space and time represented by the equations of classical mechanics, with the definitive values of both the position and the momentum of the objects assigned, as independent "elements of reality," at each point. When dealing with processes in which the velocities are slow relative to the speed of light, one need not worry about constraints imposed by relativity. (Relativity itself, however, preserves the features just described of the physical reality it considers.) S_1 and S_2 each have a history of a *continuous* motion after their interaction. This enables one to establish all measurable quantities, "elements of reality," associated with S_2 and defining its behavior, by measuring them on S_1, within *the same single measuring arrangement.* There are classical systems, such as those considered in classical statistical physics or chaos theory (or just a tossed coin), the behavior of which, due to their mechanical complexity, cannot be predicted deterministically and which may exhibit statistical correlations of some sort. This behavior is, however, classically causal and is underlain by the behavior of their individual constituents in the way just explained.

The situation is essentially different in the case of quantum phenomena, especially in RWR-type interpretations (even though the momentum conservation law still applies, of course). First, especially, again, in these interpretations, in considering any EPR pair (S_1, S_2), one cannot assume a continuous process for either system after their interaction in view of the uncertainty relations. This impossibility is correlative, by the QD postulate, to the discreteness of quantum phenomena. Secondly, now in any type of interpretation, by virtue of the irreducible role of measuring instruments and, correlatively, complementarity, one can only measure on S_1 and hence predict, *at a distance*, for S_2 only one of the two complementary quantities in question. At the same time, one also has a choice or can make a decision to predict either one of the two, which, however, irrevocably precludes the possibility of predicting the other on S_2. It follows that, in contrast to classical physics, we are "not dealing with a *single* specified experimental arrangement, but are referring to *two* different, mutually exclusive, arrangements," which fact also requires two different EPR pairs to enact each situation (Bohr 1987, v. 2, p. 57). (Technically, Bohr's statement refers to a different experiment, the photon-box experiment, proposed by Einstein, but it equally applies to the EPR experiment, clearly on Bohr's mind at this point, in 1949.) As discussed below, by assuming the first fact (that we always have a choice or decision which type of measurement to perform), while disregarding the significance of the second (that we are dealing with two different, mutually exclusive, arrangements), EPR concluded that QM is either incomplete, even Bohr-incomplete, or else Einstein-nonlocal. Bohr, by contrast, by taking into account both facts, counterargued that QM is both Bohr-complete (as complete as any theory of quantum phenomena can be, as things stand now) and Einstein-local. While possible, the predictions with probability one in question in the EPR experiment do not establish a reality. Only measurements do. But quantum measurements, in contrast to classical ones, are defined by the irreducible role of measuring instruments in them and by complementarity. At the moment in time for which the prediction is made, one can always measure a quantity complementary to a predicted one, and thus disable the possibility of establishing the reality corresponding to this prediction. By contrast, in classical physics, the reality of both elements can always be assumed independently of observation and then be measured *within the same experimental arrangement.*

It follows that the quantum-nonlocal nature of quantum predictions defines them and, *once the corresponding measurements are performed*, any possible reality involved, such that observed in quantum correlations, as essentially different from classical predictions or classical correlations. This difference is manifested in the statistical numerical data involved, such as that considered in Bell's and the Kochen–Specker theorems. It is also, by the same token, clear that, because of the irreducible role of measuring instruments in the constitution of quantum phenomena, the quantum-nonlocal nature of quantum predictions is correlative to the fundamental discreteness of quantum phenomena and complementarity.

EPR's paper considered an *individual* quantum system, S_2, prepared in a particular way, as one of the two systems (which have previously interacted with each other, but are spatially separated), forming an EPR pair (S_1, S_2). EPR argued that, by means of this preparation, it was possible to establish as real both "elements of reality," associated with two complementary variables, such that of position and that of momentum, pertaining to S_2 by performing measurements on S_1 and, thus, without "in any way disturbing" S_2, by predicting (by using QM) each the value of each variable with the probability equal to unity, in accord with their criterion of physical reality (Einstein et al. 1935, p. 138). On the other hand, QM had no means to do so because it could only predict one or the other of these variables in any given experiment. Hence, it was Bohr-incomplete, by EPR's criteria or reality and completeness, unless, EPR argued in closing, their criterion of reality would not apply (as too restricted), in which case, however, Einstein-locality would be violated (Einstein et al. 1935, p. 141). Accordingly, it could have been expected that an Einstein-complete theory, fully representing the behavior of individual quantum system, should be possible. Bohr, again, contested this argument by counterarguing that QM could, in fact, predict all that is possible to predict, all "elements of reality" that could be unambiguously defined, without violating Einstein-locality.

Einstein admitted that Einstein-nonlocality could be avoided if one assumed that QM is only a statistical theory that does not provide a representation of, nor ideally exact or deterministic predictions concerning, the behavior of the ultimate individual constituents of nature of the type found in classical mechanics or classical electromagnetic theory and relativity. Einstein was not willing to accept this alternative for a complete theory, even as only Bohr-complete, either, because it was in conflict with his conviction that a fundamental physical theory should do both, describe individual fundamental constituents considered so as to predict them ideally exactly, in other words be Einstein-complete as well as Einstein-local, thus guaranteeing local realism. For one thing, why QM was able to make its statistical predictions remained unexplained, as against, for example, classical statistical physics, where the behavior, predictable statistically, of the systems considered is underlain by the behavior of their individual constituents governed by classical mechanics. For Einstein, as noted, this aspect of QM made the theory more akin to magic than science (e.g., Born 2005, pp. 155, 205; Einstein 1949a, p. 81). More importantly, Einstein always argued, just as EPR did initially, that, if Einstein-locality is assumed, QM was still Bohr-incomplete because could not predict all that was possible to establish as real, "without in any way disturbing the system," in considering individual quantum systems.

Most of the key findings and arguments involved in more recent debates, following Bell's theorem, deal with discrete variables and Bohm's version of the EPR experiment. There are conceptual reasons for this shift, on some of which I shall comment in Chap. 9. The main reason appears to be that the thought experiment proposed by EPR cannot be performed in a laboratory. Bohm's version of the EPR experiment can and has been performed, confirming the existence of quantum correlations, which can be ascertained experimentally, apart from QM or any quantum theory.[2] Among the best known of these findings are those of Daniel M. Greenberger, Michael Horne, Anton Zeilinger, and Lucien Hardy, and, from the experimental side, Alain Aspect's experiment and related experimental work, such as that by Zeilinger and his group (Aspect et al. 1982; Greenberger et al. 1989, 1990; Hardy 1993).[3] The advent of quantum information theory, in part stimulated by these findings, gave this problematic further prominence and significance. The meaning of these findings, beginning, again, Bell's and the Kochen–Specker theorems, or indeed the EPR experiment and Einstein's and Bohr's arguments concerning it, and their implications continue to be debated. While I shall by and large bypass these debates here, some of their key aspects will be addressed in Sect. 7.5 in considering Fine's and Mermin's arguments concerning quantum correlations and Bell's theorem.[4] I shall argue that the key issues at stake in EPR-type experiments concerning discrete variables are not different from those at stake in the EPR original experiment.[5]

[2] There are experiments (those involving photon pairs produced in parametric down conversion) that statistically approximate the EPR experiment for continuous variables. I shall not consider these experiments here, but they are consistent with the present argument. They also reflect the fact that the EPR thought experiment is a manifestation of correlated events for identically prepared experiments with EPR pairs on the model of the Bell–Bohm version of the EPR experiment. The analysis of the Bell–Bohm experiment, too, involves further subtleties, such as those concerning "degrees" in which one can assess these correlations and, thus, quantum nonlocality, which are related to the strict numerical limits on Bell's and related inequalities, extensively discussed in recent literature (see, for example, Bub 2016 and references there).

[3] I only cite some key earlier experiments. There have been numerous experiments performed since, some in order to find loopholes in these and subsequent experiments.

[4] The literature dealing with these subjects is nearly as immense as that on interpretations of QM, and indeed most of the latter literature in recent decades involves and is even defined by these subjects. Among the standard treatments are (Bell 2004; Cushing and McMullin 1989; Ellis and Amati 2000). Mermin offers a particularly lucid treatment (Mermin 1990, pp. 81–185). See also (Brunner et al. 2014), for a more recent assessment of Bell's theorem. It should also be kept in mind that there are realist and classically causal views of quantum entanglement and correlations, either in realist interpretations of QM, such as the many worlds interpretation, or in alternative theories, such as Bohmian mechanics, or theories in which the reality handled by QM is underlain by a deeper reality (within the scope of QM), such as that of classical random fields (Khrennikov 2012; Plotnitsky and Khrennikov 2015). So-called superdeterminism is another realist view, which explains away the complexities discussed here by denying an independent decision of performing one or the other EPR measurements (e.g., t'Hooft 2001, 2018). These views, however, can be put aside here because they either allow for Einstein-nonlocality or, if they claim Einstein-locality, they leave no space for the concept of quantum nonlocality, because all events are assumed to be connected by continuous and classically causal processes.

[5] I have considered Bohr's reply in detail previously (Plotnitsky 2009, pp. 237–312, 2012, pp. 107–136, 2016, pp. 136–154). The present discussion, however, modifies these treatments on several key points, in part in view of some of the overall argument of this study, which offers a new interpretation of quantum phenomena and QM. The exchange and Bohr's reply has of course been extensively discussed in literature, including in most studies of Bohr cited in Chap. 6, although not from the RWR-type perspective, and some of them misread Bohr's argument by assuming that it allowed for Einstein-nonlocality, a misreading that, as discussed below, is found in Einstein's comments on Bohr's reply. I also consider and emphasize the role of probability in the EPR experiments, which is rarely done, in part because EPR predictions are with probability equal to unity, which tends to

7.3 "Can Quantum–Mechanical Description of Physical Reality be Considered Complete?": The EPR Experiment, Measurement, and Complementarity

Bohr, as noted, contested EPR's argumentation by analyzing the irreducible role of measuring instruments and complementarity in quantum phenomena, including those of the EPR type, even though the latter allow for predictions with certainty without physical, "mechanical," interference, by means of measuring instruments, with the objects concerning which these predictions were made (Bohr 1935, p. 700). This role, Bohr argued, was underappreciated by EPR. This analysis allowed him to conclude that QM "would seem to fulfill, within its scope, all rational demands for completeness," at least Bohr-completeness (Bohr 1935, pp. 696, 700n; 1987, v. 2, p. 57). Bohr also argued that QM need not entail a violation of Einstein-locality by virtue of the compatibility of his argument with "all exigencies of relativity theory," which implies Einstein-locality (Bohr 1935, p. 701n).

Bohr's interpretation in his reply was different from his ultimate interpretation, which no longer allowed for an assignment of elements of reality to quantum objects even at the time of measurement. Such elements could only be assigned to measuring instruments themselves and then only as a result of performed measurements, defining quantum phenomena, and not on the basis of prediction even with probability equal to unity. This assumption implies an outright rejection of EPR's criterion of reality. This could be questionable if one assumed this criterion to be valid in considering quantum phenomena or QM. Given, however, that Bohr, in his reply, argued that this criterion was ambiguous and, hence, not valid as such (unless significantly qualifies), this was not an issue for Bohr's ultimate view, assuming, of course, that one accepts Bohr's argument, as Bohr himself obviously did. In his reply, however, in part in counterarguing EPR's argument in terms closer to their own, the assignment of elements of reality to quantum objects was possible at the time of measurement and even on the basis of a prediction "with probability equal to unity." The second type of assignment required significant qualifications, explained below, considered by Bohr but not taken into account by EPR. Briefly, such an assignment was assumed by Bohr to be ideally *guaranteed in the future*, if and only if, which is the main qualification in question (suggested by the discussion given in Sect. 6.6. of Chap. 6), the corresponding measurement could still be performed and when it was performed, but not at the time when the prediction was made. Both types of assignment still implied that the *properties* of quantum objects could not be considered independently of their interaction with measuring instruments. Bohr also assumes, as do EPR, that quantum objects (while, for Bohr but not EPR, still beyond representation or even conception) exist independently of measurement. Accordingly, in considering Bohr's argument, I will follow this assumption, although

---

obscure this role. See (Fine 2020), for a general outline and references from a perspective different from that of this study.

I shall, when relevant, explain how the situation appears in the present interpretation, which assumes that quantum objects are defined only at the time of measurement.

Bohr, again, only argued for (along with Einstein-locality) Bohr-completeness of QM, that is, that it was able to predict all that is possible unambiguously to define as elements of reality in considering quantum phenomena. This was, however, all that he needed to do, because EPR argued that QM was not even Bohr-complete. They contended that it was possible to *establish*, in accordance with their criterion of reality, the "elements of reality" by ascribing definite values corresponding to both complementary variables pertaining to a given quantum system, S_2, of the EPR pair (S_1, S_2) by means of predictions (by using QM for predicting each such element separately) with probability equal to unity, by means of a measurement on S_1, spatially separated from S_2, and thus without "in any way disturbing" S_2. On the other hand, this ascription is never possible in considering the state of a quantum system in the formalism of QM. My emphasis on "establish" reflects a key point of EPR's argument, explained below, which has to do with counterfactual aspects of this argument, insofar as these elements of reality could only be predicted alternatively but not simultaneously. Contending that both such variables could be established for an individual quantum object would make QM incomplete, unless EPR's criterion of reality is modified and, as a result, Einstein-nonlocality is allowed, and would make it reasonable to believe that a more complete, in effects, Einstein-complete theory is possible (Einstein et al. 1935, p. 141). By considering the irreducible role of measuring instruments in the constitution of quantum phenomena, including those of the EPR type, and complementarity, Bohr argued that EPR's claim, based on their criterion of reality, that one could so establish all elements of reality considered was not sustainable in view of "the essential ambiguity" of their criterion of reality in considering quantum phenomena. On the other hand, he also argued, Einstein-locality could still be maintained (Bohr 1935, p. 701). This would also mean that an Einstein-complete local theory (local realism) may not be possible, even if not outright impossible. It is a matter of interpretation whether Bohr adopted the first, more cautious view, suggested by his phrasing in his reply that "*would seem to fulfill*, within its scope, all rational demands for completeness" (emphasis added), or the second, more prohibitive view, suggested by some of his statements elsewhere. This study, again, adopts the first view. Either way, if as Bohr argued, QM in fact predicted all that was possible to establish as real, it could have been considered as Bohr-complete, as well as, as discussed in the next section, Einstein-local. As explained there as well, Einstein had misread Bohr's argument by assuming that Bohr preserved the Bohr-completeness of QM by allowing Einstein-nonlocality, an alternative viewed by Bohr as impermissible as by Einstein, but entertained by some.

As indicated above, an EPR prediction concerning quantum object S_2 of the EPR pair (S_1, S_2) is enabled by performing a measurement on another quantum object, S_1, with which S_2 has previously been in interaction (the interaction that entangled S_1 and S_2) but from which S_2 is spatially separated at the time of the measurement on S_1. Specifically, once S_1 and S_2, are separated, QM allows one to simultaneously assign both the distance between the two objects and the sum of their momenta, because the corresponding Hilbert-space operators commute. With these quantities in hand, by measuring either the position or the momentum of S_1, one can predict exactly either the position or the momentum for S_2 without physically interfering with it, which, EPR assumed, implies that one can simultaneously assign to S_2 both quantities as elements of reality, without in any way disturbing S_2. Bohr explains the case in a note at the outset of his reply via so-called "transformation theorems," a more general feature of the formalism, which grounds the mathematics of EPR's argument (Bohr 1935, pp. 697–698, n.). According to Bohr:

> The deductions contained in the article cited [EPR's paper] may in this respect be considered as an immediate consequence of the transformation theorems of quantum mechanics, which perhaps more than any other feature of the formalism contribute to secure its mathematical completeness and its rational correspondence with classical mechanics. In fact, it is always possible in the description of a mechanical system, consisting of two partial systems (1) and (2), interacting or not, to replace any two pairs of canonically conjugate variables $(q_1 p_1)$, $(q_2 p_2)$ pertaining to systems (1) and (2), respectively, and satisfying the usual commutation rules
>
> $$[q_1 p_1] = [q_2 p_2] = ih/2\pi,$$
>
> $$[q_1 q_2] = [p_1 p_2] = [q_1 p_2] = [q_2 p_1] = 0,$$
>
> by two pairs of new conjugate variables $(Q_1 P_1)$, $(Q_2 P_2)$ related to the first variables by a simple orthogonal transformation, corresponding to a rotation of angle in the planes $(q_1 q_1)$, $(p_1 p_2)$
>
> $$q_1 = Q_1 \cos\theta - Q_2 \sin\theta \quad p_1 = P_1 \cos\theta - P_2 \sin\theta$$
>
> $$q_2 = Q_1 \cos\theta + Q_2 \sin\theta p_2 = P_1 \cos\theta + P_2 \sin\theta$$
>
> Since these variables will satisfy analogous commutation rules, in particular
>
> $$[Q_1 P_1] = ih/2\pi, \quad [Q_1 P_2] = 0$$
>
> it follows that in the description of the state of the combined system, definite numerical values may not be assigned to both Q_1 and P_1, but that we may clearly assign such values to both Q_1 and P_2. In that case, it further results from the expressions of these variables in terms of $(q_1 p_1) = (q_2 p_2)$, namely,
>
> $$Q_1 = q_1 \cos\theta + q_2 \sin\theta \, P_2 = -p_1 \sin\theta + p_2 \cos\theta$$
>
> that a subsequent measurement of either q_2 or p_2 will allow us to predict the value of q_1 or p_1, respectively. (Bohr 1935, pp. 696–697, note)

Bohr discusses an (idealized) physical model of the quantum–mechanical state of the two free particles, a state, mathematically expressed by EPR, later in his reply (Bohr 1935, p. 699). The question is that of what is possible to establish as real in the EPR experiment and whether QM predicts all that is thus possible to establish. EPR argue that QM does not, while Bohr argues that it does. When one considers a single system S, one makes measurement of a given variable, say, that of position, Q, and then associates a wave-function with it so as to be able to predict, with a given probability, the value that the variable will be within a given range, in this case, a certain area. In the EPR case, because of the entanglement of S_1 and S_2, one can, after both systems are entangled, associate a wave-function for either variable, P or Q, with S_2 so as to predict, moreover, with probability equal to unity, the value of either of the two complementary quantities associated with each variable for S_2 by measuring the same variable on S_1. As discussed in Chap. 6, however, in fact one similarly makes either prediction in the standard case of quantum measurement, because the apparatus could be set up for either measurement after the object already left the location of the apparatus. These predictions, however, will not be with probability one. The situation is subtler in the present interpretation, because one cannot speak of the interactions between two quantum objects, as defined in and by measurement, even if quantum objects are still beyond conception. As explained earlier, however, our instruments can be prepared so that such measurements, each defining the corresponding quantum object, are possible. This is better exemplified in the Bohm–Bell version of the EPR experiment for discrete variables, in which case the first step already prepares the situation in such a way that the subsequent measurements on two quantum objects, defined by these measurements, are simultaneously possible. These qualifications, however, do not affect the overall situation of the EPR experiment as considered here.

"The authors [EPR]," Bohr said in his reply, "therefore want to ascribe an element of reality to each of the quantities represented by such variables. Since, moreover, it is a well-known feature of the present formalism of quantum mechanics that it is never possible, in the description of the state of a mechanical system, to attach definite values to both of two canonically conjugate variables, [EPR] consequently deem this formalism to be incomplete, and express the belief that a more satisfactory theory can be developed" (Bohr 1935, p. 696). EPR's actual argument was more elaborate. They derived a contradiction between the assumption that QM is complete (Bohr-complete) and the assumption of the impossibility of attaching definite values to both variables in question, which, since this impossibility is inherent in QM, implies that QM, while correct, is incomplete, unless it is nonlocal. Essentially, however, this conclusion is the same as stated by Bohr. Even though one can predict (exactly) the two quantities considered only alternatively, EPR still contended that both quantities corresponded to the elements of reality jointly pertaining to S_2, according to their criterion, which did not require simultaneity of such measurements or predictions. This requirement would, in their view, imply Einstein-nonlocality (Einstein et al. 1935, p. 141). EPR disallowed this possibility, as did Bohr, although, it has been assumed by some as the way of resolving the difficulty. Later arguments by Einstein (who was unhappy with the exposition of EPR's paper) were more streamlined, but not essentially different, except for adding that, if one assumed that, if assumed to be only a statistical theory of ensembles, QM could be considered Einstein-local. That, however, would still leave it Bohr-incomplete, because Einstein continued to argue, along the lines of EPR's argument, that, assuming Einstein-locality, QM could not predict all that was possible to establish as real, "without in any way disturbing the system," for individual quantum systems.

Bohr counterargued that, contrary to EPR's assumption, the situation defined by the EPR experiment, does not allow one to dispense with the role of measuring instruments. This is not because the system in question, S_2 is in any way physical, "mechanically," disturbed, but because this role entails limitations on the *types* of measuring arrangements used in determining the quantities in question, even when our predictions concerning, S_2, are made by means of a measurement on another quantum system, S_1, with which S_2 has previously interacted, and thus, "without in any way disturbing," S_2. These limitations result from "an influence on *the very conditions which define the possible types of predictions* [concerning S_2, by measurements performed on S_1]" (Bohr 1935, p. 700). Taking this influence (which is, again, not a physical influence on S_2!) into account, which EPR failed to have done, both revealed "essential ambiguity" of their criterion of reality in the case of quantum phenomena, and allowed Bohr to argue that QM is both Bohr-complete and Einstein-local.

Both EPR and Bohr assume, then, that the EPR experiment for (S_1, S_2) can be set in two alternative ways so as to predict, with probability equal to unity, either one or the other complementary measurable quantities for S_2 on the basis of measuring the corresponding quantities for S_1. Let us call this assumption "*assumption A*."

EPR infer from this assumption that *both* of these quantities can be assigned to S_2, even though it is impossible to do so simultaneously, in view of the uncertainty relations for the corresponding measurements on S_1. This makes QM incomplete (by EPR's criterion) because it has no mechanism for this assignment, unless one assumes that EPR's criterion is too restrictive, as against assuming that such quantities can only be assumed to be real if they are measured or predicted simultaneously. The latter assumption, EPR contend, allows for Einstein-nonlocality (Einstein et al., p. 141). Let us call this inference "*inference E*" (for Einstein).

Bohr argued that, while *assumption A* is legitimate, *inference E* is unsustainable because a realization of the two situations necessary for the respective assignments of these quantities would, in effect, involve two incompatible experimental arrangements and, thus, one might add (Bohr did not expressly say this), would require two different quantum objects, and two different EPR pairs to be prepared. There is no physical situation in which this joint assignment is possible for the same object. If one has made the EPR prediction, say, for variable P with probability equal to unity, based on the measurement of the same variable on the first object, S_{11}, for the second object, S_{12}, of a given EPR pair, (S_{11}, S_{12}), an alternative EPR measurement and prediction for a complementary variable Q, can no longer be made for the same pair. In order to measure and predict the complementary variable, Q, one would need to prepare a different EPR pair (S_{21}, S_{22}), to make the corresponding measurement on S_{21} so this variable could be predicted for S_{22}. I designate this inference as "*inference B*" (for Bohr).

One can diagrammatically represent the situation as follows. Let X and Y be two complementary variables (which while formally similar to, are physically different from the conjugate variables in classical mechanics) in the Hilbert-space formalism ($XY - YX \neq 0$) and x and y the corresponding physical measurable quantities ($\Delta x \Delta y \approx h$); (S_1, S_2) of the EPR pair of quantum objects; and p the probability of prediction, via the wave function, Ψ_1 or Ψ_2, associated with S_2, of either x or y, on the basis of two alternative measurements of either x or y performed on $S_{1.}$ Then:

The EPR experiment in EPR's view:

$$\begin{array}{ll} S_1 & S_2 \\ x_1\ \psi_1\,(\text{with } p = 1) \rightarrow & x_2 \\ y_1\ \psi_2\,(\text{with } p = 1) \rightarrow & y_2 \end{array}$$

EPR admit that x_1 and x_2 are, and because of the uncertainty relations, could only be, measured alternatively by "disturbing" S_1. But, because either x_2 or y_2 can be assigned to S_2 "without in any way disturbing it," EPR argue that both x_2 or y_2 could be established as elements of reality pertaining to S_2, it follows, at the time of either prediction, and by implication in general.

The EPR experiment in Bohr's view:

$$\begin{array}{ll} S_{11} & S_{12} \\ x_{11}\ \psi_1(\text{with } p = 1) \rightarrow & x_{12} \\ S_{21} & S_{22} \\ y_{21}\ \psi_1(\text{with } p = 1) \rightarrow & y_{22} \end{array}$$

The first diagram is, I argue, impossible to realize physically. The second can be physically realized, and it is that of a complementarity, which may, against EPR's own grain, be called "the EPR complementarity." This complementarity can be described as follows. Once one type of measurement (say, that of variable x) is performed on S_{11}, enabling the corresponding prediction on S_{12}, we irrevocably cut ourselves off from any possibility of making the alternative, complementary, measurement y on S_{11}, and, thus, from the possibility of ever predicting the second variable for S_{12} (Bohr 1935, p. 700). This fact is a manifestation of both complementarity and quantum causality. As Bohr stressed in his reply, stemming from "our freedom of handling the measuring instruments, characteristic of the very idea of experiment" in all physics, our "free choice" concerning what kind of experiment we want to perform is essential to complementarity (part (b) of the main definition of complementarity given in Chap. 6) (Bohr 1935, p. 699). The expression "free choice" or its equivalents, such as "freedom of choice," are repeated throughout his reply (Bohr 1935, pp. 699–701). However, as against classical physics or relativity, implementing our decision concerning what we want to do will only allow us to make certain types of predictions and will exclude the possibility of certain other, *complementary*, types of predictions, predictions at stake in the EPR experiment. In this sense, complementarity is, again, a generalization of causality, defined in Chap. 6 as quantum causality, in the absence of classical causality and, in the first place, realism, because it defines what *may or may not* happen (which is, again not the same as will or will not happen) as a result of our decision concerning which experiment to perform. In the EPR case, according to Bohr:

> [By measuring the position of the first particle of a given EPR pair] we have by [the very procedure necessary to do so] cut ourselves off from any future possibility of applying the law of conservation of momentum to the system consisting of the diaphragm and the two particles and therefore have lost our only basis for an unambiguous application of the idea of momentum in predictions regarding the behavior of the second particle. Conversely, if we choose to measure the momentum of one of the particles, we lose ... any possibility of deducing from the behavior of this particle [its] position ... relative to the rest of the apparatus, and have thus no basis whatever for predictions regarding the location of the other particle. (Bohr 1935, p. 700)[6]

Accordingly, it is only possible to establish both quantities for two EPR pairs, (S_{11}, S_{12}) and (S_{21}, S_{22}), but never for one, and it is not possible to maintain that if we had predicted the second quantity, instead of the first one, for S_{12}, it would be the same, even ideally, as it is for S_{22}. There is simply no way to define that variable for S_{12}, except, by a measurement and thus by disturbing S_{12}, which defeats the very purpose of EPR's argument. Moreover, doing so no longer allows one to ever verify the original prediction, thus requiring one to further qualify EPR's criterion of reality, a qualification that, as will be seen, is crucial in considering locality. By prediction, this could only be done on S_{22}, that is, by preparing another EPR pair and performing a measurement of y on S_{21}, which will, however, irrevocably prevent us from establishing x for S_{22}. It is only possible to establish both quantities for two EPR pairs, (S_{11}, S_{12}) and (S_{21}, S_{22}), and never for one. This is one of the deepest aspects of complementarity, and as discussed in Chap. 6, the analogous complementary situation obtains in the case of the standard measurement, as Bohr's reply shows (Bohr 1935, pp. 697–699). If we repeat the experiment for yet another pair, (S_{31}, S_{32}), so as to make predictions concerning the position of S_{32}, we can, again, make such a prediction exactly, but the outcome of the measurement on S_{32} will not in general be the same as for S_{12} or S_{22}, because the corresponding measurements on S_{11}, S_{12}, and S_{13} will be different. Once we perform both types of measurement for a large number of pairs, we will have statistically correlated measurements, akin in spin measurements of the Bell-Bohm type, to which the preceding argument would apply as well.

Bohr does not explain the situation in terms of two different objects and EPR pairs necessary in order to make both EPR predictions. This is, however, at least an implication of his argument, given his insistence in his reply and elsewhere that "in the problem in question we are not dealing with a *single* specified experimental arrangement, but are referring to *two* different, mutually exclusive, arrangements" (Bohr 1987, v. 2, p. 57, 60, 1937, p. 86). Bohr expressly "stages" the EPR case by the double-slit arrangement for two systems considered to shows the necessity of two such mutually exclusive arrangements (Bohr 1935, pp. 699–700). This fact was underappreciated, if realized at all, by EPR. In view of the necessity of these two arrangements and their mutual exclusivity, the second quantity in question cannot in principle be assigned to the same quantum object, once the first is assigned, although

[6] For a discussion of Bohr's "staging" of the EPR experiment, as a double-slit thought experiment, in a measuring arrangement described in this passage, see (Plotnitsky 2009, pp. 294–301). See also (Bacciagaluppi 2015), for a different perspective.

we can always make an alternative choice in selecting a measuring arrangement and thus measuring or predicting the other complementary variable in question. The joint assignment is not possible even if one accepts EPR's criterion of reality, whereby such an assignment is made on the basis of a prediction, *unless we add the context of measurement to this criterion.* This is, in effect, what Bohr says, with the implication that the second, complementary, quantity can no longer be assigned because it can no longer be predicted.[7] The second prediction is not possible once an experiment enabling the first prediction is performed, because the first object S_1 is no longer available for the second prediction. The simultaneous assignment of both quantities is precluded by the uncertainty relations, as recognized by EPR. They, however, aim to circumvent this limitation by arguing that both variables could be assigned at any point, even though only one of them could be measured or predicted. Bohr counterargues that the uncertainty relations and complementarity, both defined by the irreducible role of measuring instruments, preclude one from ever assigning both quantities to any quantum object, even in the EPR case. One of the deep implications here, in part, against EPR's view of the situation, is, again, that, unlike a classical measurement, quantum measurement is an establishment of a new phenomenon and not a measurement of a property of the object considered or in the present view of the ultimate constitution of reality responsible for quantum phenomena (and at the time of measurement, quantum objects). A quantum phenomenon is a classically observable effect of the interaction between this object or this constitution and the instrument used, and all physically observable and measured properties considered are those of phenomena.

It is never possible, on experimental grounds, to coordinate two experiments measuring or predicting quantities that are subject to the uncertainty relations on two different objects so as to make it possible to consider (in the way it can be done in classical mechanics) both as identically prepared *objects*. We can only control the identical preparation of the physical state of the observable parts of the measuring apparatus involved because this state is (physically) classical, but not the state of the objects during the measurement and hence the outcome of the predictions we make on the basis of this measurement. In the EPR case, we can predict with a probability equal to unity the first quantity in question, say, the value of the position variable, for the second object, S_{12}, of a given EPR pair (S_{11}, S_{12}). We can then predict the second quantity, the value of the momentum variable, for the second object, S_{21}, of, unavoidably, another, "*identically prepared*," EPR pair (S_{21}, S_{22}). We cannot, however, coordinate these predictions in such a way that they could be considered as pertaining to two identically prepared objects in the way it could be done in classical physics. This is not possible, because the necessary intermediate measurements would, in general, give us different data. Were we to repeat the measurement and the

[7] This fact was emphasized, on realist lines by Philippe Grangier (e.g., Farouki and Grangier 2019) and, in a series of articles (discussed earlier in this chapter) arguing against the idea of quantum nonlocality, by Khrennikov (2019a, b, 2020a, b, c), which includes (Khrennikov 2020a, b) an exchange with the present author, on lines of this chapter, in (Plotnitsky 2019b, 2020). This chapter, however, significantly revises my earlier arguments.

prediction of the first pair of quantities, those of the position variables for, respectively, S_{21} and S_{22}, we could still make our prediction with the probability unity, but the outcome would, in general, not be the same as in the case of the first pair (S_{11}, S_{12}). One *can predict* the outcome of a given EPR experiment with a probability equal to unity but one *cannot repeat* such an experiment so as to predict with a probability equal to unity the same value of the corresponding outcome.[8] We can only coordinate such measurements and predictions statistically, and thus establish the EPR correlations (for continuous or discrete variables). This does not help EPR, since their argument de facto presupposes exact rather than statistical coordination of such variables as belonging to the same, or an *identically prepared*, object of the same, or an identically prepared, EPR pair. Masked as it may be by the predictions with a probability equal to unity involved in the EPR experiment, the probabilistic or statistical nature of quantum predictions is still essential. Bohr concludes his analysis of EPR's argument concerning the incompleteness of QM as follows:

> From our point of view we now see that the wording of the above mentioned criterion of physical reality proposed by Einstein, Podolsky, and Rosen contains an ambiguity as regards the meaning of the expression "without in any way disturbing a system." Of course, there is in a case like that just considered no question of a mechanical disturbance of the system under investigation [the second object of the EPR pair considered] during the last critical stage of the measuring procedure. But even at this stage there is essentially the question of an *influence on the very conditions which define the possible types of predictions regarding the future behavior of the system.* Since these conditions constitute an inherent element of the description of any phenomenon to which the term "physical reality" can be properly attached, we see that the argumentation of the mentioned authors does not justify their conclusion that quantum-mechanical description is essentially incomplete. On the contrary this description, as appears from the preceding discussion, may be characterized as a rational utilization of all possibilities of unambiguous interpretation of measurements, compatible with the finite [quantum] and uncontrollable interaction between the object and the measuring instruments in the field of quantum theory. In fact, it is only the mutual exclusion of any two experimental procedures, permitting the unambiguous definition of complementary physical quantities, which provides room for new physical laws the coexistence of which might at first sight appear irreconcilable with the basic principles of science. It is just this entirely new situation as regards the description of physical phenomena that the notion of *complementarity* aims at characterizing. (Bohr 1935, p. 700; Bohr's emphasis)

This elaboration, especially Bohr's claim that "the essential ambiguity" of EPR's criterion pertains to "the meaning of the expression 'without in any way disturbing a system,'" have presented difficulties for many readers, as did his reply overall. Einstein, Schrödinger, and Bell had problems with both, in the latter case, as discussed in Chap. 6, in expressly in dealing with this passage (Bell 2004, pp. 155–156). Bohr acknowledged these difficulties and the main reason for them, essentially related to the RWR-type epistemology of his argument: "I am deeply aware of the inefficiency of expression which must have made it very difficult to bring out the essential ambiguity involved in a reference to physical attributes of objects when dealing

[8] That the ideal experiment proposed by EPR cannot be performed does not diminish this point, which, however, will be reflected in the actual experiments statistically approximating the EPR experiment. See note 2 above.

with phenomena where no sharp distinction can be made between the behavior of the objects themselves and their interaction with the measuring instruments" (Bohr 1987, v. 2, p. 61). In other words, at least part of the difficulty is expressing efficiently Bohr's RWR view, which does not allow one to speak unambiguously, and ultimately at all, of the reality ultimately responsible for quantum phenomena apart from its effects on measuring instruments. The elaboration and Bohr's meaning in this clause, however, pose no special difficulties given the preceding analysis. Once one quantity in question is established (even on the basis of a prediction, in accordance with EPR's criterion of reality, with suitable qualifications) for S_{12} of (S_{11}, S_{12}), we cannot ever establish the second quantity involved without measuring and hence disturbing S_{12}. Only one of these quantities could be established for S_{12} without disturbing it, but once it is established, never the other quantity without disturbing it. We can establish such an alternative quantity without disturbing it only for a different quantum object, S_{22}, via a different EPR pair (S_{21}, S_{22}), by a measurement of a complementary type on S_{21}. These two determinations cannot, however, be coordinated so as to assume that both quantities could be associated with the same object of the same EPR pair. The coordination of such events could only be statistical, in which case, as Einstein acknowledged, the Einstein-locality of QM would be maintained as well. Hence, we cannot establish *both quantities* for the same system "without in any way disturbing it," as EPR claim. The only way to do so would be to perform a measurement on it and thus *disturb* it. This measurement would, however, annul the determination of the first quantity, made on the basis of a prediction on the first object of the corresponding EPR pair, because there will be no way to ever verify this prediction, which could only be done by measuring the first quantity. As will be seen, these circumstances also allow one to keep Einstein-locality intact. Thus, the ambiguity in question indeed relates to the clause "without in any way disturbing the system," which, if one wants to apply it unambiguously, requires qualifications explained here but not provided by EPR. As Bohr noted earlier in his reply:

> In the phenomena concerned we are not dealing with an incomplete description characterized by the arbitrary picking out of different elements of physical reality at the cost of [sacrificing] other such elements [for a given quantum object], but with a rational discrimination between essentially different experimental arrangements and procedures which are suited either for an *unambiguous* use of the idea of space location, or for a legitimate application of the conservation theorem of momentum. Any remaining appearance of arbitrariness concerns merely our freedom of handling the measuring instruments, characteristic of the very idea of experiment. (Bohr 1935, p. 699; emphasis added)

Any attempt to apply both elements is ambiguous, and complementarity provides a necessary disambiguation, in correspondence with the uncertainty relations. In sum, there is no way to ever establish both complementary quantities for the same quantum object, S_2, of any EPR pair, (S_1, S_2), even alternatively (by prediction) without disturbing it, which means that QM predicts all that is possible in principle to establish as real, and thus is Bohr-complete. I am not claiming that Bohr's argument in principle precludes some future theory from providing a more complete, possibly an Einstein-complete, account of the ultimate nature of physical reality responsible for quantum phenomena. While, as explained in Chap. 6, some of Bohr's statements may suggest that this is not possible, these statements are viewed here as representing Bohr's ultimate, RWR-type, interpretation of quantum phenomena and QM, a view that leaves this possibility open. I do argue, however, that Bohr's argument in his reply does successfully counter EPR's argument that QM is either not Bohr-complete, by virtue of failing to predict all "elements of reality" that are possible to establish as real, or else Einstein-nonlocal, because the determination of these elements as real proposed by EPR could not be sustained in view of "the essential ambiguity" of their criterion of reality. The qualifications, necessary to apply EPR's criteria unambiguously, amount to the fact that both quantities in question can never be predicted, even by two different measurements on S_1, for the same quantum object, S_2, or the same EPR pair. This fact, as I shall argue next, disables not only EPR's argument concerning the Bohr-incompleteness of QM, but also their claim concerning its Einstein-nonlocality as the only alternative.

Before I consider this subject, I note that the argument given thus far could be transferred, with a few easy adjustments, to Bohm's version of the EPR experiment and spin variables (such as measuring spin in a given direction). In this case, too, there is the EPR complementarity insofar as any assignment of the complementary spin-direction quantity to the same quantum object becomes impossible, once one such a quantity is assigned. An assignment of the other would require an alternative type of measurement, mutually exclusive with the first, on the first object of a given pair, and hence, another identically behaving EPR-Bohm pair, which cannot be guaranteed. Nothing other than statistical correlations between such assignments is possible, which also allows for their Einstein-locality.

7.4 Einstein-Locality and Quantum Nonlocality in the EPR Experiment

EPR noted a possible objection to their argument by requiring that "two or more physical quantities can be regarded as simultaneous elements of reality only when they can be simultaneously measured or predicted," in, they clearly imply, the same location (Einstein et al., p. 141).[9] They, however, see this as implying Einstein-nonlocality:

> One could object to this conclusion on the grounds that our criterion of reality is not sufficiently restrictive. Indeed, one would not arrive at our conclusion if one insisted that two or more physical quantities can be regarded as simultaneous elements of reality only when they can be simultaneously measured or predicted. On this point of view, since either one or the other, but not both simultaneously, of the quantities P and Q can be predicted, they

[9] As has been noted by several authors, Schrödinger arguably the first of them (Schrödinger 1935a, p. 160), one could simultaneously make the position measurement on S_1 and the momentum measurement on S_2, and thus simultaneously predict (ideally exactly) the second variable for each system, the momentum for S_1 and the position for S_2. This determination, however, is not simultaneous in the same location, and measuring the complementary variable instead of the predicted one in either location would define a different reality.

are not simultaneously real. This makes the reality of P or Q depend upon the process of measurement carried out on the first system, which does not disturb the first system in any way. No reasonable definition of reality could be expected to permit this. (Einstein et al. 1935, p. 141)

This conclusion is not sustainable either, however. Einstein-nonlocality would indeed follow if one assumes, as EPR do, that the measurement, say, of P, on S_1 fixes the *actual* physical state itself of S_2 by a spooky action at a distance, rather than, as Bohr argues, only defines a prediction, with probability equal to unity, at a distance concerning such a state, as a *possible* physical state, by fixing the conditions of this prediction. On later occasions, Einstein also argued that, under the first, EPR-type, assumption, one is left with a paradoxical situation insofar as, if QM is complete, two mutually incompatible states could be assigned to the same distant quantum object, S_2, by different measurements performed on S_1 (e.g., Born 2005, p. 205). This is why EPR contend that, if QM *is* complete by this more restricted criterion, then the physical state of S_2, could be *determined* by a measurement on a spatially separated S_1, in violation of Einstein-locality. If it is Einstein-local, their argument, based on their criterion of reality, showed, they believed, that it is incomplete, the argument challenged by Bohr, as discussed in Sect. 7.3.

Einstein thought that Bohr accepted the alternative of Einstein-locality versus completeness (Bohr-completeness), and retained completeness by allowing for Einstein-nonlocality. Einstein, however, misunderstood Bohr's argumentation, which only allows for a *prediction*, and *not action*, at a distance, and thus only for quantum nonlocality but not Einstein-nonlocality. Einstein presented Bohr's argument as, in his words, "translated into [Einstein's] own way of putting it." It is not entirely clear whether Einstein's actually referred to Bohr's reply, although this was the only place where Bohr presented his full analysis of the EPR experiments and his counterargument to EPR. According to Einstein:

Of the "orthodox" quantum theoreticians whose position I know, Niels Bohr's seems to me to come nearest to doing justice to the problem. Translated into my own way of thinking, he argues as follows.

If the partial systems A [S_1 in present notation] and B [S_2] form [after their interaction] a total system which is described by its Ψ-function $\Psi/(AB)$, there is no reason why any mutually independent existence (state of reality) should be ascribed to partial systems A and B viewed separately, *not even if the partial systems are spatially separated from each other at the particular time under consideration.* The assertion that, in the latter case, the real situation B could not be (directly) influenced by any measurement taken on A is, therefore, within the framework of quantum theory, unfounded and (as the paradox shows) unacceptable.

By this way of looking at the matter, it becomes evident that the paradox forces us to relinquish one of the following assertions:

(1) The description by means of the Ψ-function is *complete.*

(2) The real states of spatially separated objects are independent of each other.

On the other hand, it is possible to adhere to (2), if one regards the Ψ-function as the description of a (statistical) ensemble of systems (and therefore relinquishes [1]). However, this view blasts the framework of the "orthodox quantum theory." (Einstein 1949b, pp. 681–682)

This may have been the argument of some of the "orthodox" quantum theoreticians. It was, however, not Bohr's argument, certainly, as must be apparent from the preceding discussion, not in his reply, but also, to my knowledge, not anywhere else, including in his "Discussion with Einstein," originally published in the same ("Schilpp") volume. It might be that Einstein's "translation" refers to what Bohr says on the subject there. The article, however, only offers a sketch of his argument, which still offers no statement that support Einstein's "translation" (Bohr 1987, v. 2, pp. 59–61). Bohr never says anywhere in his writings, and certainly, again, not in his reply, that "there is no reason why any mutually independent existence (state of reality) should be ascribed to partial systems *A* and *B* viewed separately, *not even if the partial systems are spatially separated from each other at the particular time under consideration.*" On the contrary, their independent existence is central for his argument, including in the context of Einstein-locality, because his argument depends on the fact that one can perform an alternative complementary measurement on each. Nor does he ever say that "the real situation *B*" could ever "be (directly) influenced by any measurement taken on *A*," via an *action at a distance*. He only allows that "the real situation in *B*" could be *predicted at a distance*, which, again, does not make it real. It is true that the entanglement brings new complexity to the relationships between the Ψ-function $\Psi/(\mathrm{AB})$, and our knowledge concerning each system separately, complexities discussed in Sect. 7.6. As explained in Chap. 6, an entangled system cannot be measured as such, which, however, need not mean there is "no reason why any mutually independent existence (state of reality) should be ascribed to partial systems *A* and *B* viewed separately, *not even if the partial systems are spatially separated from each other at the particular time under consideration.*" A measurement that can always be performed on each is the reason why they should be and are by Bohr ascribed a mutually independent existence, at the time of a measurement. It is also true that in the present view, it is meaningless to speak of a quantum system apart from measurement. This, however, is also not the same as saying the ultimate constitution of the reality responsible for quantum phenomena or, at the time, of observation, quantum systems, is Einstein-nonlocal. In any event, this was not Bohr's view, according to which the independent existence is ascribed to quantum objects, even though no properties can be ascribed to them apart from measurement.

It is also true that, as indicated in Sect. 7.2, once the measurements in question (or any two measurements) on A and B are actually performed, defining the corresponding actual events, one cannot, in accord with the quantum indefinitiveness postulate, make definitive statements concerning a relationship between these two individual events, or any two individual events, or even to ascertain the existence of any such relationship. The postulate, it follows, also precludes one from ascertaining that the relationship between these two events is either Einstein-local or Einstein-nonlocal, *any two events that have already occurred*. It does not follow, however, that a prediction concerning B by means of a measurement performed on *A* determines (in an Einstein-nonlocal way, by an action at a distance) the real situation *B* at the time of this prediction or at any moment in time. Only a measurement on *B* does. Such a measurement may confirm this prediction. On the other hand, the complementary measurement on *B*, which can always be performed, may prevent and in principle disable confirming it.

In any event, nothing in Bohr's reply or elsewhere in his writings gives any support to Einstein's "translation" of Bohr's argument, the essential basis of which, the roles of measuring instruments and complementarity, is not even mentioned by Einstein. Beginning with bypassing these roles and ending with claiming that Einstein-nonlocality was acceptable to Bohr, Bohr's argument is lost in Einstein's "translation." This "translation" reads Einstein's own reasoning into Bohr's very different reasoning, which "ensures the compatibility between [his] argument and all exigencies of relativity theory" and thus Einstein-locality (Bohr 1935, p. 701n).[10]

Einstein's conclusion, based on this reasoning, which is, again, not that of Bohr, confirms that for Einstein, while consistent with Einstein-locality, the statistical alternative, as a form of Bohr-completeness, is unacceptable. The "completeness" assumed here is clearly Einstein-completeness. In Bohr or the present view, the statistical alternative does not mean relinquishing (1), if by "complete" one means Bohr-completeness. Does "this view blasts the framework of the 'orthodox quantum theory'"? It depends on what one means by the "orthodox quantum theory" or interpretation of QM. Even if one assumes that the wave-function provides a complete *account* (I shall explain my emphasis presently) of individual quantum systems, one could still see QM as Bohr-complete, while not Einstein-complete. One could do so, for example, on Bayesian, such as QBist, lines, by viewing QM as theory providing probabilistic predictions concerning quantum phenomena associated with individual quantum systems, without providing a representation or *description* (hence, I speak of "account" above) of these systems or the constitution of the reality responsible for these phenomena. The term "description" is itself a problem, even in Bohr's reply, which perhaps should have avoided the temptation of repeating EPR's title, "Can Quantum–Mechanical Description of Physical Reality be Considered Complete?" In fact, in his reply, Bohr argues that QM provides a complete, as complete as

---

[10] This type of misreading of Bohr's argument is not uncommon. For example, Fine suggests the same type of reading, even if without seeing it as definitive, in a somewhat tortured argument, not always based on careful reading of what Bohr actually said, including on the point in question (Fine 2020). The view of Bohr's statement "*an influence on the very conditions which define the possible types of predictions, regarding the future behavior of the system* [S_2]" as implying Einstein-nonlocality appears to be responsible for Fine's remarks that Bohr in commenting on the exchange in his "Discussion with Einstein" (Bohr 1987, v. 2, p. 60), "unfortunately, ... takes no notice of Einstein's later versions of the argument and merely repeats his earlier response to EPR" (Fine 2020). While Bohr did not expressly comment on these later arguments by Einstein in this article, he was undoubtedly familiar with them, and several of his elaborations, indirectly relate to these later arguments and Bohr's familiarity with them (e.g., Bohr 1987, v. 2, pp. 62–65). In fact, however, if one adopts the type of reading of Bohr's reply offered here, Bohr's counterargument would equally counter these later arguments. See note 11 below. Fine concludes that "it is difficult to know whether a coherent response can be attributed to Bohr reliably that would derail EPR," referring to others who made attempts to that effect "on Bohr's behalf" (Fine 2020). I argue here for the coherence of this response. Bohr's carefulness and caution may make his writing inefficient, as he, again, acknowledges (Bohr 1987, v. 2, p. 60), but they also give coherence to his arguments, including that of his reply to EPR.

possible or Bohr-complete, *account* insofar as it predicts all that is possible to ascertain as real. It does not, however, in Bohr's or the present view, as the RWR view, provide a description or representation of the ultimate constitution of physical reality responsible for quantum phenomena, and thus for what could be so ascertained.

There is a difference between *defining* the state of a physical object by a prediction with probability equal to unity and *establishing* such a state as *possible* on the basis of such a prediction. In Bohr's argument in his reply, physical states of quantum objects cannot be seen as *finally* determined, even when we have predicted them exactly, unless the actual measurement is made. Their predicted determination may, however, be seen as in principle assured insofar as such a measurement could in principle be performed so as to yield the predicted value. This last requirement in turn becomes a necessary qualification of EPR's criterion of reality in the case of quantum phenomena. This is because, as explained, the measurement of the alternative quantity, Q, on S_2 would irrevocably disable any possible verification of the original prediction. It is crucial, and again, central to complementarity, that it is always possible to perform this alternative measurement. This is one of the reasons why the assumption of the independent (RWR-type) reality responsible for quantum phenomena or something in nature idealized as this reality (in Bohr in terms of quantum objects existing independently) is important for Bohr's analysis of the EPR experiment and of the question of Einstein-locality there. That the present interpretation assumes that a quantum object is an idealization only applicable at the time of measurement does not change this assumption as such and does not affect my argument at the moment. This independence ensures the possibility of this measurement. Unless either measurement that corresponding to the prediction or the alternative one is performed, and the corresponding quantum object is established, it is meaningless to speak of the physical reality, that of P or that of Q, associated with S_2. However, once this alternative measurement is performed, the original prediction becomes meaningless as in principle unverifiable. Hence, QM could not be shown to be Einstein-nonlocal by EPR's logic, any more than it can be shown to be Bohr-incomplete by their logic, which does not of course mean that either case cannot be made by a different argument. As noted earlier and as discussed in the next section, once this measurement is performed and becomes an *event*, one cannot definitively say that, in a given individual case, the relationship between the two events, defined by the two measurements in question, is either (Einstein) local or nonlocal. This is in accord with the quantum indefinitiveness postulate, which precludes any definitive statements concerning the relationships between any two individual quantum events. All definitive claims concerning quantum events are statistical. These considerations, however, do not affect my argument, which only concerns *predictions* at a distance, because they do not imply that a prediction concerning S_2 by means of a measurement performed on S_1 determines (by an action at a distance) any element of reality associated with S_2 at the time of this prediction or at any moment in time.

The argument here offered complicates speaking, as is common, of entangled objects as forming "an indivisible whole." Bohr never does so, although his reply to EPR has, again, been misread in this way, by confusing Bohr's use of this language for describing a phenomenon in his sense as forming an indivisible whole with the quantum object considered. Bohr's concept of phenomenon was introduced later. But, if one applies this concept to Bohr's argument in his reply to EPR, a measurement performed on S_1 forms an indivisible whole composed of S_1 and the measuring instrument used, but this wholeness does not in any way involve S_2. It only enables one to make a prediction concerning S_2 and a corresponding *possible* future phenomenon. A predicted measurement outcome is, however, not a phenomenon, only a measurement confirming this prediction is. Hence, a measurement on S_2 would establish its own phenomenon, with its own indivisible wholeness between S_2 and the measuring instrument. Accordingly, measuring on S_2 the variable, Q, complementary to the predicted one, P, will define a reality alternative to the predicted one and will disable this prediction even if it is made with a probability equal to unity.

This is why Bohr argued that it was not a question of a physical, "mechanical," influence of the measurement on S_1 upon the physical state of affairs concerning S_2, which is another indication that Einstein-locality is on his mind throughout. As he says: "of course there is in [the EPR] case no question of a mechanical disturbance of the system [object] [S_2] under investigation during the last critical stage of the measuring procedure" (Bohr 1935, p. 700). Contrary to a common misunderstanding (including, again, by Einstein) of Bohr's argument, this is not a physical influence, *at a distance*, of the measurement performed on S_1 on the spatially separated situation of S_2, although this measurement defines our predictions, "spooky predictions at a distance," concerning the corresponding future measurement on S_2. Bohr makes this point very clear: Even though there is no physical or mechanical disturbance, "even at this stage [when one makes a measurement on S_1 in order to make a prediction concerning S_2] there is essentially the *question of an influence on the very conditions which define the possible types of predictions, regarding the future behavior of the system* [S_2]" (Bohr 1935, p. 700). The influence in question is, thus, defined, locally, by fixing the conditions concerning the type of possible predictions concerning S_2 by making the corresponding measurement on S_1, which *in principle excludes* an alternative EPR measurement on S_1 and thus the conditions necessary for making an alternative EPR prediction for S_2. This influence, thus, concerns the *local* conditions of the measurement on S_1 and, correspondingly, the prediction, the only possible prediction by virtue of this measurement, concerning S_2. ("Local" merits an excessive emphasis here.) It concerns the determination of one possible experimental setup, as opposed to the other possible setup, and it is never possible to combine both setups. Once one of the two possible setups is in place and defines the measurement on the first object, any determination of the second quantity becomes impossible. An alternative arrangement, which would make this type of determination possible, would inevitably involve a different quantum object. One's decision *locally*, in the location of S_1, *influences* what kind of predictions, in the location of S_1, concerning S_2 are possible, even though we do not interfere with S_2. Bohr takes the Einstein-locality requirement as given, just as Einstein does.[11] Rather than revealing QM to be (Bohr) incomplete

[11] Einstein's subsequent arguments of the EPR type are sometimes viewed as offering a stronger or even different case by focusing more sharply on the question of locality (e.g., Einstein 1936; Born 2005, pp. 166–170, 204–205, 210–211; Einstein 1949a, pp. 77–85; and, in commenting on

or (Einstein) nonlocal, as EPR argue, in Bohr's view, the EPR *experiment* confirms that "the finite interaction between object and measuring agencies conditioned by the very existence of the quantum of action entails—because of the impossibility of controlling the reaction of the object on the measuring instruments if these are to serve their purpose—the necessity of a final renunciation of the classical ideal of causality and a radical revision of our attitude towards the problem of physical reality" (Bohr 1935, p. 697). EPR's argument and related arguments by Einstein deserve much credit for revealing that, as manifested in quantum phenomena, physical reality is, while Einstein-local, quantum-nonlocal. It is the reality of quantum correlations.

7.5 Quantum Nonlocality and Quantum Correlations

It is instructive to compare the present, strong RWR-type, view of quantum nonlocality and quantum correlations with two approaches to the subject, via Bell's theorem, that by Fine (1989) and that by Mermin (1990). I shall mostly focus on Fine's argument because of important differences, along with proximities, between his and the present view. By contrast, Mermin's view is primarily significant because it confirms the present concept of quantum nonlocality defined by predictions at a distance without an action at a distance, along with the fact that the relationships between two events of any individual pair of events involved cannot be ascertained to be either Einstein-local or Einstein-nonlocal. This is, however, in accord with the quantum indefinitiveness postulate, which precludes any statements concerning, or even assuming the existence of, any relationship between any two individual quantum events, events, again, that have already occurred.

Fine does not speak in terms of realism or the lack thereof. His primary focus is on "indeterminism": the "undetermined" nature of individual quantum events, including those comprising correlated multiplicities of them. Fine's concept of "undetermined events" is, thus, in accord with this study's definition of *indeterminacy* (given in Chap. 2) as a more general category, with randomness defined as a specific form of indeterminacy, when no probability is assigned to a possible future event. Both concepts thus only refer to possible future events, rather than events that have already happened, which makes them determined. Although, as I said, I prefer the view of individual quantum events as strictly random and hence the statistical (RWR)

Bohr's view, Einstein 1949b, pp. 681–682). I would argue, however, that, while these arguments may have streamlined EPR's argument, they still make their case in terms of the alternative between Einstein-locality and Bohr-completeness along the lines of EPR's argument. The above discussion of Einstein's misunderstanding of Bohr's argument in the Schilpp volume confirms this view (Einstein 1949b, pp. 681–682). In other words, these arguments do not make an essentially different case. Einstein never claimed otherwise. Instead, he thought that these arguments brought into a sharper focus the essential features of his view. This may be. The point remains, however, that these arguments never consider—any more than EPR's argument does—that the alternative determinations of one or the other complementary variable can never be established for the same (distant) quantum object. A new EPR pair is required to do so.

interpretation of correlations, my argument in this section equally applies to probabilistic, such as Bayesian, interpretations of correlations. Coupling indeterminism to "nonessentialism," as juxtaposed to a certain, realist, form of "explanationism," brings Fine's view closer to the RWR-view, including in Bohr's reply to EPR, to which Fine refers (Fine 1989, pp 184–185). While, however, Fine sees Bohr's analysis of measurement in the EPR case, in Bohr's reply as a "whole different topic" (Fine 1989, p. 184n2), it is, as my discussion here suggests, fundamentally related to the question of correlations.[12] According to Fine:

> If we adopt an indeterminist attitude to outcomes of a single, repeated measurement, we see each outcome as undetermined by any factor whatsoever. Nevertheless we are comfortable with the idea that, as the measurements go on, the outcomes will satisfy a strict probabilistic law. For instance, they may be half positive and half negative. How does this happen? What makes a long run of positives, for example, get balanced off by the accumulation of nearly the very same number of negatives? If each outcome is really undetermined, how can we get any strict probabilistic order? Such questions can seem acute, deriving their urgency from the apparent necessity to provide an explanation for the [statistically!] strict order of the pattern, and the background indeterminist premise according to which there seems to be nothing available on which to base an explanation (Fine 1989, p. 191).

My added brackets highlights the essentially statistical, in contrast to classically causal, nature of this order, defined by the informational structure of the observed data. Fine then contests the explanationist attitude, which is essentially realist in the present definition:

> Once we accept the premise of indeterminism, we open up the idea that sequences of individually undetermined events *can* nevertheless display strict probabilistic patterns. When we go on to wed indeterminism to rich probabilistic theory, like the quantum theory, we expect the theory to fill in the details of under what circumstances particular probabilistic patters will arise. The state/observable formalism of the quantum theory, as is well known, discharges these expectations admirably. The indeterminism opens up the space of possibilities. It makes room for the quantum theory to work. The theory specifies the circumstances under which patterns of outcomes will arise and which particular ones to expect. It simply bypasses the question of how any pattern could arise out of undetermined events, in effect presupposing that this possibility is among the natural order of things. …What then of correlations?
>
> Correlations are just probabilistic patterns between two sequences of events. If we treat the individual events as undetermined and withdraw the burden of explaining why a pattern arises from each of two sequences, why not adopt the same attitudes towards the emerging pattern between the pairs of outcomes, the pattern that constituted the correlation? Why, from an indeterminist perspective, should the fact that there is a pattern *between* random sequences require any more explaining than the fact there is a pattern internal to the sequences themselves? (Fine 1989, pp. 191–192)

As Fine's formulation suggests, "the quantum theory," insofar as it "specifies the circumstances under which patterns of outcomes will arise," must also involve the

---

[12] In his survey of the EPR argument, cited above (Fine 2020) and his related writings (e.g., Fine 1996, 2007), Fine separates Bohr's argument in his reply from the subject of correlations as well. As I argue here, implicit as the subject is there, Bohr's reply tells us a great deal about correlations or entanglement, by virtue of his RWR-type understanding of the EPR experiment. In contrast to Fine, this point is, as will be seen, realized by Mermin.

specification of the corresponding experimental arrangements, which in the present definition of a quantum theory, referring to its mathematical formalism, is part of an interpretation. In any event, the existence of correlations is independent of any theory. In Fine's view, then, the quantum theory is "bypass[ing] the question of how any pattern [including any pattern defining correlations] *could* arise out of undetermined events, in effect presupposing that this possibility is among the natural order of things." This view is thus that of *renouncing* the demand for an explanation or knowledge of how correlations or any quantum data, and thus quantum phenomena, come about. This is different from the RWR view, especially, the strong RWR view, and the corresponding interpretation of quantum phenomena, according to which an explanation or knowledge, or even a conception of how correlations or quantum phenomena in the first place come about, is not possible or, in Bohr's words, is "*in principle* excluded," at least as things stand now (Bohr 1987, v. 2, p. 62). This is, again, keeping in mind that here this statement and, thus, the difference of the present and Fine's view, only refers to the present interpretation or related interpretations rather than implying (as it may in Bohr), the impossibility in principle in any such explanation or conception, now or ever. The nature of future individual quantum events, which then become correlated events, as undetermined, is part of this view as well. For, while the combination of individual indeterminacy and collective statistical correlations may be especially striking, it is equally legitimate to ask why such individual events are undetermined, for example, in view of correlations, which suggest strict indeterminacy of individual events as little as this indeterminacy suggests correlations.

Fine's and the present position agree on two key counts. First, as this study has done from the outset, Fine takes for granted both the nature and structure, the *informational* structure, of quantum phenomena and that QM predicts this structure, correlations included, in accord with the available experimental evidence. Secondly, as this study does, Fine rejects the essentialist (realist) attitude toward explanation (Fine 1989, p. 193). He also states reasons for this rejection and for taking correlations for granted:

> There was a time when we did not know this [that correlational patterns may arise between the matched events in EPR-type sequences], when the question of whether the theory was truly indeterminist at all was alive and subject to real conjecture. Foundational work over the parts fifty years, however, has pretty much settled that issue (although, of course, never beyond *any* doubt). The more recent work related to EPR and Bell's theorem has shown, specifically (although again, not beyond *all* doubt), that the correlations too are fundamental and irreducible, so that the indeterministic ideal extends to them as well. It is time, I think, to accept the ideals of order required by the theory. It is time to see patterns between sequences as part of the same natural order as patterns *internal* to sequences themselves.
>
> A nonessentialist attitude toward explanation can help us make this transition, for it leads us to accept that what requires explanation is a function of the context of inquiry. (Fine 1989, p. 193)

As Fine explains:

> The search for "influences" or for common causes is an enterprise external to the quantum theory. It is a project that stands on the outside and asks whether we can supplement the theory in such a way as to satisfy certain *a priori* demands on explanatory adequacy. Among

> these demands is that stable correlations require explaining, that there must be some detailed account for how they are built up, or sustained, over time and space. In the face of this demand, the [en]tangled correlations of the quantum theory [which resist or even defy such explanations] can seem anomalous, even mysterious. However, this demand represents an explanatory ideal rooted outside the quantum theory, one learned and taught in the context of a different kind of physical thinking. (Fine 1989, p. 192)

Both Heisenberg and Bohr rejected such classical demands already at the time of the discovery of QM. In fact, correlational patterns, too, entered the theory virtually from its emergence. The interference pattern observed in the double-slit experiment, "the greatest of all quantum conundrums," as Mermin calls it, is a correlational pattern, and Mermin expressly relates it to EPR-type correlations (Mermin 1990, p. 108). Fine is prudent to note that there are still doubts that "foundational work over the parts fifty years, however, has pretty much settled that issue" or that "the more recent work related to EPR and Bell's theorem has shown … that the indeterministic ideal extends to [correlations] as well." His elegant formulation, however, also reflects the fact that he underestimated these doubts. They have never subsided in the way Fine appears to believe and are still wide spread now, 20 years after Fine stated his position.

According to Fine, the quantum theory should "specify the circumstances under which patterns of outcomes will arise and which particular ones to expect," even as it "simply bypasses the question of how any pattern could arise out of undetermined events" (Fine 1989, p. 191). As noted, in the present definition of quantum theory, this specification would belong to an interpretation rather than the theory, as they appear to do, along with the formalism, in Fine's view. This difference does not affect the fundamentals of the situation, however. In Bohr's and, following Bohr, the present view, these circumstances define quantum phenomena, which are independent of any particular theory, especially as far as its mathematics is concerned, although not of theoretical considerations in general, for example, insofar as specifying these circumstances involves classical physics. These are the circumstances that compel those who adopt the RWR view, beginning with Bohr, to go beyond merely "*bypass[sing]* the question of how any pattern *could* arise out of undetermined events," and to adopt interpretations in which an explanation or even conception of how correlations or quantum phenomena, in the first place, come about is "*in principle* excluded." These interpretations respond to the quantum-nonlocal nature of quantum correlations between spatially separated quantum events and to the capacity of QM to predict them, while avoiding Einstein-nonlocality. Fine clearly sees correlations as Einstein-local. On the other hand, he appears to dismiss "nonlocality," perhaps too easily, although it is possible that he is only rejecting essentialist attempts at explaining the quantum-nonlocal aspects of correlations (Fine 1989, pp. 184–190, 194). He extensively commented on the subject elsewhere (e.g., Fine 1996, 2020).

Mermin's conclusion of his analysis of correlations (an analysis that also offers an elegant proof of Bell's theorem) sheds an additional light on the situation. It reveals a subtle nuance of quantum nonlocality, a nuance consistent with the strong RWR view and the quantum indefinitiveness postulate. According to Mermin:

> [It] is wrong to apply to individual runs of the experiment the principle that what happens at A does not depend on how the switch is set at B. Many people want to conclude from this that what happens at A *does* depend on how the switch is set at B, which is disquieting in view of the absence of any connections between the detectors. The conclusion can be avoided, if one renounces the Strong Baseball Principle, maintaining that indeed what happens at A does not depend on how the switch in set at B, but that this [independence] is only to be understood in its statistical sense, and most emphatically cannot be applied to individual runs of the experiment. To me this alternative conclusion is every bit as wonderful as the assertion of mysterious [spooky] action at a distance. I find it quite exquisite that, setting quantum metaphysics entirely aside, one can demonstrate directly from the data and the assumption that there are no mysterious actions at a distance, that there is no conceivable way consistently to apply the Baseball Principle [what happens at A does not depend on how the switch in set at B] to individual events. (Mermin 1990, p. 109)

This view is consistent or even equivalent to the difference between the Einstein-nonlocality of spooky *action* at a distance and quantum nonlocality of spooky *predictions* at a distance. While this type of claim concerning individual quantum events may appear "disquieting," it should not be surprising. The impossibility of definitively claiming the relationships between any two single events at A and B to be either independent or dependent in this way would be a consequence of the quantum indefinitiveness postulate, which precludes any claims concerning the relationships between any two individual quantum events, including either the existence or absence of Einstein-nonlocal connections between them. Quantum nonlocality is a statistical concept.

Similarly, in the double-slit experiment, that "greatest of all quantum conundrums" (Mermin 1990, p. 108), in the noninterference setup, with the counters installed, a single detection of, say, an electron near a given slit does not guarantee that this electron passed through that slit, as opposed to the other slit or, in principle, both slits, although if the slits are sufficiently far apart, the last assumption would imply a violation of relativity and thus Einstein-nonlocality. As just explained, however, one cannot, by the quantum indefinitiveness postulate, ascertain the locality or nonlocality of the connections between two individual events, in this case the first being an emission of the electron from the source before it reaches the diaphragm with the slit and the second the detection of the electron near the slit. Statistically, however, passing of electrons through one or another slit could be ascertained, with all individual traces on the screen with which electron collide still discrete in both setups. On the other hand, one cannot similarly statistically claim that the electrons are passing through both slits in the alternative (interference-pattern) setup, in parallel with Mermin's argument under discussion. For one thing, we have no means for such a detection; we only make this *assumption* on the basis of the interference or correlational pattern observed. More importantly, if the distance between the source and the diaphragm is sufficiently long or the slits are sufficiently far apart, this assumption would imply a definitive violation of relativity and thus of Einstein-locality statistically, in multiple events. Accordingly, it is more suitable to assume that electrons, statistically, pass through either one slit or the other, without our being able to know through which in each case. All the traces detected on the screen are discrete: they are detections of single collisions. In the present interpretation, there are no electrons passing through the slits, but only when they are detected by a counter, or collide with the screen. The language of electrons passing through the slits or whatever else happens between what is *actually* observed in the instruments is the remnant of classical physics, based on our general phenomenal intuition of how things behave around us, no longer applicable to what "happens" between observations. An electron cannot be assigned a classical motion, suggested by the word "passing," or "happenings." What one instead has are specifiable, now *classically*, experimental settings and, when observed, corresponding specifiable phenomena and the rule for predicting them, different for these settings (especially, when they are complementary) for predicting them probabilistically by means of QM or QFT, cum Born's rule, or possibly some alternative theories. The EPR-type experiments for discrete variables, such as spin (not found in classical physics), are arguably more suited to exemplify this situation, but it defines all quantum phenomena and how QM or QFT works in predicting them.

The RWR view, then, especially the strong RWR view, equally handles the indeterminate nature of individual quantum events and correlations between certain sequences of quantum events. One cannot predict these correlations correctly on the basis of the data observed in one detector: "There is no way to infer from the data at one detector how the switch was set in the other. Regardless of what is going on in detector B, the data for a great many runs at detector A is simply a random string of R's [red signals] and G's [green signals]" (Mermin 1990, p. 107). One can only predict them correctly if we know both settings. If, however, somebody, unbeknown to us, will change the setting in one detector, for example, and in particular, to registering a complementary spin direction, our predictions will no longer correspond to what is actually observed, and there would be no way to confirm them. As discussed earlier, this circumstance is important for understanding the role of complementarity in EPR-type experiments and is central to Bohr's reply to EPR. The violation of Bell and related inequalities can also be linked to these circumstances and thus to complementarity, as is clear, for example, from Fine's argument in the article discussed here (Fine 1989, pp. 177–180). Accordingly, there is no experimental basis to ascertain that any quantum object can be assigned both "elements of reality," one found in one setting and the other in the other, before the detector flashes. As Mermin notes, the EPR-Bohr exchange "could be stated quite clearly" in terms of his thought experiment, "a direct descendant of the rather more intricate but conceptually similar" EPR experiment, and the analysis given here confirms (Mermin 1990, pp. 90–91).

7.6 Complementarity and Entanglement: Quantum Knowledge and Quantum Ignorance, with Bohr and Schrödinger

I have argued in this chapter that the EPR experiment, defined by the entanglement of two objects considered, manifests not only one of the deepest aspects of quantum phenomena or QM, which is commonly recognized, but also of complementarity, *as a quantum-theoretical concept*, which is rarely noted. Entanglement is a characteristic feature of quantum physics, and some, including Schrödinger, who introduced the term have argued that it is *the* single feature of quantum physics defining its difference from classical physics. Technically, entanglement is a feature of the formalism of QM that reflects the feature of quantum phenomena that fundamentally (if perhaps not uniquely) distinguishes them from classical ones. While keeping this mathematical aspect of the concept in mind throughout, I shall, as I did in Chap. 6, by the term entanglement understand, as Schrödinger did, this situation as a whole and continue to speak, as Schrödinger did as well, of the entanglement between two or more quantum objects or the object and the instrument in quantum measurement. In other words, I shall, following Schrödinger, understand by "entanglement" the concept that represents this mathematical-experimental situation.

I would like now to consider the relationships between complementarity and entanglement in more detail, also, as the title of this section suggests, as the relationships between knowledge and ignorance in quantum theory. Complementarity and Bohr's analysis of the EPR experiment or quantum phenomena in general may be primarily linked to knowledge, and entanglement and Schrödinger's analysis of it primarily to ignorance. Primarily does not means exclusively. It is instead a matter of relative and sometimes delicate balance of knowledge and ignorance, reflected in each concept and the relationships between them. As discussed earlier, Schrödinger remained unhappy with QM, which he still saw in his cat-paradox paper as "perhaps after all only a convenient calculational trick" (Schrödinger 1935a, p. 167). Unlike Bohr, he saw EPR's paper as making more apparent (even if not decisively proving) that an alternative, better theory could be developed. He was wise, as ever, to say "perhaps." His discontent, however, did not prevent him from offering a profound analysis of entanglement and of QM in these papers.

As discussed in Chap. 6, the entanglement between the object under investigation and the measuring instrument is essential to the standard case of quantum measurement and is connected to the role of complementarity there. This interaction allows one, by decision, to measure and then predict the outcome of a future possible measurement of *either* of the two complimentary variables (such as the position or the momentum) *after* this interaction had already taken place and the object left the location of the instrument. This fact shows that all quantum predictions are in effect predictions at a distance and, hence, are quantum-nonlocal, without, however, entailing a violation of Einstein-locality. Before returning to these connections in my conclusion, I would like to discuss Schrödinger's analysis of entanglement in his response to EPR's paper in three papers, including his cat-paradox paper (Schrödinger 1935a, b, 1936).

As noted, Schrödinger introduced the term entanglement in the first place, in both German [*Verschränkung*] and English, with the German term, arguably conveying his meaning better. Just as is complementarity, however, entanglement is an "artificial word which does not belong to our daily concepts," to adopt Bohr's characterization of complementarity cited earlier (Bohr 1937, p. 87). Accordingly, the connection of the quantum-theoretical meaning of either term, English or German, to its meaning in daily language is essentially irrelevant. Entanglement is a quantum-theoretical concept in the sense defined in Chap. 2. In this case, however, the mathematics of QM is more essential than in the case of complementarity, where, however, mathematics is more significant than it is commonly assumed, because it is necessary for complementary predictions.

Schrödinger's papers also suggested yet another quantum postulate, inherent in EPR-type situations or even in quantum phenomena in general, in part by virtue of complementarity, but not expressly defined thus far. This postulate may be called the *irreducible ignorance* (II) *postulate*: it states that any possible knowledge of quantum phenomena entails an essential element of ignorance concerning it in relationships very different from the way knowledge and ignorance are related (as they are) in classical physics. In the classical theory, as classically causal, the recourse to probability, while reflecting our ignorance concerning the systems considered, is only a practical expedient due to the insufficient, but in principle possible, knowledge concerning these systems. By contrast, in quantum physics, at least in RWR-type interpretations, to return to Bohr's formulation, "the recourse to probability laws is [due to] the inability of the classical frame of concepts to comprise the peculiar feature of indivisibility, or 'individuality,' characterizing the elementary processes" (Bohr 1987, v. 2, p. 34). In other words, our ignorance is fundamental. It has to do not with the, in practice, insufficient knowledge concerning the state of the physical reality responsible for quantum phenomena, but is an effect of this reality itself as ultimately unavailable to knowledge or even conception. This situation, however, and thus the II postulate allow for and are correlative to quantum causality, which is fundamentally (rather than only practically) probabilistic.

According to Schrödinger, in the case of a quantum entanglement between two quantum systems, the "best possible knowledge of a whole does not necessarily include the same for its parts. ... The whole is in a definite state, the parts taken individually are not" (Schrödinger 1935a, p. 161). "Knowledge" here refers to expectation-catalogs we can form, to "our questions about the future":

> Whenever one has a complete expectation-catalog—a maximal total knowledge—of a ψ-function for two completely separate bodies, or, in better terms, for each of them singly, then one obviously has it also for the two bodies together, i.e., if one imagines that neither of them singly but rather the two bodies together make up the object of interest, of our questions about the future. ... But the converse is not true. *Maximal knowledge of a total system does not necessarily include total knowledge of all its parts, not even when these are fully separated from each other and at the moment are not influencing each other at all.* (Schrödinger 1935a, p. 160)

Hence, an entanglement makes our ignorance irreducible as well. "Ignorance" is, too, no longer due to our insufficient possible knowledge of the ultimate constitution of the reality considered but rather is due to the impossibility, in principle, of such a knowledge. This is true, as things stand now, even in some realist interpretations of QM, although not in such alternative theories as Bohmian mechanics in which the classical epistemology of knowledge and ignorance it retained (the mathematics is different from that of classical mechanics), in Bohmian mechanics at the cost of Einstein-nonlocality. RWR-type interpretations make this nature of quantum ignorance automatic. Indeed, as *interpretations*, they respond to this nature. Quantum correlations are linked to our ignorance, and conversely our ignorance is defined by quantum correlations, which are quantum-informational in nature. Consider, the double-slit experiment "the greatest of all quantum conundrums," as Mermin, again, calls it, as a manifestation of a quantum correlational pattern (Mermin 1990, p. 108). There must be a lack of knowledge, an ignorance, concerning through which slit each particle passes, for the correlational (interference) pattern to emerge. Although the language of "loss of knowledge" is used sometimes, this lack of knowledge is, at least in RWR-type interpretations, not a loss in knowledge that one could in principle have. All knowledge one can possibly have is that concerning what happened in each initial measurement (which prepares each object considered) and what may happen in the future when the object hit the screen, the future probabilistically estimated by QM. If we have or even could in principle have the knowledge through which slit each objects passed, we have a new measurement situation in which the correlation pattern disappears, although the (always multiple) phenomena considered are still quantum in their physical nature and could only be predicted by means of a quantum and not classical theory.

Any actual information already available (through measurement) or obtainable in the future (as reflected in our predictions) that we can in principle have concerning quantum phenomena is always classical, but its structure—organization—and the way we obtain it is different in quantum physics. There are things that we don't know in advance in classical physics, but these are things that could in principle be known, in the present or in the future. Not so in quantum physics, at least in certain interpretations, where our ignorance is not about such things, but about things that cannot be known or conceived of, which makes the term "things" inapplicable, even as "things-in-themselves" in Kant's sense. The latter, as discussed earlier, cannot be known but still could be conceived of: they could still be *thought*. Quantum ignorance is more radical; it is the ignorance of the unthinkable. According to Schrödinger, then:

> When two systems, of which we know the states by their respective representatives, enter into temporary physical interaction due to known forces between them, and when after a time of mutual influence the systems separate again, then they can no longer be described in the same way as before, viz. by endowing each of them with a representative of its own. I would not call that *one* but rather *the* characteristic trait of quantum mechanics, the one that enforces its entire departure from classical lines of thought. By the interaction the two representatives [the quantum states] have become entangled. Another way of expressing the peculiar situation is: the best possible knowledge of a *whole* does not necessarily include the best possible knowledge of all its *parts,* even though they may be entirely separate and therefore virtually capable of being 'best possibly known,' i.e., of possessing, each of them, a representative of its own. The lack of knowledge is by no means due to the interaction being insufficiently known—at least not in the way that it could possibly be known more completely—it is due to the interaction itself. (Schrödinger 1935b, p. 555)

Schrödinger's point that "the lack of knowledge is by no means due to the interaction being insufficiently known—at least not in the way that it could possibly be known more completely—it is due to the interaction itself" represents the situation concerning the nature of knowledge and information in quantum theory as just outlined. At least in the present view (which or something akin to which is assumed by Schrödinger here, even if with a negative attitude toward it), this lack of knowledge is a reflection of the fundamental nature of our ignorance concerning the ultimate nature of reality responsible for quantum phenomena and, thus, for this entanglement. As just stated, this is not ignorance due to the lack of knowledge that could at least in principle be obtained, a secret that could in principle be discovered one day, and so forth. It cannot because it may not even be possible to conceive of what we are ignorant about. In strong RWR interpretations, even a conception of how the situation that gives rise to entanglement comes about is not possible, because the conception of how quantum phenomena come about is not possible. But then, as I argue here, all quantum phenomena involve and are defined by entanglement. Our ignorance of the *ultimate* nature of the physical reality responsible for quantum phenomena and all physical phenomena (if one adopts the U-RWR view, as I do here) is insurmountable in principle. It precludes us from assuming that it is in principle possible to know or conceive of this reality. This is what makes this reality a reality without realism. The qualification "ultimate" is, again, essential. We can think and know a great deal about the reality formed by effects of the ultimate reality that is beyond the reach of thought. It is worth recapitulating how the situation appears in the present interpretation, in which one can only speak of quantum objects at the time of measurement.

First, it, again, follows that one cannot rigorously speak of the *interaction* between quantum objects as an event that occurred before an experiment, because one cannot rigorously speak of what happens between experiments in general. This includes the case when one of these entities is the quantum part of a measuring instrument that has "interacted" with the quantum object under investigation, which interactions precedes the measurement as such, defined by the amplification of the quantum state of this part to the level of observation. Any rigorous statements can only refer to observable phenomena, which define all possible events and with which the corresponding quantum objects are associated, as suitable idealizations. In this view, in the situation considered by Schrödinger, there are only two quantum objects associated with the measurement or measurements performed initially and then two quantum objects associated with two measurements performed subsequently. Quantum entanglement can be defined in terms of such measurements. The entanglement between two quantum objects, S_1 and S_2, say, forming an EPR pair (or, again, the interaction between them, handled by the mathematics of entanglement), allows one by means of a measurement performed on S_1 to make predictions, with probability one, concerning S_2. S_2 is only defined once the corresponding measurement is performed, but not when the prediction concerning it is made, which makes it impossible to speak of any independent properties of S_2, however predicted, because there is no S_2, in the first place, until the corresponding measurement is performed. There is only the independent, RWR-type, reality which is responsible for the appearance of S_2 when this measurement is performed, and then, S_2 is still an RWR-type entity, which means that no physical properties can be attributed to it as such even at the time of measurement. These properties could only be attributed to the measuring instruments. As predictions at a distance, these predictions are quantum-nonlocal, while remaining Einstein-local.

As explained in Chap. 6, *quantum phenomena* are never entangled. Nor, in the present view, are quantum objects, as a concept or idealization only applicable at the time of measurement and thus always linked to quantum phenomena. So, one cannot speak even of their existence between experiments. Rigorously, two initial measurements associated with S_1 and S_2 lead to the situation of the state of the ultimate, RWR-type, reality or the quantum field, in which possible future measurements can be handled by the mathematics of entangled states and expectation catalogs they enable. One can speak of this reality in the cases of entanglement as an "entangled" reality insofar as our measurements of its effects are handled by the mathematics of entanglement, which thus also gives this mathematics a fundamental role, on which I shall comment below. *Quantum states*, ψ-functions, are rigorously defined as entangled, again, making our ignorance irreducible. We cannot assume that it is in principle possible to know the ultimate character of reality one or another effect of which we can predict. We are irreducibly ignorant of it. If one assumes, as Schrödinger (or Bohr) does, an independent existence of quantum objects between measurements, then one could say that they become entangled. In RWR-type interpretations, however, such as that of Bohr, the nature of the reality defining this entanglement is still beyond representation or even conception.

As Bohr said in commenting on EPR's paper, "the issue is of a very subtle character and is suited to emphasize how far, in quantum theory, we are beyond the reach of pictorial visualization" and ultimately any representation or even conception (Bohr 1987, v. 2, p. 59). Bohr and Schrödinger exchanged letters on some aspects of EPR's article and Bohr's reply (published in October 1935), while Schrödinger was finishing his cat-paradox paper, published in November (Schrödinger, Letter to Bohr, 13 October 1935; Bohr, Letter to Schrödinger, 26 October 1935, in [Bohr 1972–1996, v. 7, pp. 503–509, 511–512]). This exchange, however, concerned primarily Bohr's views rather than those of Schrödinger. Bohr does not appear to have expressly commented on Schrödinger's papers or addressed entanglement as a concept. I surmise, however, that Bohr would have agreed that these papers imply that entanglement and complementarity are essentially connected.

As stated above, as related to "questions about the future," any *possible* knowledge at stake in Schrödinger's argument is probabilistic or statistical (Schrödinger 1935a, p.160). This is the only knowledge the wave function provides, by way of expectation-catalogs, and when it comes to predictions, any particular complementarity is, too, a complementarity of expectation-catalogs. (Complementary *measurements* do give us definitive knowledge, but generally not concerning the same quantum object.) I am, of course, not saying that entanglement and complementarity are the same but only that they are subtly related to each other. Conceptually, they both radically put into question the relationships the whole and the parts, as we understand them in classical physics or daily life. Indeed, analogously (but, again, not identically) to complementarity it is not clear in what sense the combined system in question here is the whole of its two parts. Schrödinger explains this in terms of expectation catalogs in his cat-paradox paper:

> This is the point. Whenever one has a complete expectation-catalog—a maximal total knowledge—of a ψ-function for two completely separate bodies, or, in better terms, for each of them singly, then one obviously has it also for the two bodies together, i.e., if one imagines that neither of them singly but rather the two bodies together make up the object of interest, of our questions about the future.
>
> But the converse is not true. Maximal knowledge of a total system does not necessarily include total knowledge of all its parts, not even when these are fully separated from each other and at the moment are not influencing each other at all. Thus it may be that some part of what one knows may pertain to relations or stipulations between the two subsystems (we shall limit ourselves to two), as follows: if a particular measurement on the first system yields this result, then a particular measurement on the second the valid expectation statistics are such and such, but if the measurement in question on the first system should have that result, then some other expectation holds for that on the second; should a third result occur for the first, then still another expectation applies to the second; and so on, in the manner of a complete disjunction of all possible measurement results which the one specifically contemplated measurement on the first system can yield. In this way, any measurement process at all or, what amounts to the same, any variable at all of the second system can be tied to the not yet known value of any variable at all on the first, and of course vice versa also. If that is the case, if such conditional statements occur in the combined catalog, then it cannot possibly be maximal with regard to the individual systems. For the content of two maximal individual catalogs would by itself suffice for a maximal combined catalog; the conditional statement could not be added on. (Schrödinger 1935a, p. 160)

One can see here some elements of Bohr's reply to EPR insofar as no more complete knowledge than provided by QM is possible. I shall consider Schrödinger's comments, thus grounded, on the EPR experiment below. For the moment, Schrödinger continues:

> These conditional predictions, moreover, are not something that has suddenly fallen in here from the blue. They are in every expectation-catalog. If one knows the ψ-function and makes a particular measurement and this makes a particular result, then one again knows the ψ-function, *voilà tout*. It's just that for the case under discussion, because the combined system is supposed to consist of two fully separated parts, the matter stands out as a bit strange. For thus, it becomes meaningful to distinguish between measurements on the one and measurements on the other subsystem. This provides to each full title to a private maximal catalog; on the other hand it remains possible that a portion of the attainable combined knowledge is, so to say, squandered on conditional statements, that operate between the subsystems, so that the private expectancies are left unfulfilled—even though the combined catalog is maximal, that is even though the ψ-function of the combined system is known.
>
> Let us pause for a moment. This result in its abstractness actually says it all: Best possible knowledge of a whole does not necessarily include the same for its parts. Let us translate this into terms of Sect. 9 [describing "the ψ-function as description of state"]. The whole is in a definite state [ψ-function], the parts taken individually are not.
>
> "How so? Surely a system must be in some sort of state." "No. State is ψ-function, is maximal sum of knowledge. I didn't necessarily provide myself with this, I may have been lazy. Then the system is in no state."
>
> "Fine, but then too the agnostic prohibition of questions is not yet in force and in our case I can tell myself: the subsystem is already in some state, I just don't know which."
>
> "Wait. Unfortunately no. There is no 'I just don't know.' For as to the total system, maximal knowledge is at hand ..."
>
> *The insufficiency of the ψ-function as model replacement* [for classical theory] *rests solely on the fact that one doesn't always have it.* If one does have it, then by all means let it serve

> as description of the state. But sometimes one does not have it, in cases where one might reasonably expect to. And in that case, one dare not postulate that it "is actually a particular one, one just doesn't know it"; the above-chosen standpoint forbids this. "It" is namely a sum of knowledge; and knowledge, that no one knows, is none—. (Schrödinger 1935a, p. 161)

The last case is, of course, decisive in conveying the fact that the ultimate nature of the reality responsible for quantum phenomena is defined by ignorance and not knowledge, and thus is subject to the II postulate, to which the concept of reality without realism responds, as an interpretation of the quantum–mechanical situation in general. Schrödinger says next:

> That a portion of the knowledge should float in the form of disjunctive conditional statements *between* the two systems can certainly not happen if we bring up the two from opposite ends of the world and juxtapose them without interaction. For then indeed the two "know" nothing about each other. A measurement on one cannot possibly furnish any grasp of what is to be expected of the other. Any "entanglement of predictions" that takes place can obviously only go back to the fact that the two bodies at some earlier time formed in a true sense *one* system, that is were interacting, and have left behind *traces* on each other. If two separated bodies, each by itself known maximally, enter a situation in which they influence each other, and separate again, then there occurs regularly that which I have just called *entanglement* of our knowledge of the two bodies. The combined expectation-catalog consists initially of a logical sum of the individual catalogs; during the process it develops [classical] causally in accord with known law (there is no question whatever of measurement here). The knowledge remains maximal, but at its end, if the two bodies have again separated, it is not again split into a logical sum of knowledges about the individual bodies. What still remains *of that* may have becomes less than maximal, even very strongly so. — One notes the great difference over against the classical model theory, where of course from known initial states and with known interaction the individual end states would be exactly known. (Schrödinger 1935a, p. 161)

The passage confirms and amplifies my argument as concerns the difference, "the great difference," of the lack of knowledge or the nature of our ignorance in quantum and classical physics, as reflected, perhaps most dramatically, in the concept of entanglement. In quantum theory, our ignorance of the ultimate nature of reality is insurmountable *in principle*, at least in RWR-type interpretations. Complementarity and entanglement are manifestations of this insurmountable "it" or, as the case may be, "not even it" or, as I said, "it without bit," from the structure of the "bits" of information (classical in character) found in measuring instruments. *How* the situation in which "the maximal knowledge of a total system does not necessarily include the total knowledge of its parts" emerges, the ultimate nature of the reality responsible for it, is something that is ultimately impossible to know or even conceive of. The idea that this is not "*one* but rather *the* characteristic trait of quantum mechanics, the one that enforces its entire departure from classical lines of thought" literally "haunts" Schrödinger's articles on entanglement: "best possible knowledge of the whole *does not* include best possible knowledge of its parts—and that is what keeps coming back to *haunt* us" (Schrödinger 1935a, p. 166, b, p. 555; emphasis added). I end my discussion of Schrödinger with his elegant commentary of the EPR experiment, as "a simple 'pointed' example" of his argumentation, which also shows the mathematics of entanglement at work. He says:

For simplicity, we consider two systems with just one degree of freedom. That is, each of them shall be specified through a *single* coordinate q and its canonically conjugate momentum p. The classical picture would be a point mass that could move only along a straight line, like the spheres of those play things on which small children learn to calculate. p is the product of mass by velocity. For the second system we denote the two determining parts by capital Q and P. As to whether the two are "threaded on the same wire" we shall not be at all concerned, in our abstract consideration. But even if they are, it may in that case be convenient not to reckon q and Q from the same reference point. The equation $q = Q$ then does not necessarily mean coincidence. The two systems may in spite of this be fully separated.

In [EPR's] paper, it is shown that between these two systems, an entanglement can arise, which at a particular moment, can be compactly shown in the two equations: $q = Q$ and $p = -P$. That means: *I know, if* a measurement of q on the system yields a certain value, that a Q-measurement performed immediately thereafter on the second will give the *same* value, and vice versa; *and I know*, if a p-measurement on the first system yields a certain value, that a P-measurement performed immediately thereafter will give the opposite value, and vice versa.

A single measurement of q or p or Q or P resolves the entanglement and makes both systems maximally known. A second measurement on the same system modifies only the statements about it, but teaches nothing more about the other. So one cannot check both equations in a single experiment. But one can repeat the experiment *ab ovo* a thousand times; each time set up the same entanglement; according to whim check one or the other of the equations; and find confirmed that one which one is momentarily pleased to check. We assume that all this has been done.

If for the thousand-and-first experiment one is then seized by the desire to give up further checking and then measure q on the first system and P on the second, and one obtains

$$q = 4; \ P = 7;$$

can one then doubt that

$$q = 4; \ p = -7$$

would have been a correct prediction for the first system, or

$$Q = 4; \ P = 7;$$

a correct prediction for the second? Quantum predictions are indeed not subject to test as to their full content, ever, in a single experiment; yet they are correct, in that whoever possessed them suffered no disillusion, whichever half he decided to check. There's no doubt about it. Every measurement is for its system the first. Measurements on separated systems cannot directly influence each other—that would be magic. Neither can it be by chance, if from a thousand experiments it is established that virginal measurements agree.

The prediction catalog $q = 4$, $p = -7$ would of course be hypermaximal. (Schrödinger 1935a, pp. 163–164)

This is the essence of quantum predictions: "Quantum predictions are indeed not subject to test as to their full content, ever, in a single experiment; yet they are correct, in that whoever possessed them suffered no disillusion, whichever half he decided to check. There's no doubt about it." We can bet on them, just as we can bet with them. But we need the mathematics of QM to do so. Indeed, Schrödinger also notes here a crucial point discussed above but missed by EPR: "So one cannot check both equations in a single experiment." One needs two separate experiments and two EPR pairs to check and in the first place to make both predictions. The calculations involved, which include the entangled state constructed by EPR

$$|\psi\rangle = \iint_{-\infty -\infty}^{\infty \infty} A(x_1, x_2)|x_1\rangle|x_2\rangle dx_1 dx_2$$

(with A, which cannot be separated into a function of x_1 and x_2, assumed to be Dirac's delta function $\delta(x_1 - x_2)$), and other mathematics used in considering the EPR experiment (which, again, cannot be performed in a laboratory because the above state is not normalizable) were not as simple at the time as Schrödinger might have them appeared to be in this passage. Obviously, they would now be a routine exercise in an undergraduate course on QM, and of course nowhere near the complexities of the present-day quantum theory, such as QFT, or mindboggling calculations related to the entangled entropy of black holes, using just about the most complex mathematics found in fundamental physics.

My point here is that the calculations necessary for handling the EPR experiment return us to the irreducible role of mathematics in quantum theory, for one thing, given the ψ-function, even if were only mentioned, and it was far from merely mentioned by Schrödinger or here. Every time it is even mentioned, the whole mathematical architecture of QM enters the argument. The ψ-function, Schrödinger's great invention, especially the complex ψ-function is the part of genealogy of both QED and QCD, and thus of the standard model. For the moment, there is no entanglement without it, and it is through the ψ-function, the main vehicle of knowledge, the irreducible probabilistic knowledge, and ignorance in QM (if also through the fact we cannot always have the ψ-function) that we *know* that in quantum physics "the maximal knowledge of a total system does not necessarily include the total knowledge of its parts."

As discussed in Chap. 5, Bohr eventually came to realize that the standard situation of quantum measurement is essentially connected to the EPR experiment because the quantum object under investigation and the quantum part of the measuring instrument, interacting with this object, are entangled. It is true that in the standard case, probabilities of our predictions are, in general, not equal to one. This, however, does not affect the essential aspects of the situation. Besides, as Bohr also came to realize, one can reproduce the EPR situation by making the quantum object and the quantum part of the instrument as forming the EPR pair, by the set of simple additional measurements for $q_1 - q_2$ and $p_1 + p_2$, both of which can be accurately measured, and by using these measurements to make the EPR-type predictions, which are ideally exact (Bohr 1938, pp. 101–104, 1987, v. 2, pp. 59–60).

Most crucial, however, is the fact of the initial entanglement in question between the object considered and the measuring instrument involved. It is this entanglement that enables any possible predictions concerning the outcome of a future experiment on the basis of the data obtained in this measurement, defined by this entanglement, which entanglement always preceded the outcome registered in the observed part of the instrument. We can decide on the arrangement for registering a given measurement, say, either that of the position or that of the momentum, *before* this outcome is registered, but we can also do this *after* the interaction entangling the object and the instrument had taken place and the object has left, after this interaction, the location of the instrument. This, as explained, means that we actually register the outcome associated with the quantum part of the instrument and not with the object which is by this point elsewhere, while our prediction will pertain to the interaction (within the same structure) between the object and another instrument. This is strictly parallel to the EPR experiment in which, after the interaction between two objects, S_1 and S_2, that entangles them, we can decide to perform either the position or the momentum measurement on S_1 and make a prediction, with probability equal to unity, concerning the outcome of the corresponding measurement on S_2, "without in any way disturbing" it. The same, however, is in effect true in the standard measurement because if we make, as we always can, one or the other arrangement *after* the object left the location of the initial interaction, we make our prediction concerning the object, at a distance, without in any way disturbing it. EPR concluded on this basis that one can assign both quantities to S_2 or else that we must assume Einstein-nonlocality, a conclusion challenged by Bohr in the way considered in this chapter.

All quantum predictions, even the standard ones, are, then, predictions at a distance without in any way disturbing the system, without, however, need to assume any action at a distance and thus Einstein-nonlocality. On the other hand, unlike in classical physics (where we can predict distance events within a classically causal process), we can never either measure or, accordingly, predict both quantities simultaneously, by virtue of the uncertainty relations, a fact acknowledged by EPR. This, again, does not in principle preclude classical causality (found, for example, in Bohmian mechanics, which is, however, Einstein-nonlocal), but it does not require it or realism in the first place, thus allowing for RWR-type interpretations of the situation, while preserving Einstein-locality. There is only quantum nonlocality defined by the prediction at a distance, with probability one, but with the qualifications discussed earlier, insofar as an alternative measurement on S_2, which is always possible, will disable the possibility of verifying this prediction and thus the assignment of the quantity predicted. In classical physics, these complexities do not arise because both quantities can be assigned to the object considered as possessed by this object independently, as a permissible idealization even in the U-RWR view.

Accordingly, although ingenious, EPR's contrivance in designing the EPR experiment to make predictions on the basis of performing a measurement on a different quantum object so as not to interfere with the object in question merely reflects the way quantum measurements and predictions always work. It took, however, the EPR experiment and EPR's paper to make Bohr realize this fact, even though Bohr sometimes claimed that EPR's argument contained nothing essentially new, possibly for the reasons just outlined. On the other hand, the features of quantum measurement just discussed were only considered in Bohr's writings and were given a deeper analysis following EPR's paper. Bohr's thinking concerning quantum phenomena and QM was given a new impetus by EPR's paper, an impetus that also brought with it revisions of his interpretation, leading him to his ultimate interpretation. This change in Bohr's thinking is a reflection of the fact that EPR's paper compelled physicists and philosophers to ask deeper questions concerning the relationships between Einstein-locality and quantum nonlocality, complementarity and entanglement, and knowledge and ignorance, and thus between physics and mathematics and among physics, mathematics, and philosophy, in quantum physics.

7.7 Conclusion

The historical trajectory engendered by EPR's paper, from Bohr's and Schrödinger's responses to it to Bell's theorem and beyond, transformed our understanding of nature and our interactions with it in quantum physics as a mathematical-experimental science, with mathematics assuming the primary role in this conjunction. As I have argued, following Heidegger, from the outset of this study, from Descartes and Galileo on, modern mathematics "is experimental because of its mathematical project" (Heidegger 1967, p. 93). As I have also argued, however, QM and then QFT, in RWR-type interpretations, gave mathematics a hitherto unprecedented role in physics. Quantum entanglement is a remarkable manifestation of this role. As such, it was yet another unintended gift to QM and the RWR view by Einstein, who was no friend of either QM or the RWR view. It took Schrödinger, no friend of the QM or the RWR view either, to give quantum entanglement its name and a fitting mathematical expression, and to realize its full conceptual significance: "best possible knowledge of the whole *does not* include best possible knowledge of its parts." In the RWR view, this classical expression or even the mathematical representation (which is rigorous) of entanglement in the formalism of QM cannot capture the ultimate character of the reality responsible for the phenomena associated with entanglement, such as correlations, or for quantum phenomena in general, in the first place. The mathematics of QM, however, enables us to predict the outcomes of quantum experiments, which also differentiate quantum phenomena in terms of the effects observed, such as those of correlations.

While, at one level, a tremendous reduction of our phenomenal thinking from which it was born, mathematics enables us to *relate* to things in nature and mind which are beyond the reach of all thinking, including mathematical ones. Mathematics allows our thinking to reach beyond mathematics and, again, beyond anything that we can in principle conceive or think about, except as that which we cannot conceive or think about, but to which we can, by mathematics, relate otherwise. In quantum theory, mathematics allows us to estimate probabilities or statistics of quantum events, and no other estimates are, again, empirically possible, as things stand now. This, however, makes QM, a Bohr-complete theory of (nonrelativistic) quantum phenomena, as complete as such a theory can be, as things stand now. This is no small achievement. Indeed, QM is one of the greatest mathematical-experimental achievements in the history of modern physics, an achievement in part made possible by the RWR-type thinking that led to Bohr to his 1913 theory and Heisenberg to QM.

Schrödinger had always resisted the RWR view, and he never warmed up to complementarity either. He hoped that entanglement would expunge the RWR view, by that time, along with QM itself, as "perhaps after all only a convenient calculational trick," from our understanding of quantum phenomena and would rekindle a search for an alternative theory (Schrödinger 1935a, p. 167). While this search was indeed rekindled later, in the 1960s, by related developments, such as Bell's and the Kochen-Specker theorem, QM or, in high-energy quantum regimes, QFT, has remained our standard theory of quantum phenomena. For Bohr, by contrast, the EPR experiment and, by implication, quantum entanglement were manifestations of, as he said in his reply to EPR, the "entirely new situation as regards the description of physical phenomena that the notion of *complementarity* aims at characterizing" (Bohr 1935, p. 700). It is a rigorous response to this situation that, he argued, made QM a complete theory, as complete as a theory of (nonrelativistic or low-energy) quantum phenomena, including those of the EPR-type, can be, as things stand now. With Dirac's introduction of QED in 1927, QFT took on high-energy quantum phenomena. Remarkably, Yukawa published his theory of mesons in 1935, the same year that EPR's paper, Bohr's reply, and Schrödinger's cat-paradox paper were published. 1935 was, especially in retrospect, a great year for quantum theory.

References

Aspect, A., Dalibard, J., Roger, G.: Experimental test of Bell's inequalities using time varying analyzers. Phys. Rev. Lett. **49**, 1804–1807 (1982)

Bacciagaluppi, G.: Did Bohr understand EPR?. In: Aaserud, F., Kragh H. (eds.) One Hundred Years of the Bohr Atom (Scientia Danica, Series M, Mathematica et physica) vol. 1, pp. 377–396. Royal Danish Academy of Sciences and Letters: Copenhagen, Denmark

Bell, J.S.: Speakable and Unspeakable in Quantum Mechanics. Cambridge University Press, Cambridge, UK (2004)

Bohr, N.: Can quantum-mechanical description of physical reality be considered complete? Phys. Rev. **48**, 696–702 (1935)

Bohr, N.: Causality and complementarity. In: Faye, J., Folse, H.J. (eds.) The Philosophical Writings of Niels Bohr, Volume 4: Causality and Complementarity, Supplementary Papers, 1999, pp. 83–91. Ox Bow Press, Woodbridge, CT, USA (1937)

Bohr, N.: The causality problem in atomic physics. In Faye, J., Folse, H. J., (eds.) The Philosophical Writings of Niels Bohr, Volume 4: Causality and Complementarity, Supplementary Papers, 1999, pp. 94–121. Ox Bow Press, Woodbridge, CT, USA (1938)

Bohr, N.: Niels Bohr: Collected Works, vol. 10. Elsevier, Amsterdam, Netherland (1972–1996)

Bohr, N.: The Philosophical Writings of Niels Bohr, vol. 3. Ox Bow Press, Woodbridge, CT, USA (1987)

Born, M.: The Einstein-Born Letters (tr. Born, I.). Walker, New York, NY, USA (2005)

Brunner, N., Gühne, O., Huber, M. (eds.): Special issue on 50 years of Bell's theorem. J. Phys. A **42**, 424024 (2014)

Bub, J.: The Banana World: Quantum Mechanics for Primates. Oxford University Press, Oxford, UK (2016)

Cushing, J.T., McMullin, E. (eds.): Philosophical Consequences of Quantum Theory: reflections on Bell's Theorem. Notre Dame University Press, Notre Dame, IN, US (1989)

Einstein, A.: Physics and reality. J. Franklin Inst. **221**, 349–382 (1936)

Einstein, A.: Autobiographical Notes (tr. Schilpp, P. A.). Open Court, La Salle, IL, USA (1949a)

Einstein, A.: Remarks to the essays appearing in this collective volume. In: Schilpp, P.A. (ed.) Albert Einstein: Philosopher–Scientist, 1949, pp. 663–688. Tudor, New York, NY, USA (1949b)

Einstein, A., Podolsky, B., Rosen, N.: Can quantum-mechanical description of physical reality be considered complete? In: Wheeler, J.A., Zurek, W.H. (eds.) Quantum Theory and Measurement, 1983, pp. 138–141. Princeton University Press, Princeton, NJ, USA (1935)

Ellis, J., Amati, D. (eds.): Quantum Reflections. Cambridge University Press, Cambridge, UK (2000)

Farouki, N., Grangier, F. The Einstein-Bohr debate: finding a common ground of understanding, arXiv:1907.11267 (2019)

Fine, A.: Do correlations need to be explained? In: Cushing, J.T., McMullin, E. (eds.) Philosophical Consequences of Quantum Theory: Reflections on Bell's Theorem, 1989, pp. 175–194. Notre Dame University Press, Notre Dame, IN, USA (1989)

Fine, A.: The Shaky Game: Einstein, Realism and the Quantum Theory, 2nd edn. University of Chicago Press, Chicago, IL, USA (1996)

Fine, A.: Bohr's response to EPR: Criticism and defense. Iyyun Jerus. Philos. Quart. **56**, 31–56 (2007)

Fine, A.: The Einstein-Podolsky-Rosen argument in quantum theory. In: Zalta, E.N. (ed.) Stanford Encyclopedia of Philosophy (Summer 2020 Edition) (2020). https://plato.stanford.edu/archives/sum2020/entries/qt-epr/

Greenberger, D.M., Horne, M.A., Zeilinger, A.: Going beyond Bell's theorem. In: Kafatos, M. (ed.) Bell's Theorem, Quantum Theory and Conceptions of the Universe, pp. 69–72. Kluwer, Dordrecht, Netherlands (1989)

Greenberger, D.M., Horne, M.A., Shimony, A., Zeilinger, A.: Bell's theorem without inequalities. Am. J. Phys. **58**, 1131–1142 (1990)

Hardy, L.: Nonlocality for two particles without inequalities for almost all entangled states. Phys. Rev. Lett. **71**, 1665–1668 (1993)

Heidegger, M.: What Is a Thing? (tr. Barton, WB, Jr., Deutsch, V). Gateway, South Bend, IN, USA (1967)

Khrennikov, A.: Quantum probabilities and violation of CHSH-inequality from classical random signals and threshold type detection scheme. Prog. Theor. Phys. **128**, 31–58 (2012). https://doi.org/10.1143/PTP.128.31

Khrennikov, A.: Get rid of nonlocality from quantum physics. Entropy **21**(8), 806 (2019a). https://doi.org/10.3390/e21080806

Khrennikov, A.: Quantum versus classical entanglement: eliminating the issue of quantum nonlocality. arXiv:1909.00267 (2019b)

Khrennikov, A.: Two faced Janus of quantum nonlocality. arXiv:2002.01977v1 [quantum-ph] (2020a)

Khrennikov, A.: Quantum versus classical entanglement: eliminating the issue of quantum nonlocality. Found. Phys. **50**, 1762–1780 (2020b). https://doi.org/10.1007/s10701-020-00319-7

Khrennikov, A.: Quantum postulate vs. quantum nonlocality: is [the] devil in h? arXiv:2003.05718v1 [quantum-ph] (2020c)

Mermin, N.D.: Boojums All the Way Through: Communicating Science in a Prosaic Age. Cambridge University Press, Cambridge UK (1990)

Plotnitsky, A.: Epistemology and Probability: Bohr, Heisenberg, Schrödinger and the Nature of Quantum-Theoretical Thinking. Springer, New York, NY, USA (2009)

Plotnitsky, A.: Bohr and Complementarity: An Introduction. Springer, New York, NY, USA (2012)

Plotnitsky, A.: The Principles of Quantum Theory, from Planck's Quanta to the Higgs Boson: The Nature of Quantum Reality and the Spirit of Copenhagen. Springer/Nature, New York, NY, USA (2016)

Plotnitsky, A.: Spooky predictions at a distance: reality, complementarity and contextuality in quantum theory. Philos. Trans. R. Soc. A **377**, 20190089 (2019a). https://doi.org/10.1098/rsta.2019.0089

Plotnitsky, A.: "Without in any way disturbing the system:" illuminating the issue of quantum nonlocality. arXiv:1912.03842v1 [quant-ph] (2019b)

Plotnitsky, A.: "The unavoidable interaction between the object and the measuring instrument": reality, probability, and nonlocality in quantum physics. Found. Phys. **50**, 1824–1858 (2020). https://doi.org/10.1007/s10701-020-00353-5

Plotnitsky, A., Khrennikov, A.: Reality without realism: on the ontological and epistemological architecture of quantum mechanics. Found. Phys. **25**(10), 1269–1300 (2015). https://doi.org/10.1007/s10701-015-9942-1

Schrödinger, E.: The present situation in quantum mechanics. In: Wheeler, J.A., Zurek, W.H. (eds.) Quantum Theory and Measurement, 1983, pp. 152–167. Princeton University Press, Princeton, Princeton, NJ, USA (1935a)

Schrödinger, E.: Discussion of probability relations between separated systems. Proc. Camb. Philos. Soc. **31**, 555–563 (1935b)

Schrödinger, E.: Discussion of probability relations between separated systems. Proc. Camb. Philos. Soc. **32**, 446–452 (1936)

t'Hooft, G.: Quantum mechanics and determinism. In: Frampton, P., Ng, J., (eds.) Particles, Strings, and Cosmology, pp. 275–285. Rinton Press, Princeton, NJ, USA (2001)

t'Hooft, G.: Time, the arrow of time, and quantum mechanics (2018). arXiv:quant-ph/1804.01383

Chapter 8
"Something Happened": The Real and the Virtual in Elementary Particle Physics

Something must have happened ...
—Joseph Heller, *Something Happened* (Heller 1974)

Abstract This chapter addresses the question of elementary particles. As testified by the persistent title, "What is an elementary particle?," "an elementary particle" has been a problem to which only fragments of a possible solution could be offered. At a certain point, "a virtual particle" has become part of this problem. This chapter does not aim to do more than consider this problem as a problem. In contrast, however, to most approaches to this problem, realist in nature, it pursues an approach based on the RWR view. QFT extends the RWR view on two counts. First, it does so by joining the irreducibly unknowable or, in the strong RWR view, the irreducibly unthinkable to the irreducibly multiple. Second, QFT requires a rethinking of what happens between quantum experiments in high-energy regimes, which leads to the concept of virtual particles. The first subject is addressed in Sect. 8.2, via the concept of a quantum field, and the second in Sect. 8.3, which also considers asymptotic freedom in QCD. I conclude with some philosophical reflections on elementary-particle physics, initiated by Dirac's discovery of antimatter, which Heisenberg saw as "perhaps the biggest change of all the big changes in physics of our [twentieth] century."

Keywords Asymptotic freedom · Elementary particles · Feynman diagrams · Hadron jets · Quantum field · Real particles · Symmetry groups · Virtual particles

8.1 Introduction

This chapter addresses the thorny subject of elementary particles from the RWR perspective. As testified to by the persistent title, "What is an elementary particle?" used by, among others, Heisenberg (1989, pp. 71–88) and Weinberg (1996), "an elementary particle" has been and remains a problem to which only fragments of possible solutions could be offered. At a certain point of this history, "a virtual

A. Plotnitsky, *Reality Without Realism*,
https://doi.org/10.1007/978-3-030-84578-0_8

particle" has become part of this problem, with an even greater uncertainty as to what virtual particles might be or whether they exist at all. The existence of, as they came to be called in contrast to virtual particles, "real particles," including elementary ones, is less in doubt, although whether one could know what they are is. Because I shall primarily deal with elementary particles, by particles I shall, unless qualified, refer to elementary particles, and when using the term without a further specification, refer to real particles, while specifying virtual particles. Of course, a "particle," in the first place, is a problem, too, a problem that underlies that of an elementary particle and has been around for much longer than that of elementary particles, indeed much longer than quantum theory.

This chapter does not aim to do more than consider this problem as a problem. In contrast, however, to most approaches to this problem, which are realist in nature, this chapter pursues a nonrealist one, based on the strong RWR view, developed in the preceding chapters in the case of low-energy quantum phenomena and QM (or low-energy, nonrelativistic QFT), and extended here to high-energy regimes and QFT.[1] As stated from the outset of this study, there is no single form of nonrealism, even of the strong RWR-type, any more than a single form of realism. In particular, in contrast to other RWR-type interpretations, including that of Bohr, the one adopted here is characterized by a tripartite stratified view of the physical reality defining quantum phenomena. This stratification acquires a new significance in considering elementary particles, especially in high-energy QFT regimes, in which the particle identity cannot be maintained within a single experiment. I briefly reprise this stratification here.

While still beyond knowledge or conception, quantum objects, including elementary particles, are considered as idealizations defined only at the time of measurement, and not as refering to something existing independently, in contrast to the ultimate constitution of the reality responsible for quantum phenomena, or for quantum objects, both at the time of measurement. This reality is assumed to exist independently, and it will be considered in this chapter in terms of the concept quantum field, which is, it follows, not a quantum object, as it is some or even most definitions of a quantum field. Quantum phenomena are, forming an "idealization of observation,"

---

[1] As a result, my engagement with literature on the subject will be more limited than one might ideally prefer. On the other hand, limiting this engagement allows me to focus more sharply on my key points, shaped by my RWR-type approach to the problem of elementary particles and related issues of QFT. Also, while it offers a concept of quantum field, this chapter is not, strictly speaking, on QFT, another vast subject with extensive literature dealing with it, which, too, will only be engaged in a limited fashion for the same set of reasons. Most sources on QFT cited here, such as (Jaeger 2019, 2021), (Kuhlmann 2020), and (Ruetsche 2011), contain extensive bibliographies. Among the standard technical textbooks are (Peskin and Schroeder 1995), (Weinberg 2005), and (Zee 2010). Most works on QFT in physics and the philosophy of physics, too, adopt and primarily consider realist views. For an extensive philosophical treatment of the concept of a particle, see, for example Falkenburg (2007). The book traverses several territories in common with the present study, including Bohr's views, which are, however, given an interpretation different from the *ones* (plural) considered here, in part because the present book focuses primarily on Bohr's ultimate interpretation, bypassed by Falkenburg, along with changes in Bohr's view. Falkenburg appears to read Bohr primarily in terms of his Como lecture, some of the key aspects of which were, as discussed earlier, abandoned by Bohr.

assumed to be representable, indeed by means of classical physics (Bohr 1987, v. 1, p. 55). It is important to keep in mind the concept of quantum measurement as understood in this study, following Bohr. A quantum measurement does not measure any properties of quantum objects or, in this stratification, of the ultimate constitution of reality responsible for quantum phenomena, or quantum objects, at the time of measurement. A quantum measurement (technically, an observation) creates a quantum phenomenon, defined by what is observed in the measuring instrument used. With this phenomenon in place, it is possible to measure the physical property of this instrument, a property that is an effect, which defines this phenomenon, of the interaction between the ultimate constitutions of the physical reality considered (and, at the time of this interaction, the quantum object considered) and the instrument. This property is measured in the same way as are the properties of *objects* in classical physics or relativity. This is possible because measuring instruments are assumed to be described classically, and, thus, considered as classical objects, in contrast to quantum objects associated with each phenomenon. (Rigorously speaking, one also measures the properties of observed phenomena in classical physics or relativity, but their objects can be identified with observed phenomena for all practical purposes.) Measuring instruments, however, also have a quantum stratum through which they interact with the ultimate constitution of the reality responsible for, at the time of measurement, both quantum objects, or what are so idealized, and quantum phenomena.[2]

The strong RWR view and the corresponding interpretations (which may, again, be different) of QFT imply that the question "What is an elementary particle?" has no answer, at least as things stand now, because this view allows for no specifiable concept of elementary particles or quantum objects in general, apart from in terms of their effects on measuring instruments, which are, as just explained, rigorously specifiable by means of classical physics. As noted from the outset of this study, however, it does not follow that the ultimate nature of reality responsible for quantum phenomena is constant or uniform. While each time unknowable or, in the strong RWR view, unthinkable, this reality is assumed to be each time different in its ultimate character as well as in its manifested effects, making each quantum phenomenon or event, as such an effect, unique in turn, by the QI (quantum individuality) postulate, as well as discrete in relation to any other quantum phenomena, by the QD (quantum discreteness) postulate. The circumstance that one can no longer deal, as in low-energy regimes, with particles of the same type even in the same experiment adds the

[2] This last assumption poses additional complexities in high-energy quantum regimes because, while QFT thus presupposes that the constitution of measuring instruments has quantum aspects, it does not take these aspects into account, as Bohr noted already in 1937 (Bohr 1937, p. 88), following his work with Léon Rosenfeld on measurement in QFT (Bohr and Rosenfeld 1933). It is sometimes argued, via Bohr and Rosenfeld's analysis, that this fact may be responsible for the infinities of QFT and the necessity of renormalization there. Bohr and Rosenfeld's 1950 update of the article, in the wake of the work of Schwinger, Feynman, Tomonaga, and Dyson, briefly addresses renormalization (Bohr and Rosenfeld 1950). I shall put these complexities aside because they do not affect my argument here. See, however, (Plotnitsky 2016, pp. 239–246).

dimension of the irreducibly multiple to that of the irreducibly unthinkable, defining (strong) RWR-type interpretations in all energy regimes.

In the interpretation of quantum phenomena and QFT in high-energy regimes adopted here, the status of real and virtual particles as idealizations (as both are assumed to be in this interpretation) is different. The idealization defined by the concept of real particles, as quantum objects, is required by these interpretations in all energy regimes. The idealization defined by the concept of virtual particles, which are never detected by experiments and thus cannot be considered as quantum objects in the present definition, is allowed but is not required. The circumstances that led to the rise of the concept of virtual particles do play a major role in the present interpretation and in most RWR-type interpretations of high-energy quantum phenomena and QFT. The experimental data found in high-energy quantum regimes and the character of quantum phenomena tell us that certain workings of nature that take over in these regimes make them essentially different from low-energy quantum regimes. This subject is addressed in Sect. 8.2, via the concept of a quantum field. High-energy quantum regimes and QFT also require, in part correlatively, rethinking what "happens" between quantum experiments in these regimes, which lead to the concept of virtual particles, keeping in mind that, as discussed in Chap. 2, the expression "something happens" only applies provisionally in RWR-type interpretations in any energy regimes. This subject is considered in Sect. 8.3.

The first subject is more straightforward, although it reflects a radical change in quantum physics, beginning with Dirac's equation and the resulting discovery of antiparticles. As stated above, in contrast to low-energy quantum regimes (QM or QFT), in high-energy quantum regimes, an investigation of a particular type of elementary particle unavoidably involves not only other particles of *the same type*, say, electrons, but also *other types* of particles, such as, in QED, positrons, photons, or electron–positron pairs, that is, dealing with the corresponding effects, even in the same experiment. By the same token, it becomes meaningless to speak of the same electron detected in any two successive measurements, which lends further support to the tripartite structure of quantum measurement adopted here. While, in the present view, assuming the identity of two successively detected quantum objects is only a statistically permissible idealization in even low-energy quantum regimes, this assumption is no longer possible in high-energy quantum regimes. The identity of particles of the same type is strictly maintained, as it is in the case in QM or low-energy QFT. Although it is also applicable even in QM or in low-energy QFT, the concept of a quantum field introduced here is designed to handle these new types of effects observed in high-energy quantum regimes. As noted, rather than, as is more common, as a quantum object, a quantum field is defined here as a form of an independent RWR-type reality responsible for quantum phenomena and quantum objects, in particular, elementary particles, at the time of measurement.

The second subject, that of virtual particles, is more delicate. The concept of virtual particles is used widely, including in realist interpretations of QFT, although some realist interpretations of QFT also deny the existence of virtual particles.[3] In the RWR view, the situation is as follows. Because of certain observable effects in high-energy (QFT) quantum regimes, the RWR view assumes that there exists a stratum of the ultimate, RWR-type, reality responsible for these effects that *allows* for treating this stratum by using the concept of virtual particles. In other words, while, in RWR-type interpretations, still beyond knowledge or even conception, the nature of the ultimate constitution of reality in high-energy regimes is different from its character in low-energy regimes. This difference is responsible for new types of effects of the interactions between measuring instruments and high-energy quantum objects, such as real elementary particles, which is an idealization assumed, as defined by these interactions, in both low- and high-energy quantum regimes in RWR-type interpretations, in the one adopted here, again, at the time of measurements. By contrast, the idealization based on the concept of virtual particles is *not required*. That based on the concept of real particles is required because, as explained in detail, Chap. 2, in RWR-type interpretations in any quantum regime, the character of quantum phenomena observed in measuring instruments implies the existence of something idealized as, among other quantum objects, real particles.

To briefly recapitulate the situation, while discriminating between the object and the measuring instrument is essential in each case, and while what is actually observed, as a quantum phenomenon, it is uniquely defined, what is considered as the measuring instrument and what is considered as the object under investigation is not uniquely defined. This definition depends on the position of the cut, which allows for considerable latitude, even if not an unlimited one. In particular, the ultimate, RWR-type, reality responsible for quantum phenomena is always on the other side of the cut. It follows, that this reality, which is assumed to exist independently, and the quantum object under investigation, which, or again, something that is so idealized, is assumed to exist only when an experiment is performed, or the corresponding idealizations, are in general different, even though some quantum objects, such as and in particular, real elementary particles, are always on the other side of the cut. Nevertheless, real elementary particles are still quantum objects and, as such, in the present view, refer, as concepts or idealizations, to something that is ascribed existence only at the time of experiment and are defined (as electrons, photons, quarks, and so forth) in terms of effects observed in measuring instruments. In low-energy regimes and QM, it is, again, permissible, at least as statistical idealization, to assume that the same quantum object, including an elementary particle, can be detected in two or more successive measurements. This assumption no longer holds in high-energy (QFT) regimes. The situation in low-energy regimes (QM or low-energy QFT) becomes, however, better understood as well by using the concept of a quantum field, as a limiting case.

[3] See Jaeger (2019, 2021) for a realist perspective on the subject and helpful references. Both articles also offer effective critiques of several arguments, primarily realist in nature, against the existence of virtual particles, arguments, thus, very different from the present, RWR-type, argument. See also Fox (2008), promptly cited by Jaeger's article (Jaeger 2019, Ref. 1), for the debates on the subject, again, however, not addressing the type of argument offered here.

The case is different in the case of virtual particles. The observed (high-energy) effects of those workings of the ultimate, RWR-type, reality that are commonly accounted for by using the concept of virtual particles are only associated with measurements performed on real particles, assumed to be affected, prior to these measurements, by virtual particles. On the other hand, neither these workings themselves (by definition, in RWR-type interpretations) nor anything handled or idealized by the concept of virtual particles (in any interpretation) is ever observed. Hence, again, in the present definition of a quantum object, a virtual particle can never be viewed as a quantum object.[4] Virtual particles are idealizations that only refer to something that happens between experiments and that is never registered in experiments in the way real particles are, keeping in mind the provisional nature of the expression "something happens" in the RWR view. This something has effects on the outcomes of experiments, effects arising from the interaction between this something and the workings of the ultimate, RWR-type, reality that give rise to real elementary particles, which come into existence in quantum measurements. While, however, assuming the existence of this something is necessary in view of these effects in, arguably, any interpretation, this assumption does not require the concept of virtual particles in order to account for these effects. This something could be and has been accounted for otherwise, in contrast to real elementary particles as quantum objects defined by measurements.

It is true that using the formalism of QFT (or QM) does not strictly require the assumption of real particles or quantum objects in general either. For, one can by using this formalism predict, probabilistically or statistically, what will happen in future experiments on the basis of the data obtained in previously performed experiments, while merely assuming some form of reality as responsible for quantum phenomena, without assuming that this reality is that of particles.[5] The present interpretation

[4] One might argue that we do not always register real particles either, specifically in the case quarks and gluons, which only exist as confined inside nuclei and, thus, cannot be registered as independent quantum objects. Their reality, or again, the reality thus idealized, could, however, still be registered inside nuclei, and it has been in famous LEP (Large Electron–Positron Collider) experiments that established the existence of this reality and justified its idealization in terms of quarks and gluons, via confinement and asymptotic freedom. It is true that understanding and even performing these experiments relied on the concept of virtual quarks and gluons. As will be seen, however, this does not undermine the present argument juxtaposing real and virtual particles, because virtual quarks and gluons have never been registered in any experiments, see also Falkenburg (2000).

[5] See, for example, Ulfbeck and [Aage] Bohr (2001) and [Aage] Bohr et al. (2004). These articles target the use of the idea of particles in quantum theory, including (Niels) Bohr, mistakenly in my view. They disregard Bohr's understanding, already in the Como lecture, of quantum objects, including elementary particles, as "abstractions," with "their properties being definable and observable only through their interactions with other systems [measuring instruments]" (Bohr 1987, v. 1, p. 57). By contrast, Bohr's concept of "atomicity," essentially equivalent to his concept of phenomena, reflects the fact that phenomena have features, specifically, discreteness, individuality, and indivisibility, generally associated with atomic properties, but now no longer applied to quantum objects, or the reality thus idealized, to which no properties of any kind are assigned. Phenomena are not atomic, particle-like, in the physical sense because they consist of a very large number of atoms and hence of elementary particles. It is true that, in contrast to the present view, Bohr adopted the concept of a quantum object, including as an "elementary particle," as an idealization applicable

makes this type of assumption, while placing this reality beyond any representation or conception. However, it also assumes the concepts of particles and elementary particles, still as RWR-type entities, as applicable at the time of measurement. While QM and QFT qua theories are falsifiable by the data observed in measuring instruments, the assumption of the existence of real particles is not falsifiable, any more than that of virtual particles. An interpretation of quantum phenomena is, however, a different matter. Importantly, RWR-type interpretations do not assume a nonfalsifiable ontology as a representation or even conception of the ultimate character of physical reality responsible for quantum phenomena. They only assume a (classical) ontology of quantum phenomena, while quantum objects, including real elementary particles, still as RWR-type entities, are only assumed to be an idealization applicable at the time of measurement.

While, however, the change in the ultimate character of reality responsible for quantum phenomena in high-energy quantum regimes does not, in RWR-type interpretations, require the idealization defined by the concept of virtual particles, this idealization may be provisionally used in these interpretations because of its practical effectiveness. I shall adopt this provisional use, while strictly distinguishing this *provisional* idealization, never corresponding to quantum objects, from that defined, as an *actual* idealization, by the concept of real elementary particles, as quantum objects, at the time of measurement. As explained in Sect. 8.3, in the present view, the role of virtual particles is that of a heuristic device, akin to (which is not to say, the same as) Feynman diagrams, which helps calculations in QFT.

8.2 Elementary Particles and Quantum Fields

Low-energy quantum regimes permit and most interpretations of quantum phenomena and QM (or QFT in these regimes) adopt a conception of (real) elementary particles, as quantum objects. The same conception is also applicable in, although generally not sufficient for, high-energy regimes, beginning with the circumstance that elementary particles of the same type cannot be distinguished from each other, while these types themselves are rigorously distinguishable. In the present interpretation, a particle, again, is only defined as a quantum object by a measurement. Hence, in this interpretation, one cannot, rigorously, speak of the same quantum object, such as an electron, in two successive measurements, including those that may define a given experiment. As discussed in Chap. 2, however, the assumption that it is the

independently of measurement. Even so, however, he would not ascribe to any quantum object any physical properties, including those associated with any concept of particle, any more than that of wave. The "elementary" and "particle" aspects of "elementary particles" only apply to certain effects of the interaction between quantum objects (or in the present view, the ultimate constitution of the reality responsible for quantum phenomena) and measuring instruments. See also note 4 above.

same is a statistically permissible idealization in low energy (QM) regimes. A statistical dimension of the situation is present even if one adopts a more conventional view, in which quantum objects have an independent existence. One cannot be certain that one encounters the same electron in an experiment designed to detect it after it was assumed to be emitted from a source even in low-energy (QM or QFT) regimes, although the probability that it would be a different electron is generally low. In high-energy QFT regimes, speaking of the "same" electron detected in the course of a given experiment involving several measurements loses its meaning altogether.

For the moment, in any quantum regime, two electrons could be distinguished by changeable properties associated with them, such as their positions in space or time, momentums, energy, or the directions of spins, but, in the RWR view, only as properties manifested in measuring instruments and only at the time of measurement. Such properties are subject to the uncertainty relations and complementarity. It is possible to locate (and in the present view, establish) by measuring two different electrons, as quantum objects, in separate regions in space. It is not possible to distinguish them from each other on the basis of their mass, charge or spin. These quantities are not subject to the uncertainty relations or complementarity. As Weyl noted long ago, "the possibility that one of the identical twins Mike and Ike is in the quantum state E1 and the other in the quantum state E2 does not include two differentiable cases which are permuted on permuting Mike and Ike; it is impossible for either of these individuals to retain his identity so that one of them will always be able to say 'I'm Mike' and the other 'I'm Ike.' Even in principle one cannot demand an alibi of an electron!" (Weyl 1931, p. 241).[6] In RWR-type interpretations, properties defining electrons or other elementary particles within each type could only be *associated* with quantum objects (even if one assumes their independent existence between measurements, as opposed to, as here, only at the time of measurement) by means of the corresponding effects observed in measuring instruments and are not attributable to these objects themselves. It is possible, however, to maintain both the indistinguishability of particles of the same type and the strict distinguishability of the types themselves in RWR-type interpretations because both features can be consistently defined by the corresponding sets of effects manifested in measuring instruments.

This view is, thus, in accord with the assumption, defining RWR-type interpretations that the *character* of elementary particles and their behavior, or of the reality thus idealized, is beyond representation or even conception, just as is the ultimate, RWR-type, character of the reality itself responsible for quantum phenomena. In the present interpretation, this reality is, again, assumed to exist independently, while elementary particles are assumed to idealize the (RWR-type) reality that exists only at the time of measurement. An elementary particle of a given type, say, again, an electron, is specified by a discrete set of possible phenomena or events (the same for all electrons), observable in measuring instruments in the experiments associated with particles of this type. An elementary particle can thus only be idealized as part of

[6] The statement is cited in French (2014), which is a useful source on the subject, primarily, again, considered from a realist perspective.

a composite system, consisting of the RWR-type reality it idealizes and a measuring instrument, which system has a registered effect upon the observable, classically describable, part of this instrument. The elementary character of a particle is defined by the fact that there is no experiment that allows one to associate the corresponding effects on measuring instruments with more elementary individual quantum objects. Once such an experiment becomes conceivable or performed the status of a quantum object as an elementary particle could be challenged or disproven, as it happened when hadrons and mesons were discovered to be composed of quarks and gluons. If so, in the present view, this composite nature will manifest itself in a new set of effects observed in each corresponding expriment. It is possible to treat hadrons and mesons as elementary particles on an appropriately larger scale, where their interior constitution can be disregarded, which makes the concept of an elementary particle scale dependent. This dependence would be reflected mathematically in the corresponding symmetry group and its irreducible Hilbert-space representation and physically, in the set of effects observed in experiments.

It should be reiterated that the present concept of an elementary particle, defined in terms of such effects, does not imply that "elementary particles," even if never shown to be composite, are, or idealize, fundamental elementary constituents, "building blocks," of nature. This assumption is impossible in RWR-type interpretations, as is any assumption concerning this constitution, as is applying the concepts of "elementary" or "constituents." Nor, by the same token, is it possible to apply to elementary particles any specifiable concept of a particle, any more than any other specifiable concept, such as wave or field, although, as will be explained presently, the concept of a quantum field could be defined otherwise, as a mode of RWR-type independent reality (rather than a quantum object) beyond the reach of all specifiable concepts.

In their recent book, Laurent Baulieu, John Iliopoulos, and Roland Seneor open the chapter, entitled "What is an Elementary Particle?," which is, as noted, a persistent title, used by Heisenberg (1989, pp. 71–87) and Weinberg (1996), with the following revealing remarks: "A chapter on this subject should normally begin with a definition of what is an elementary particle. The trouble is that we have no such definition." They offer a sensible alternative: "We can only give a table [of elementary particles] whose entries evolve with time and represent at any given moment, the current state of our knowledge (or ignorance) of the structure of matter" (Baulieu et al. 2017, pp. 625–626). Given the experimental data in question and the character of QM or especially QFT, it indeed appears to be difficult to ascertain that anything in the ultimate constitution of the reality responsible for quantum phenomena conforms to any physical concept of elementary particle that we can form. If one adopts an RWR-type interpretation, nothing does, at least as things stand now. All that is assumed in such interpretations is that something, a form of reality, exists in nature that has certain effects observed in measuring instruments, which represent our *knowledge* and which compel us, even in low-energy (QM) regimes, to introduce, as an idealization of this reality, the concept of elementary particles (in the present view, defined only at the time of and by measurement) as specifiable in terms of these effects but not in terms of independent attributes of or entities belonging to this reality. As just explained, invariant characteristics associated with elementary particles, such as mass, charge, or spin, could also be understood in terms of such effects. An entry into the table of elementary particles and the knowledge or ignorance associated with it would, then, only refer to effects observed on measuring instruments, while nothing can be known or, in the strong RWR view, even thought about the nature of such entities, or, in the present interpretation, about the ultimate, RWR-type, reality that gives rise to them through its interaction with measuring instruments.

While most QFT conceptions of an elementary particle are transferred from QM to high-energy quantum regimes, they are insufficient in these regimes and need to be adjusted or supplemented by additional concepts, most commonly that of a quantum field. The present approach follows this pattern by defining the concept of quantum field in RWR terms. First, however, I shall explain why the concept of an elementary particle operative in QM is insufficient in high-energy regimes. This insufficiency arises in view of the following situation, not found in QM (or low-energy QFT, even though the latter introduces new features into quantum phenomena), to which the mathematical architecture of QFT responds.[7] In fact, with Dirac's equation, it was this architecture, discovered first, that led to the discovery of this situation. This discovery is a manifestation of both the general significance of mathematics in quantum theory, stated at the outset of this study and discussed throughout it, as an invention of an abstract mathematical machinery not tied, as it is in classical physics or relativity, to a physical description of reality, and specifically, Dirac's particular way of thinking, proceeding from mathematics to physics, more pointedly so than that of Heisenberg, or especially (given his attachment to realist thinking) that of Schrödinger. Still, as discussed in Chap. 5, even Schrödinger, driven by what his theory had to predict, was manipulating mathematics by playing with its abstract structure, akin to the way Heisenberg and Dirac did.

Speaking for the moment in classical-like terms, suppose that one arranges for an emission of an electron, at a given high energy, from a source and then performs a measurement at a certain distance from that source, say, by placing a photographic plate there. The probability or, if we repeat the experiment with the same initial conditions (defined by the state of the emitting device), the statistics of the outcomes would be properly predicted by QED. But what will be the outcome? The answer is not what our classical or even quantum–mechanical intuition would expect. This answer was a revolutionary discovery made by Dirac through his equation, although it took a few years to realize what this equation implied in terms of physics.

Let us consider first what happens if one deals with a classical object analogous to an electron and then if one considers a nonrelativistic QM electron in the same type of arrangement. I speak of a classical object analogous to an electron because the "game of small marbles" for electrons was finished well before QM. An electron, say, a Lorentz electron, of a small finite radius, would be torn apart by the force of its negative electricity. This led to treating the electron mathematically as a dimensionless point, without giving it any physical structure, while still assigning it

[7] Low-energy QFT is essential for explaining some quantum phenomena, such as the non-zero energy of the vacuum, not explicable by QM.

measurable physical quantities, permanent (such as mass, charge, or spin), or variables (such as position, time, momentum, or energy). A point electron in quantum theory is, however, a mathematical idealization, different from a point-like idealization in classical mechanics, where the body thus idealized could still be assumed to have spatial dimensions physically. One can take as an example of the classical situation a small ball that hits a metal plate. The place of the collision could be predicted (ideally) exactly by classical mechanics, and we can repeat the experiment with the same outcome on an identical or even the same object. Regardless of where we place the plate, we always find the same object. (It is assumed that the situation is shielded from outside interferences.)

By contrast, if one considers an electron in the QM regime, it is, first of all, impossible, because of the uncertainty relations, to predict the place of collision exactly or with the degree (ideally unlimited) of approximation possible in classical physics. An (ideally) exact prediction of the position (or other variables) of a quantum object is also possible, specifically in EPR-type experiments by means of a measurement performed on the other particle of the (EPR) pair considered. As discussed earlier, however, even a prediction with probability one is still a probabilistic prediction and not a guarantee of the reality of what is predicted, because one can always perform a complementary measurement, thus disabling any possibility of verifying such a prediction and hence assigning the corresponding quantity. In addition, a single emitted electron could, in principle, be found anywhere in a given area or not found at all. Nor can an emission of an electron be guaranteed. There is a small but nonzero probability that such a collision will not be observed or that the observed trace is not that of the emitted electron. Finally, assuming that one observes the same electron in two successive measurements is still an idealization in the present interpretation, given that it defines any quantum object as an idealization applicable only at the time of measurement. This idealization is, however, permissible in low-energy regimes. QM gives correct probabilities or statistics for all low-energy quantum events thus far.

Once one moves to high-energy quantum phenomena, beginning with those governed by QED, the situation is, again, different, even radically different. In a subsequent measurement, one can find in the corresponding region, not only an electron (or nothing), as in QM regimes, but also other particles: a positron, a photon, an electron–positron pair. That is, in RWR-type interpretations, one can register the events or phenomena (observed in measuring instruments) that we associate with such entities. QED predicts which among such events can occur, and with what probability or statistics, and just as QM, QED, in RWR-type interpretations, does so without representing or, in the strong RWR view, allowing one to conceive of how these events come about. The corresponding Hilbert-space machinery becomes more complex, in the case of Dirac's equation making the wave function a four-component Hilbert-space vector, as opposed to a one-component or, if one considers spin, two-component Hilbert-space vector, as in quantum mechanics, keeping in mind that each component is infinite dimensional. These four components represent the fact that Dirac's equation

$$\left(\beta mc^2 + \sum_{k=1}^{3} \alpha_k p_k c\right)\psi(x,t) = i\hbar\frac{\partial\psi(x,t)}{\partial t}$$

$$\alpha_i^2 = \beta^2 = I_4$$

(I_4 is the identity matrix).

$$\alpha_i\beta + \beta\alpha_i = 0$$

$$\alpha_i\alpha_j + \alpha_j\alpha_i = 0$$

is an equation for both the (free) electron and the (free) positron, including their spins, and they can transform into each other or other particles, such as photons, in the corresponding high-energy processes, again, transformations, in the RWR view, only manifested in measuring instruments. By the same token, one can no longer speak of the same electron, positron, and so forth as detected in two successive measurements in low energy quantum regimes. In the current standard versions of QFT, the wave functions are commonly replaced by operators (the procedure sometimes known, for historical reasons, as "second quantization"): to every point x a Hilbert-space operator acting on this space is associated, rather than a state-vector as in QM. As Kuhlman notes: "both in QM and QFT states *and* observables are equally important. However, to some extent their roles are switched. While states in QM can have a concrete spatio-temporal meaning in terms of probabilities for position measurements, in QFT states are abstract entities and it is the quantum field operators that seem to allow for a spatio-temporal interpretation" (Kuhlman 2020). As he qualifies, however, "since 'quantum fields' are operator valued it is not clear in which sense QFT should be describing physical fields, i.e., as ascribing physical properties to points in space. In order to get determinate physical properties, or even just probabilities, one needs a quantum state. However, since quantum states as such are not spatio-temporally defined, it is questionable whether field values calculated with their help can still be viewed as local properties" (Kuhlman 2020). As will be seen, while the present RWR-type concept of a quantum field is physical, all calculated values are only probabilities and all local properties are only those of the observable parts of measuring instruments, just as in QM.

Once one moves to still higher energies governed by QFT, the panoply of possible outcomes becomes much greater. Correspondingly, the Hilbert spaces and operator algebras involved have still more complex structures, linked to the appropriate Lie groups and their representations, defining (when these representations are irreducible) different elementary particles. In the case of QED, we only have electrons, positrons, and photons, single or paired; in QFT, depending on how high the energy is, one can literally find any known and possibly as yet unknown elementary particle or combination. It is as if instead of identifiable moving objects and motions of the type studied in classical physics, we encounter a continuous emergence and disappearance, creation and annihilation, of particles, further complicated by the role of virtual particles, or again, something in nature which compels some to introduce the latter concept. This is still a classical-like and thus metaphoric picture, which is ultimately inapplicable. But we have no other pictures. If one wants to convey anything that "happens" between experiments my means of a picture or phenomenal concepts (such as that signified by the word "happen"), rather than only use mathematics to predict the outcome of experiments, classical-like pictures are our only recourse. I shall consider this aspect of the situation in the context of QFT in closing this chapter. In the present interpretation, again, only the ultimate, RWR-type, reality is assumed to exist independently, while each real particle only comes into existence in a measurement. Although, like anything quantum, these transformations can only be handled probabilistically or statistically, they also have a complex ordering to them. In particular, in addition to various correlational patters akin to those found in low-energy (QM) regimes, they obey various symmetry principles, especially local symmetries. The latter have been central to QFT, not the least in leading to discoveries of new particles, such as quarks and gluons inside the nucleus, and then various types of them, eventually establishing the standard model of particle physics. Thus, QED is an abelian gauge theory with the symmetry group U(1) and has one gauge field, with the photon being the gauge boson. The standard model is a non-abelian gauge theory governed by the tensor product of three symmetry groups U(1)⊗ SU(2)⊗ SU(3) and broken symmetries, and it has 12 gauge bosons: the photon, three weak bosons, and eight gluons.

The role of symmetry in particle physics merits a brief detour. There are extensive technical treatments of symmetries in QFT (e.g., Duncan 2012, pp. 414–569). The mathematics is extraordinary and is prohibitive for nonspecialists.[8] It is difficult to give a rigorous philosophical treatment of the role of symmetry in QM and QFT or relativity, or even in classical physics, without an engagement with high-level mathematics, such as, in quantum theory, that of complex Hilbert-space representations of both finite and infinite dimensions of Lie groups. There are cases in which it is possible to circumvent these difficulties and to convey the essential points. Consider Noether's celebrated theorems, which states that every differentiable symmetry of the action of a physical system has a corresponding conservation law. There are, however, subtle mathematical features and nuances involved, beginning with fact that these theorems only apply to differential symmetries over physical space. The action of a physical system is an integral over time of a Lagrangian, which allows one to determine the system behavior by using the principle of least action (the quantity defined as the integral of the momentum over a given distance traveled by a body). The theorems are a great application of the principle, arguably, stated first by Pierre Louis Maupertuis in 1774 and mathematically developed by Leonhard Euler around the same time, and then by Joseph Louis Lagrange and Sir William

[8] Wilczek's book, discussed below, offers a helpful nontechnical treatment of symmetry in QCD (Wilczek 2009, pp. 57–69). See Kuhlman (2020) for a philosophical introduction to the subject which (rightly) emphasizes not only a crucial role of symmetries but also the difficulties just stated.

Rowan Hamilton. Leibniz had related ideas earlier. Philosophically, the idea that nature follows the most efficient way possible can be traced to much earlier. The proof of the theorems, while elegant, is not nearly as prohibitive as some of the mathematics of QFT (which uses Noether's theorems, as considered, for example, in Duncan's discussion of symmetries in QFT just cited). Still, one needs to be at least an advance mathematics or physics undergraduate major to follow it. According to Yvette Kosmann-Schwarzbach's historically oriented account of the theorems, technically, Noether's first theorem, from which the theorems related to different conservation laws follow:

> Noether's first theorem, a generalization of several conservation theorems that were already known in mechanics .... She considers a multiple integral,
>
> $$I = \int \cdots \int f(x, u, \frac{\partial y}{\partial x}, \frac{\partial^2 y}{\partial x^2}, \ldots)dx,$$
>
> of a higher order Lagrangian f that is a function of n independent variables, $x_1, \ldots, x_\lambda, \ldots, x_n$, and of μ dependent variables, $u_1, \ldots, u_i, \ldots, u_\mu$, as well as of their derivatives up to a fixed but arbitrary order, κ. She then considers a variation of u, $\delta u = (\delta u_i)$, and derives identity (3),
>
> $$\sum_{i=1}^{\mu} \psi_i \delta u_i = \delta f + \mathrm{Div} A,$$
>
> where the ψ_i are the Lagrangian expressions, which is to say the components of the variational derivative (Euler–Lagrange derivative) of f, and where the components A_λ of A are linear in the variation δu and in its derivatives. The opposite of the quantity A is now called the Legendre transform of the Lagrangian f. Here Div is the ordinary divergence, $DivA = \sum_{\lambda=1}^{n} \frac{\partial A_\lambda}{\partial x_\lambda}$, of $A = (A_1, \ldots, A_n)$ considered as a vector in n-dimensional space, and δf is the variation of f corresponding to the variation δu of u, while the variation of x is assumed to vanish. Identity (3) is obtained by an integration of parts. In the case where $n = 1$, the case of a simple integral, Noether gives an expression for A for an arbitrary μ, first for $\kappa = 1$, which yields what she calls Heun's "central Lagrangian equation," then for an arbitrary κ, and then she states her theorem:
>
> **I.** *If the integral I is invariant under a [group] $\mathfrak{G}_\rho$, then there are ρ linearly independent combinations among the Lagrangian expressions which become divergences—and conversely, this implies the invariance of I under a [group] $\mathfrak{G}_\rho$. The theorem remains valid in the limiting case of an infinite number of parameters.*
>
> Noether explains that "in the one-dimensional case," that is, when $n = 1$, one obtains first integrals, while, "in higher dimensions," i.e., when $n > 1$, "one obtains the divergence equations which, recently, have often been referred to as conservation laws." By the "limiting case" included in the statement of Theorem I is meant the case in which the elements of the group depend on an infinite but denumerable set of parameters, as opposed to the case dealt with in her Theorem II. (Kosmann-Schwarzbach 2011, p. 57)

One could write a book on the history, or on the philosophy of the concept of Lagrangian from analytical mechanics, which is part of the genealogy of Noether's theorem, to relativity (also part of this genealogy) to QM and QFT and beyond (string theory, for example). I shall comment on the Lagrangian of QCD in closing this chapter. For the moment, central as this mathematics, including, the Lagrangian considered, may be for Noether's proof and, I would surmise, her thinking, the physical and philosophical significance of the theorems in connecting symmetries, and conservation laws could be understood apart from their mathematical content. Thus, it follows from them that, for systems with suitable Lagrangian (and most classical, relativistic, and, with some qualifications given below, quantum systems are), the symmetry under continuous translations in time implies the conservation of energy, the symmetry under continuous translations in space implies the conservation of linear momentum, and the symmetry under continuous rotations implies the conservation of the angular momentum. There are further nuances concerning the relationships between symmetry and conservation principles, for example, the fact that not all physical systems allow for a Lagrangian formulation (e.g., Wigner 1954). These nuances can, however, be put aside, because they do not affect my point concerning the difference between symmetries in classical (or relativistic) and quantum physics.

The description of Noether's theorems just given is essentially linked to classical phenomenal intuitions, some of which are refined into features of these theorems, just as they are into features of classical physics and (with limitations) relativity. The situation is entirely different in the case of the mathematical symmetries of QM and QFT, from the mathematical unitary transformations, to the invariance under the Heisenberg group in QM, to the symmetry groups and their irreducible representations (mathematically) defining "elementary particles." These symmetries and related conservation laws are not directly manifested in the physically observed quantum phenomena, like those related to classical and relativistic conservations laws covered by Noether's theorems, although they are related to quantum phenomena, either probabilistically or otherwise. For example, all systems with a given spin (a discrete variable) are handled by Hilbert spaces with the same finite dimension (over $\mathbb{C}$), while Hilbert spaces for continuous variables will be infinite-dimensional, possibly uncountably infinite dimensional in QFT. In RWR-type interpretations, however, these relations are between the mathematics of the theory, which does not represent the ultimate constitution of the reality responsible for quantum phenomena, and these phenomena. They are not defined, as they would be in classical physics and relativity, by mathematically idealized continuous, indeed differential, representations of classical causal processes connecting phenomena, which could, for all practical purposes be identified with the corresponding objects. This difference, acutely realized by Heisenberg, was in part responsible for his later view, discussed in Chap. 2, which allowed for the possibility of representing the ultimate constitution of nature, including elementary particles, in mathematical terms, in particular, symmetries, apart from physical concepts, at least as we understand such concepts in classical physics or relativity. The dynamical conservation laws and Noether's theorems, in their standard forms, still apply in QM or QFT in RWR-type interpretations. They are, however, only manifested in measuring instruments. Indeed, they expressly apply to quantum phenomena observed in measuring instruments (and thus can be expressed in their physical content in classical terms), but not to the reality that connect quantum phenomena. The situation is different from many mathematical symmetries of QM or QFT and particle physics, or analog of Noether's theorem there, such as the Ward-Takahashi identity in QFT, which implies conservation laws such as the conservation of electric changes relative to a change of the phase factor in the complex field of a particle and the associated gauge of the electric and vector potential. These symmetries could, in the RWR view, only be related to quantum phenomena via the probabilities or statistics of the outcomes of quantum experiments. Thus, the Ward-Takahashi identity imposes symmetries at the level of quantum (probability) amplitudes. These circumstances, again, make the role of symmetries in QM and QFT difficult to explain apart from the mathematics of these theories.

The concepts (there have been quite a few) of a relativistic quantum field respond to the situation in QFT here outlined. Most concepts of nonrelativistic quantum fields can be seen as limit cases of relativistic ones. Accordingly, by quantum fields I shall, unless qualified, refer to relativistic quantum fields. These concepts were initially developed as forms of quantization of the electromagnetic field, again, necessary even in low energy quantum regimes. The character and even the very possibility of such concepts, especially as physical concepts, is a subject of seemingly interminable debates, just as and often correlatively is the concept of elementary particles. While there is a strong general sense concerning the mathematics involved (although the range of specific mathematical tools offers one several choices) and while there is a large consensus, although not a uniform one, that viable physical concepts of a quantum field are necessary, most of the proposals concerning such concepts are realist.[9] By contrast, I suggest a nonrealist *physical* concept of quantum field defined by the strong RWR view, which is (interpretively) consistent with the mathematics of QFT and most currently available *mathematical* concepts of quantum field, such as those, those based on the Lagrangian formulation and canonical commutation or anticommutation relations for fields, for, respectively, bosonic and fermionic fields, analogous to those of QM, say, for bosonic field, ϕ and π:

$$\left[\phi(x,t),\pi(\mathbf{y},t)\right]=i\delta^3(\mathbf{x}-\mathbf{y})$$

$$\left[\phi(x,t),\phi(\mathbf{y},t)\right]=\left[\pi(x,t),\pi(\mathbf{y},t)\right]=0$$

(Some of these mathematical concepts are associated or combined with physical ones.)

As understood here, a quantum field is *not a quantum object* but a particular *mode* of the RWR-type reality, which, as any such mode in the present interpretation, is

[9] This assessment is confirmed by Kuhlmann (2020), keeping in mind that the term "nonrealist" is sometimes used for interpretations that would qualify as realist in the present definition. Among alternative proposals (which leave the mathematical formalism of QFT intact) is Wilczek's concept of "grid," derived from the idea of the ether, which, Wilczek argues, has underappreciated features he used in defining his concept (Wilczek 2009, pp. 73–75).

assumed to exist independently and is manifested only by its effects on measuring instruments, via quantum objects, such as elementary particles. A quantum field is independent of measurement, while quantum objects are, as concepts or idealizations, always defined by measurements. These effects are more multiple than those observed, via the corresponding quantum objects, in low energy regimes. This multiplicity is defined by the fact that these effects correspond to elementary particles, to which a quantum field gives rise and which can be of various types even in a single experiment, consisting of two or more successive measurements, with the first one performed on a given particle. The initial quantum object could also be a set of elementary particles of the same or different types, with a different such set, possibly consisting of entirely different types of particles, appearing in each new measurement. As a mode of the RWR-type reality assumed to exist independently, a quantum field is responsible for transforming effects associated with elementary particles at the time of measurement. These effects may be either invariant (as concerns a given particle type), such as those associated with mass, charge, or spin, or variable, such as those associated with position, momentum, or energy. As concerns this association, always via real particles, as quantum objects, there is no difference from low energy regimes; the difference is in what kind of effects are observed. These effects have a kind of multiplicity in high-energy regimes, not found in the case of effects observed in low-energy regimes. The multiplicities of types of elementary particles become progressively greater in higher energy regimes. Hence, the concept of a quantum field just defined brings together the irreducibly unthinkable, discovered by QM, and the irreducibly multiple, discovered by QFT. It may be useful to briefly comment, by way of a contrast, on the concept of a classical field.

A classical field is represented, in a realist way, by a *continuous* (technically, differential) manifold with a set of scalar (a scalar field), vector (a vector field), or tensor (a tensor field) variables associated with each point and the rules for transforming these variables by means of differential functions, from point to point of this manifold. One can also define it as a fiber bundle over a manifold with a connection. The concept of a fiber bundle is used in QFT, where it is associated with local gauge symmetry, in the RWR view, without representing, any more than any part of the mathematical formalism of QFT, any quantum physical process but only being part of the probabilistically or statistically predictive machinery of QFT. In classical physics or relativity, the variables in question map measurable quantities associated with the field, thus providing a field ontology, which also allows for (ideally) exact predictions concerning future events associated with this field via measurable field quantities. In quantum physics, in RWR-type interpretations, this type of ontology is impossible. One deals with a discrete manifold of phenomena and discrete sets of quantities associated with each phenomenon, without assuming any continuous process that would connect them. As does QM, QFT (in any regime) relates, in terms of probabilistic or statistical predictions, the continuous, technically, differential, mathematics to the discontinuous configurations of the observed phenomena and data.

As discussed in Chap. 2, in considering quantum events, say, again, two successive measurements, which register different outcomes, it is humanly natural, to assume that something "happened" or that there was a "change" in the physical reality responsible for these events between them. However, in any interpretation, one cannot give this happening a determined location in space and time, and in RWR-type interpretations, there is nothing we can say or even think about the character of this change, including as a "happening" or "change," apart from its effects, which are more multiple in high-energy regimes than in low-energy regimes. Nor can one assume (again, in any interpretation), as one can, ideally, in low-energy regimes, that we observe the same quantum objects in two successive measurements. For example, it is no longer possible to think of a single electron in the hydrogen atom, as the same electron detected (and in RWR-type interpretations, defined) by different measurements. Each measurement is assumed to detect a different electron, if one makes a measurement between two measurements each of which detects an electron, this measurement can detect a positron or a photon, or an electron–positron pair. One could also speak of quantum fields, including in the sense defined here, in QM (or in low-energy QFT), but in a reduced form that preserves the particle identities, as representatives of a given particle type—each photon always remains the "same" photon (or disappears), each electron the "same" electron (or disappears), and so forth, in the present view, again, strictly as a statistically permissible idealization because each is only defined by a measurement. In high-energy regimes, particles transform into one another, within and beyond a given particle type. An electron could reappear in this process, that is, be detected by a measurement, after a positron appeared in it after the previous appearance of an electron, but these two electrons are not the same.

In this understanding of the concept, speaking, as is common, of the quantum field of a particle, say, again, an electron, entails new complexities. Mathematically, the formalism of, say, QED, allows one to make predictions concerning the electron, which, mathematically, invited one to speak of the electron as a quantum field. Physically, in the present understanding of a quantum field, this only means that the RWR-type reality defining the quantum field considered in a given experiment has strata that enables the corresponding measurements detecting electrons or, again, what is so idealized by us. It is not possible, however, to separate these strata from those similarly associated with the possibility of detecting a positron or a photon in the same experiment (in the sense of being defined by the same initial measurement or preparation), because neither of these strata as such is observed in measurement. Only electrons, positrons, or photons are, as quantum objects, and only in terms of the corresponding effects (defining each particle or set of particles as the corresponding idealization) observed in measuring instruments. On the other hand, it is possible, to specify quantum fields as associated with fundamental forces and the corresponding types of particles, field bosons, electromagnetic (photons), weak W^+, W^-, and Z, or strong (gluons), or gravitation (gravitons), in the latter case, hypothetically, given that there is no theory of gravity as a form of QFT.

These circumstances reflect the fact that the present concept of a quantum field is a physical, rather than a mathematical, concept, which defines a quantum field as part of the independent reality ultimately responsible for quantum phenomena and not as a quantum object, always dependent on an experiment. This concept can, as noted, be associated with most current standard versions of a mathematical concept of a quantum field, defined in terms of a predictive Hilbert space formalism with a particular vector and operator structure, enabling the proper probabilistic predictions of the QFT phenomena concerned. The operators enabling one to predict the probabilities for the "annihilation" of some particles and "creation" of others, that is, for the corresponding measurable quantities observed in measuring instruments, are called annihilation and creation operators or also lowering and raising operators, commonly designated as $\hat{a}$ and $\hat{a}^\dagger$, each lowering or increasing the number of particles in a given state by one. In RWR-type interpretations, these operators do not represent any physical reality: they only enable one to calculate the probabilities or statistics of the outcomes of experiments, just as the wave functions do in quantum mechanics. Both, in Schrödinger's language, provide expectation-catalogs for the outcomes of possible experiments. Those provided by QFT give probabilities or statistics of the appearance of quantities associated with other types of particles even in experiments initially defined by a particle of a given type. In QFT regimes, it is, again, meaningless to ever speak of a single electron even in the hydrogen atom.[10] QFT correctly predicts these effects, probabilistically or statistically, which is, however, fully in accord with quantum experiments in all regimes. As noted, QED is the best confirmed physical theory ever.[11]

The concept of a quantum field here defined does not introduce any new mathematics. It is not designed to do so. This concept is part of the (strong RWR-type) interpretation of quantum phenomena and quantum theory proposed by this study, and thus of how mathematics works in quantum theory, especially in the case of QFT as different from QM, in particular, given the concept of quantum measurement and the tripartite scheme of physical reality in this interpretation. This scheme—consisting of (1) the ultimate constitution of physical reality, as an RWR-type reality, responsible for quantum phenomena; (2) quantum objects, including elementary particles, defined by measurement as RWR-type concepts, and (3) quantum phenomena, also defined by measurement but represented classically—is redefined by configuring (1) in terms of quantum fields. Indeed, while especially important to this understanding in high-energy quantum regimes and QFT, the concept of a quantum field proposed

[10] The description just outlined is in accord with, and provides an RWR-type interpretation of, the view, held even in realist interpretations, that the application of creation operators to quantum states (in the mathematical sense of vectors in a Hilbert space), does not represent a physical process of particle creation out of nothing, which would entail a violation of conservation laws. There are various ways of handling this situation in realist interpretations, for example, in terms of collisions (e.g., Schwinger 1970, p. 73). In the present view, this procedure and thus creation and annihilation operators are just part of the mathematical machinery that allows us to predict the transition probability between two quantum events, which would be associated with different particles, in the present interpretation, each defined by the corresponding measurement. Any measurement performed before the second measurement could reveal a different particle or set of particles than that found either in the first or in the second measurement.

[11] One must keep in mind the mathematical complexities involved (those of QCD are especially difficult) and conceptual difficulties, some of which, such as renormalization, are still not entirely resolved. Nevertheless, these theories work very well, not the least with the help of renormalization, considered from the RWR perspective in (Plotnitsky 2016, pp. 239–246).

here is applicable in all quantum regimes, to the point of compelling one to conclude that, in quantum theory, reality without realism is the reality of quantum fields.

8.3 Virtual Particles

As noted from the outset of this chapter, while not required in RWR-type interpretations, the concept of virtual particles may be adopted by these interpretations because of its usefulness as a heuristic device. Conventionally, the role of virtual particles is defined by their interactions, *between experiments*, with real particles. In the present view, virtual particles and these interactions would represent, one might say, symbolically, the ultimate, RWR-type, constitution of physical reality, prior to experiments, which, by means of the interaction between this constitution and measuring instruments, give rise to real particles as quantum objects (which virtual particles are not), and via them, effects observed in measuring instruments. "Events" only refer to the observations of these effects and not to what happens between experiments, as anything associated with virtual particles does. It is true that, in certain cases, virtual particles, if assumed to exist, can become real particles. Thus, while virtual particles are assumed often to appear in pairs of a particle and an anti-particle, which exist for a very short time and then mutually annihilate, there are situations, such as vacuum decay, and Hawking radiation, and the Unruh effect when it is possible to separate the pair by external energy so that they avoid annihilation and become real particles, which may have direct effects on measuring instruments. Such situations make adopting the concept of virtual particles compelling, but still not imperative because the particles thus emerging are still real particles and their emergence, manifested in the corresponding effects observed in measurements (which and only which allow one to speak of them as quantum objects in the present interpretation), could be explained by means of other mechanisms. Thus, although Stephen Hawking used the concept of virtual particle formation in his explanations of Hawking radiation, the phenomenon itself, which is more complex and multifaceted (and has open problems associated with it), could be interpreted as arising from the production of pairs of real particles.

Conventionally, when assumed, virtual particles are entities that are born and disappear very quickly but exist long enough to affect quantum fields between experiments and, as a result, what is subsequently observed in experiments. The concept emerged in the perturbative QFT, which considers the birth and disappearance of real particles, a characteristic of QFT regimes, as involving virtual particles. These interactions are represented by Feynman diagrams, with virtual particles represented by internal lines. I discussed below, from the present perspective, Feynman diagrams are as only heuristic devices, helping calculations. In the present view, however, virtual particles are heuristic devices as well. While virtual particles conserve energy and momentum, they do not necessarily have the same mass as their real counterparts.[12] They are considered "off-shell" in the standard terminology because they do not satisfy certain standard equations (as "on-shell" objects do), for example, insofar as they do not strictly obey the energy–momentum relation $m^2c^4 = E^2 - p^2c^2$ and their kinetic energy may not have the standard relationships to their velocity. The probability amplitude for the existence of a virtual particle interferes with that for its nonexistence, while for a real particle the cases of existence and nonexistence are not coherent and do not interfere with each other. What is, again, significant is that, even though they are not observed, virtual particles serve as effective, but, again, in RWR-type interpretations, not required, idealizations (although some among these interpretations may adopt the concept), because of statistically ascertainable effects on real particles and our measurements concerning the latter. Among them are (in addition to those mentioned above as more specifically dealing with virtual particles becoming real particles) the Lamb shift, the Casimir effect, the role of virtual gluons in establishing the constitution of protons as comprised by quarks and gluons and in giving the proton its mass, and asymptotic freedom. All these cases would illustrate the difference between two idealizations, real and virtual particles, in RWR-type interpretations. Here, I would like to focus on the case of asymptotic freedom.

Asymptotic freedom was discovered by Gross and Wilczek (1973, 1974) and independently Politzer (1973), eventually bringing them their Nobel Prize in 2004. Asymptotic freedom enables quantum chromodynamics (QCD), the QFT of the strong interactions between quarks and gluons, to account for the behavior of quarks and gluons, specifically their confinement insider nuclei, thus, the impossibility of observing them outside nuclei. This statement conveys the situation in classical terms, as there is no other way to *speak* of anything in the sense of representing by means of language what "happens." Mathematically, asymptotic freedom is a property of some gauge theories that reflects, if one, again, speaks conventionally and, hence, classically, the "fact" that interactions between particles become asymptotically weaker with an increase of the energy scale and a corresponding decrease of the length scale. While quarks interact weakly at high energy scales (allowing perturbative calculations by QCD), at low energy scales, the interactions become strong, ultimately too strong for quarks to break out of nuclei, thus confining them inside nuclei. This leads to the impossibility of observing quarks or gluons as such, as real particles, at least not in the way other (unconfined) elementary particles, such as electrons, photons, or

[12] There are further subtleties as concerns conservation laws, which are handled by using the uncertainty relations, including those for time and energy variables, which involve subtleties of their own because, in contrast to coordinate and momentum variables, time is not an operator but a parameter (e.g., Shilladay and Bush 2006). In considering virtual particles, coordinates are parameters as well. According to Weinberg: "virtual particles … are not directly observable while they are being exchanged, because their creation as real particles (e.g., a free electron turning into a photon and an electron) would violate the law of conservation of energy. However, the quantum–mechanical uncertainty principle dictates that the energy of a system that survives for only a short time must be correspondingly highly uncertain, so these virtual particles can be created in intermediate states of physical processes but must be reabsorbed again very quickly" (Weinberg 1977, p. 24).

neutrinos, can be. Quark and gluons, can, however, be "observed" as real particles, indirectly, or so it is assumed, in so-called hadron jets.

The meaning of the question (as it remains a question) of so "observing" even *real* quarks and gluons either inside the nuclei or in jets remains debated on several counts (e.g., Falkenburg 2000; Fox 2009, and further references there). I shall put this debate by and large aside, given that my main concerns at the moment are virtual particles and how the concept of observation is defined in the RWR view, vs. more realist views. I shall, accordingly, restrict myself to a few comments on the issue shaped by this difference, in particular by considering Tobias Fox's conclusion of the article just mentioned, contrasting the case of quarks with Carl D. Anderson's discovery of the positron, vis-à-vis the present view. According to Fox:

> Obviously, images of jet-events give an idea of a good, working particle detector but they are not crucial for the confirmation of theoretically-determined measuring values. By comparison, think of the discovery of the positron by Carl D. Anderson in 1932. In his publication, we see images of positron tracks. Since these images show that a single track is caused by a single positron, they were key to his claim that positrons are indirectly observable.... Getting even one image of a certain particle track means the discovery of a positron by indirect observation. And the charge-mass ratio typical for a positron can be determined by evaluating the curvature of that track. Remember our definition [an object is indirectly observed if the physical phenomenon created by the object is observed directly]. Anderson observed indirectly positrons in the same way as we observe indirectly an aeroplane by watching a vapour trail in the sky.
>
> This type of investigation is impossible for quarks, although quark physics and quantum chromodynamics is none the worse for this. That quarks cannot be observed has little impact on the successful verification of the physical entity "quarks," if the term "entity" is understood in its most basic sense. Many entities are not observable—forces, natural constants, fields, virtual particles, resonance particles, as well as natural laws or terms of natural laws. ...
>
> There is a strong consensus that some objects, which are not directly observable, might be indirectly observable. There are additionally entities that to us do not exist because, among other reasons, they are not indirectly observable.
>
> In the case of quarks, we have seen that they cannot be indirectly observed, even in the comparatively evidential context of jet-events. This should be a strong hint that quarks are in principle unobservable, although, this is not yet proved. This brings us to the debate of scientific realism. And the indications are that quarks can only be treated as entities, similar to other unobservable entities in science. (Fox 2009, pp. 186–187)

While Fox is right to invoke "the debate of scientific realism," all the possibilities here listed concerning the observability of quarks or lack thereof would still fall within realism in the present understanding of the term. By contrast, in the present, strong RWR, view, while there are differences between quarks or gluons and unconfined objects as concerns their observability or, in present terms, the emergence of the corresponding quantum phenomena, no objects of either type are assumed ultimately to exist as such. Observed quantum phenomena are assumed to exist as, to return to Bohr's language, "idealization of observation," thus akin to Fox's "direct observation" (Bohr 1987, v. 1, p. 55). In the RWR view, however, reversing Fox's definition, a physical phenomenon constructed by means of our observational technology and observed directly, or, again, as an idealization of observation, establishes a connection to an object, observed indirectly. Or rather this object is assumed, as an idealization, to exist on the basis of the effect of the interaction between it and the measuring instrument, an effect manifested in the observed parts of the instrument and as such observed in this instrument directly. In the present view, moreover, the concept of a quantum object only applies at the time of observation. The difference, then, between, say, electrons or positrons and quarks or gluons are that of the structure of effects, traces, of the interactions between our measuring instruments and the ultimate constitution of the reality considered, which is assumed (still an unfalsifiable assumption which is only practically justified) to exist independently, as a quantum field. In the case of electrons or positrons, the corresponding structure of effects allows us to associate it with (the idealization defined as) real electrons or positrons. On the other hand, given the debate considered by Fox, in the case of quarks or gluons, the corresponding structure of effects may or may not allow us to associate it with (the idealization defined as) of quarks or gluons. Or rather this structure of effects may or may not allow us to do so *in the same way*. But, "the confirmation of theoretically-determined measuring values," still allows one to speak of *real* quarks and gluons as a suitable idealization, which is not possible in the same way in the case of virtual quarks and gluons. But then, this is not possible in the same way for virtual electrons or positrons either. In dealing with virtual particles there is no difference: this idealization is possible but not required in the present view, or at least does not have the same status as that of real particles.

It is still possible, however, that we don't need (the idealization of) even real quarks or gluons, or need them less than (the idealization of) unconfined particles. All we might need, as Gell-Mann thought initially, may be the corresponding symmetry groups and their irreducible representation (associated with quarks) or some other mathematical technology that enable us to predict what is observed in high-energy QCD regimes. This is, however, the kind of mathematical technology that we always need, along with the technology of experiments, if we want to have physics as a mathematical-experimental science.

With these considerations in mind, I would like to discuss Wilczek's vivid *ordinary language* account, via asymptotic freedom, of "why quarks and gluons appear (only) as jets," by "explain[ing] why soft radiation [which makes hadron jets out of a quark and antiquark] is common but hard radiation [which involves intermediate gluons and leads to an extra jet] is rare." He says:

> The two central ideas of asymptotic freedom are: first, that the intrinsic color charge of the fundamental particle—whether quark, antiquark, or gluon—is small and not very powerful; second, that the cloud of virtual particles that surrounds the fundamental particle is thin nearby but grows thicker far away. It's the surrounding cloud [of virtual particles], not the particle's core charge, that makes the strong interaction strong.
>
> Radiation occurs when a particle gets out of equilibrium with its cloud. Then rearrangements that restore the equilibrium in color fields cause radiation of gluons or quark-antiquark pairs, much as rearrangements in atmospheric electric fields cause lightening, or rearrangements in tectonic plates cause earthquakes and volcanos. How can a quark (or antiquark or gluon) get out of equilibrium with its cloud? One way is if it suddenly pops out from a virtual photon, as happened in the experiments at LEP [Large Electron-Proton Collider] we've been discussing [experiments that "showed" quarks and gluons inside nuclei]. To reach equilibrium, the newborn quark has to build his cloud, starting from the center—where the

> small color charge initiates the process—and working its way out. The changes involved are small and graded, so they require only a small flow of energy and momentum—that is, soft radiation. The other way a quark can get out of equilibrium with its cloud is if it's jostled by quantum fluctuations of the gluon field. If the jostling is violent, it can cause hard radiation. But because the quark's intrinsic color charge is small, the quark's response to quantum fluctuations in the gluon field tends to be limited, thus hard radiation is rare. That's why three jets are less likely than two. (Wilczek 2009, p. 56)

A more standard, less vibrant, qualitative description of asymptotic freedom would go roughly as follows. Virtual quark-antiquark pairs tend to screen the color charge, similarly to the Landau pole screening of virtual charged electron–positron pairs in vacuum in QED. In the close vicinity of a charge, the vacuum becomes polarized because virtual particles of opposing charge are attracted to the charge and those of the same charge are repelled, with the effect of partially canceling out the field at any finite distance. Approaching the central charge, the effect of the vacuum diminishes and the effective charge increases. QCD, however, has a crucial new property: gluons carry a color charge as well. Moreover, they do so differently because each gluon carries both a color charge and an anti-color magnetic moment. The total effect of polarization of virtual gluons in a vacuum is not to screen the field but to amplify it and change its color, the phenomenon known as antiscreening. Approaching a quark decreases the antiscreening effect of the surrounding virtual gluons, which weakens the effective charge with decreasing distance. Because the virtual quarks and the virtual gluons, thus, contribute opposite effects, which effect wins out depends on the number of different flavors of quarks. For standard QCD with three colors, if there are no more than 16 flavors of quarks (including anti-quarks), antiscreening dominates and the theory is asymptotically free. There are only six known flavors of quarks.

This description obviously bypasses the extraordinary and extraordinarily difficult mathematics, without which there would be no QCD. The following account gives a sense of this mathematics, still mixing in some verbal and, thus, classical descriptions. This mixing may be unavoidable even in technical accounts, even though such verbal descriptions may be misleading and, at least in the RWR view, are ultimately inapplicable. Mathematics, at least in the present view, does not describe or represent what "happens" anymore than anything else does. It only provides correct probabilistic predictions of the outcome of quantum experiments. Asymptotic freedom, as a mathematical property of QCD, is derived by means of calculating the so-called beta-function that represents the transformations of the coupling parameter, g, (dependent on the energy-scale, μ) of the theory under the renormalization group:

$$\beta(g) = \frac{\partial g}{\partial \log(\mu)}.$$

The renormalization group is one of the most extraordinary concepts of QFT, introduced by Kenneth G. Wilson, the technical description of which is formidable and will be bypassed here. Conceptually, it is a mathematical technology, part of the so-called effective QFTs, that enables one to properly investigate the changes of a quantum system as represented ("viewed") at different scales (e.g., Kulhmann 2020, Sect. 2.4.). It reflects the changes in the underlying force laws of QFT, changes due to the fact that the energy scale at which physical processes occur varies, with energy–momentum and resolution distance scales obeying the uncertainty relations. Because the underlying renormalization group, β only depends on μ implicitly through g. This dependence on the energy scale is called "running" the coupling parameter. Its explicit computation is difficult but possible by means of various mathematical techniques.

For sufficiently short distances or large values of momentum, an asymptotically free theory allows for calculations by a perturbation theory, calculations guided (in the present view heuristically) by Feynman diagrams. These cases are, accordingly, more manageable by calculations than those that could deal with long-distance, strong-coupling behavior, equally found in such theories, behavior that is assumed to correspond to confinement. The beta-function, then, describes how the coupling constant changes with scaling of the system. In order to calculate the beta-functions, one needs to evaluate Feynman diagrams, which is to say, the transformations in the formalism guided by these diagrams (which in the present view, are, again, only heuristic guides to such calculations) contributing to the event of the emission or the absorption of a gluon by a quark. The calculation is performed by using rescaling in position-space or momentum-space. In non-abelian gauge theories, such as QCD, the existence of asymptotic freedom depends on the gauge groups and number of flavors of interacting particles. To the lowest nontrivial order, the beta-function in an SU (N) (the Lie group of $n \times n$ unitary matrices with the determinate one) with n_f kind of quark-like particles is

$$\beta_1(\alpha) = \frac{\alpha^2}{\pi}\left(-\frac{11N}{6} + \frac{n_f}{3}\right).$$

Here α is the equivalent of the fine-structure constant in the theory, $\frac{g^2}{(4\pi)}$. If$\beta_1 < 0$, the theory is asymptotically free. For SU (3), the color charge gauge group of QCD, $N = 3$, and $\beta_1 < 0$

$$n_f < \frac{33}{2}$$

Hence, for SU(3), the theory is asymptotically free if there are 16 or fewer flavors of quarks. There are, again, only six flavors of quarks.

This account is, of course, still far from representing the actual mathematics of asymptotic freedom or QCD, which is necessary to arrive at these numbers, the mathematics exemplified and discovered in Gross and Wilczek's and Politzer's papers and their calculations, like those of the renormalization group parameters. That these papers get these numbers right is almost a miracle. On the other hand, quantum theory or all modern physics, from Galileo on, is about getting "numbers" right. The numbers in question here are those observed strictly in measuring instruments, or inferred from those observed in measuring instruments, such as the numbers of colors (and colors here are only numbers) or the number of quarks, never observed as such. No classical-like language or concepts, such as those used by Wilczek in the above verbal account, or Feynman diagrams can represent or, in the RWR view, allow one to even imagine anything physically corresponding to the mathematics of asymptotic freedom. We have no representational concepts or language other than classical-like to represent what happens in nature, with only mathematics giving us a freedom from them, as Heisenberg argued in the case of the mathematics of QM (Heisenberg 1930, p. 11). As discussed earlier, this mathematics itself was a product of the historical development of mathematics defined by its divorce from its representational functioning in classical physics, which was a mathematized refinement of our general phenomenal thinking. This freedom may not be complete, as we still use verbal means to help us to describe our mathematical concepts (as for example, in Gross and Wilczek's or Politzer's papers on asymptotic freedom). It is, however, sufficient to circumvent the constraints of our daily phenomenal intuition and concepts or language either in mathematics itself or in certain areas of physics, such as relativity, where mathematics can still serve as a means of representing physical processes, or quantum theory, where it may no longer serve this purpose. It does not in RWR-type interpretations, where it only serves to predict probabilistically the numerical outcomes of quantum experiments, which could also allow one to estimate other numbers, such as that of possible quarks or their flavors, as just explained.

Virtual particles help such estimates but are not required for them, and no numbers associated strictly with them are registered in any experiments, let alone any “clouds,” so evocatively used by Wilczek's verbal description of asymptotic freedom. Virtual particles remain unfalsifiable constructs. This is the case not only because anything that can be said or thought about them and their behavior could only be classical and thus cannot apply to them, for this is also true about real particles or the ultimate, RWR-type, reality of quantum fields, which are sometimes described by evocative metaphors as well. This is also, and primarily, the case because virtual particles neither leave any traces in measuring instruments nor are strictly necessary for the mathematical formalism of QCD. Although, as discussed earlier, the lack of this necessity is sometimes claimed in the case of real quarks and gluons or any real particles, real quarks and gluons are assumed to leave such traces, even of a more complex and qualified nature, as discussed above. Even if, in view of this more complex nature, this assumption is questioned or even denied in some interpretations, the ontological status of real quarks and gluons, including as concerns their impact on observed traces, is still different from that of virtual quarks and gluons.

The irreducibly statistical nature of QCD as a quantum theory, a feature central to this study, remains crucial in dealing with these phenomena, and it enters Wilczek's account, cited above. According to Wilczek (referring by “quantum mechanics” to all quantum theory as a probabilistic or statistical theory):

> The connection of our photographs [of jets] to the profundities of quantum mechanics is even more direct. … It is simply that once more, we find that doing the same thing over and over again gives different results each time. We saw that before with the ultrastroboscopic nanomicroscope that takes pictures of protons [as consisting of quarks and gluons]; we're seeing it now with the creative destruction machine that takes pictures of empty space [false

> vacuum]. If the world behaves classically and [deterministically] predictably, the billion euros invested in LEP would have underwritten a very boring machine: every collision would just reproduce the result of the first one, and there'd be only one photograph to look at. Instead, our quantum-mechanical theories predict that many results can emerge from the same cause. We can predict the relative probabilities of different results. Through many repetitions, we can check those predictions in detail. In that way, short-term unpredictability can be tamed. Short-term unpredictability is, in the end, perfectly compatible with the long-term precision. (Wilczek 2009, p. 56)

The expression "the same cause" could, in the present view, only refer to the identical preparation of the measuring instruments used, rather than to a cause defining classical causality. By referring to our capacity to predict these probabilities, this conclusion brings the mathematics of QCD, in its probabilistically or statistically predictive role, into Wilczek's verbal and, thus, conceptually classical description, again, unavoidable in nontechnical and even technical accounts of QCD or quantum theory in general. In the RWR view, however, such a description cannot correspond to what "happens," which word itself is, again, ultimately inapplicable, as Heisenberg argued (Heisenberg 1962, p. 57). While it may be helpful, any such description is ultimately inadequate metaphor, even in dealing with real rather than only virtual quantum objects. In the RWR view, even a purely mathematical representation of what happens between experiments is not possible, although Heisenberg in the same book (Heisenberg 1962, pp. 145, 167–186) and other later works assumed that it is which assumption does not require the term or concept of "something happens." Wilczek is aware of these complexities, even if without adopting the RWR view. His position is similar to that of Heisenberg or (ontic) structural realism, with, as in Heisenberg, symmetry as its core. Ontic structural realism has some currency in interpretations of QFT, including with the help of category theory, discussed in a different context in Chap. 9 (e.g., Bain 2013). Wilczek prefaces the account of quantum randomness and probability just cited as follow: "There is direct connection between quarks and gluons appearing in jets and asymptotic freedom. It's easy to explain the connection using Fourier transforms, but unfortunately Fourier transforms themselves aren't easy to explain [for a lay reader], so we won't go there. Here's a word explanation that is less precise but calls for more imagination (and less preparation)" (Wilczek 2009, p. 54). Although a beautiful piece of mathematics and technically complex in this case (it is measured by the moments of the scaling structure function, which is no simple matter, as Wilczek rightly states here [e.g., Gross and Wilczek 1973, p. 3643]), the concept is only a small part of the mathematics of asymptotic freedom.

In the RWR view, especially the strong RWR view, Wilczek's verbal "explanation" is not only less precise but is also rigorously inapplicable, however imaginative it might be descriptively, because this imagination can only create a classical picture or a proto-classical one (that is, refinable into a classical physical representation). Our mathematical imagination can be free from such pictures, even if, again, not free from pictures or daily language altogether. Thus, as discussed earlier, when one thinks of a Hilbert space, one tends to use some geometrically thinking, to form some, unavoidably classical, picture of spatiality, which helps our thinking, physical and mathematical, but is not required for proofs or calculation. In this case, however, our geometrical intuitions are of a very limited help, while the technical aspects of this geometry (defined in Chap. 4 as spatial algebra) are essential.

The situation just discussed in connection with asymptotic freedom, thus, reflects two related tensions that preoccupied this study from the outset and acquired new dimensions in the context of the question of elementary particles. The first, more general, is between the abstract mathematical architecture of quantum theory and our general phenomenal representation or, or accompanied by, a verbal description, always classical in nature, also applicable to the observable parts of measuring instruments. The description of these observable parts is, in RWR-type interpretations, entirely divorced from the abstract mathematics of the formalism, nevertheless, able to predict what is so observed as well as possible thus far. The second tension is between algebra and geometry, which are more interactive, although geometry still play a greater role in both our, ultimately inapplicable, verbal account of what "happens" between experiments and in our representation of observed quantum phenomena. (Their representation by means of classical physics would of course also involve the algebra of classical physics.) Finally, as I argue, it is not clear that, as still a human way of thinking, mathematics should be able to represent, however ideally, the ultimate constitution of nature, and in the U-RWR view adopted here it cannot. Nothing can. We are fortunate enough that it enables us to predict, in probabilistic or statistical terms at that, quantum events.

In QCD, or QFT in general, then, the concept of virtual particles helps us to make these predictions. It would, however, be difficult to argue that it is strictly necessary for doing so and in the present view it is not. This assessment is corroborated by examining the nature and role of Feynman diagrams, arguably the main vehicle for working with the concept of virtual particles. No virtual particle found in Feynman diagrams is ever observed, unless it becomes a real particle, as it can in certain circumstances. Consider the following simple, routinely used, Feynman diagram (Fig. 8.1) with an electron and a positron annihilating, producing a virtual photon that then disappears, giving birth to a quark and anti-quark, with an antiquark then emitting a gluon.

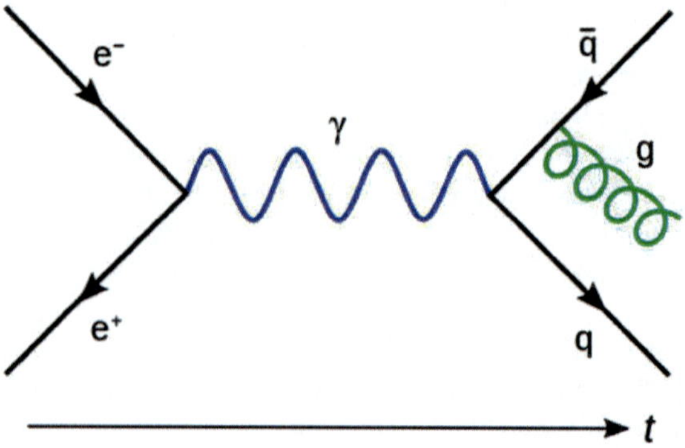

Fig. 8.1 An electron and a positron annihilate, producing a virtual photon that then disappears, giving a birth to a quark and anti-quark, with an antiquark then emitting a gluon

Only the electron and positron, and quark and antiquark (and then gluon) involved are observable: only they could leave traces in measuring instruments from which we infer their existence, in the case of quarks and gluons, again, with qualifications given earlier. The virtual photon is never observed, unless it becomes a real photon, as it can in certain circumstances, as discussed above, keeping in mind that, in the RWR view, all observations only pertain to quantum phenomena as defined by the observable parts of measuring instruments. No real particles as such are ever observed, and in the present interpretation, even as unobservable as such, they are defined, as an idealization, only at the time of observation. Ultimately, however, Feynman's diagrams are only diagrams, especially in the RWR view. They do not represent not only, obviously, any trajectories in space and time, a concept not applicable to quantum behavior in the first place, but also any physical connections between these events, as only events involving real particles could, in principle, be observed. Feynman diagrams provide a guidance to which of such events can be observed and which probabilities can be calculated. This guidance is helpful, sometimes exceptionally so even to the point of being indispensable *in practice*, as for example, in Gross and Wilczek's papers on asymptotic freedom, which might not have been possible apart from them. In principle, however, one can do without them. Schwinger, for whom, admittedly, the idea of virtual particles was crucial, did not use Feynman diagrams in his renormalization of QED.[13] The concept of antimatter emerged from the mathematics of Dirac's equation, without using anything like virtual particles either. This view of Feynman diagrams is not restricted to RWR-type interpretations and is found elsewhere, including in realist interpretations of QFT. Some of the latter, however, often assume the existence of virtual particles (e.g., Kaiser 2014; Meynell 2018; Jaeger 2019). RWR-type interpretations, by contrast, deprive of reality not only Feynman diagrams but also the concept of virtual particles and behavior. Both are heuristic tools, helping QFT calculations, admittedly, tools are different in nature.

When it comes to QFT calculations, one needs all the help one can get. The mathematics of QFT, especially QCD, could, again, be extremely difficult. Most calculations in QCD require powerful approximation methods, such as lattice QCD, and tremendous computer resources. The QCD Lagrangian governing the interaction of quarks and gluons, including quark confinement, may be elegant on the surface:

$$L = -\frac{1}{4} F^{\alpha}_{\mu\upsilon} F^{\mu\upsilon}_{\alpha} - \sum_{n} \bar{\psi}_n \gamma^{\mu} \left[\partial_{\mu} - igA^{\alpha}_{\mu} t_{\alpha}\right] \psi_n - \sum_{n} m_n \bar{\psi}_n \psi_n$$

[13] Dyson's proof of the equivalence of Schwinger's and Feynman's schemes offers support for this view of Feynman's diagrams (Dyson 1949). This view, however, could be justified from within Feynman's own scheme. See Schweber (1994, pp. 474–575) for an excellent discussion of Dyson's contribution, although the book, intriguingly, pays next to no attention to virtual particles, either in this discussion or in general.

Once, however, one begins calculating, unfolding its symbols, as, in Gross and Wilczek's or Politzer's remarkable papers on asymptotic freedom (Gross and Wilczek 1973, 1974; Politzer 1973), this elegance retreats, and, guided as one might be, as these authors were, by Feynman's diagram, one enters the calculational nightmares for which QFT is famous, but from which, dreams of new theories are born.

8.4 Conclusion

In RWR-type interpretations, in high-energy (QFT) regimes, the ultimate, RWR-type, reality, responsible for the quantum phenomena in these regimes, produces multiple and multiply transforming effects, a form of the Heraclitean many. But, in these interpretations, it does so in a way that we cannot know or even conceive of, which may be seen as a radical form of the Heraclitean hidden, which "nature loves" (Kahn 1981, Fragment X, p. 33).[14] The mathematics of QFT is, just as that of QM, a Pythagorean mixture of algebra and geometry, through which we can probabilistically or statistically predict the effects of the unthinkable on what can be thought and experienced as quantum phenomena. This, as I said, need not mean that the ultimate level of reality is defined by permanence or oneness, the One, as Parmenides, another key pre-Socratic thinker, and Plato thought: the concept of permanence is no more applicable to this "it" than that of change or any other concept, including "something happened," or "it," "hidden," or "many." Thinking in terms of the Heraclitean vision of the hidden, becoming, and the many are more suited to QFT because the effects of the hidden (idealized as unrepresentable or even inconceivable), effects observed in measuring instruments, are defined by multiplicity and becoming. Parmenides and, following him, Plato would have seen this multiplicity and becoming as illusions of our senses rather than the characteristics of the ultimate nature of reality, hidden from our senses, if for them, not, at least not entirely, from thought. In the present, strong RWR, view one would accept their first claim, when considering the ultimate constitution of the reality responsible for quantum phenomena, but not their second claim. For it does not follow that this reality is defined by absolute permanence, any more than by change, or as anything hidden because this word implies that what is hidden might be revealed one day. Nor may it be accessible to thought. Hence, rather than referring to something, some "secret," that can be revealed one day, "reality" is a name without a concept associated with it, or even a name without a name, because it is difficult for a name to avoid a concept associated with it.

[14] An appeal to the Heraclitean many in QFT is found in literature, as in Redhead (1990, p. 22), but not, to my knowledge, along the RWR lines of bringing together the hidden, becoming, and the multiple of QFT. Admittedly, these are English terms. Hence, I use the adjective Heraclitean, rather than referring to anything in Heraclitus's own thought. To what degree these terms or concepts thus designated correspond to Heraclitus's meaning is a separate issue, complicated by the fact we only have famously enigmatic fragments of his writings quoted by others. My claims here strictly concern the present view of QFT in these (English) terms, as we use them now.

Beginning with Dirac's equation, QED and eventually QFT became far-reaching extensions of the revolution ushered by Heisenberg's discovery of QM, which also initiated the RWR view of quantum phenomena. This was acutely realized by Bohr and Heisenberg in assessing Dirac's discovery of his equation and, as a consequence of his equation, antimatter. Bohr spoke of "Dirac's ingenious quantum theory of the electron," as "a most striking illustration of the power and fertility of the general quantum–mechanical way of description," which was, at this point, defined by Bohr's ultimate, RWR-type, interpretation (Bohr 1987, v. 2, p. 64). Heisenberg, who made major contributions to QFT, including that of nuclear forces, saw Dirac's discovery as even more significant. He saw it "as perhaps the biggest change of all the big changes in physics of our [twentieth] century. It was a discovery of utmost importance because it changed our whole picture of matter. … It was one of the most spectacular consequences of Dirac's discovery that the old concept of the elementary particle collapsed completely" (Heisenberg 1989, pp. 31–33). But it also brought with it new ways of thinking about the problem of the elementary particles and quantum fields.

Heisenberg's own concept of an elementary particle by this point became mathematically realist, with a Platonist flavor, although not strictly Platonist insofar as he still maintained the idea of matter as defining the ultimate constitution of physical reality. Perhaps, ultimately more Pythagorean that Platonist, this concept was based on the idea that the ultimate constitution of nature could, in principle, be represented mathematically in the absence of physical concepts, at least as the latter was previously understood. In Bohr's or the present view, the traces observed in measuring instruments and associated with elementary particles, in the present view, idealized as such (as a particle type of quantum objects), only at the time of measurement, are messengers of the reality beyond thought, mathematical thought included. This reality, however, has effects on that part of the world to which we can give concepts and which we can know, because we can stage quantum experiments enabling us to observe quantum phenomena and use the mathematics of quantum theory to predict the outcomes of quantum experiments thus observes. That we can only do so probabilistically or statistically need not be viewed as detrimental; quite the contrary, as Wilczek suggested above, in speaking of "short-term unpredictability as, in the end, perfectly compatible with the long-term precision," probabilistic or statistical in nature (Wilczek 2009, p. 56). Quantum phenomena and the mathematics of probability are great gifts to our thought, the first a gift of nature and the second a gift of thought to itself. Yet another gift of thought is our ability to create experimental technology that, in its interaction with nature, brings about quantum phenomena. It is true that all these are ultimately created by nature or matter, dead and living, with living matter giving us our brains and thus our capacity for thought. Still, working together, matter, thought, and technology enable us to discover and connect to something in nature that our thought may not be able to reach.

References

Bain, J.: Category-theoretic structure and radical ontic structural realism. Synthese **190**, 1621–1635 (2013)

Baulieu, L., Iliopoulos, J., Seneor, R.: From Classical to Quantum Fields. Oxford University Press, Oxford (2017)

Bohr, N.: Causality and complementarity. In: Faye, J., Folse, H.J. (eds.) The Philosophical Writings of Niels Bohr, Volume 4: Causality and Complementarity, 1999, pp. 83–91. Ox Bow Press, Supplementary Papers, Woodbridge, CT, USA (1937)

Bohr, N.: The Philosophical Writings of Niels Bohr, 3 vols. Ox Bow Press, Woodbridge, CT (1987)

Bohr, N., Rosenfeld, L.: On the question of the measurability of electromagnetic field quantities. In: Wheeler, J.A., Zurek, W.H. (eds.) Quantum Theory and Measurement, 1983, pp. 479–522. Princeton University Press, Princeton, NJ (1933)

Bohr, N., Rosenfeld, L.: Field and charge measurements in quantum electrodynamics. In: Wheeler, J.A., Zurek, W.H. (eds.) Quantum Theory and Measurement, 1983, pp. 523–534. Princeton University Press, Princeton, NJ, USA (1950)

Bohr, A., Mottelson, B.R., Ulfbeck, O.: The principles underlying quantum mechanics. Found. Phys. **34**, 405–417 (2004)

Duncan, A.: The Conceptual Framework of Quantum Field Theory. Oxford University Press, Oxford (2012)

Dyson, F.J.: The S-matrix in quantum electrodynamics. Phys. Rev. **75**, 1736–1755 (1949)

Falkenburg, B.: How to observe quarks. In: Agazzi, E., Pauri, M. (eds.) The Reality of the Unobservable, 2020, pp. 329–341. Kluwer, Dordrecht, Netherland (2000)

Falkenburg, B.: Particle Metaphysics. Springer, Heidelberg, Germany (2007)

Fox, T.: Haunted by the specter of virtual particles: A philosophical reconsideration. J. Gener. Philos. Sci. **39**, 35–51 (2008)

Fox, T.: Why quarks are unobservable. Philosophia Scientiae **13**(2), 167–189 (2009). https://doi.org/10.4000/philosophiascientiae.303

French, S.: Identity and individuality in quantum theory. In: Zalta, E. N. (ed.) Stanford Encyclopedia of Philosophy (2014). http://plato.stanford.edu/entries/quantum-field-theory/

Gross, D.J., Wilczek, F.: Asymptotically free gauge theories II. Phys. Rev. D **8**, 3633–3651 (1973). https://doi.org/10.1103/PhysRevD.8.3633

Gross, D.J., Wilczek, F.: Asymptotically free gauge theories II. Phys. Rev. D **9**, 980–993 (1974). https://doi.org/10.1103/PhysRevD.9.980

Heisenberg, W.: The Physical Principles of the Quantum Theory, Trans. Eckhart, K., Hoyt, F.C, rpt. 1949. Dover, New York, NY (1930)

Heisenberg, W.: Physics and Philosophy: The Revolution in Modern Science. Harper & Row, New York, NY (1962)

Heisenberg, W.: Encounters with Einstein, and other Essays on People, Places, and Particles. Princeton University Press, Princeton, NJ (1989)

Heller, J.: Something Happened. Knopf, New York, NY (1974)

Jaeger, G.: Are virtual particles less real? Entropy **21**(2), 141 (2019). https://doi.org/10.3390/e21020141

Jaeger, G.: Exchange forces in particle physics. Found. Phys. **51**(1), 1–31 (2021). https://doi.org/10.1007/s10701-021-00425-0

Kahn, C. (ed.): The Art and Thought of Heraclitus: A New Arrangement and Translation of the Fragments with Literary and Philosophical Commentary. Cambridge University Press, Cambridge (1981)

Kaiser, D.: Drawing Theories Apart: The Dispersion of Feynman Diagrams in Postwar Physics. University of Chicago Press, Chicago, IL (2014)

Kosmann-Schwarzbach, Y.: The Noether Theorems: Invariance and Conservation Laws in the Twentieth Century (Trans. Schwarzbach, B.). Berlin, Germany, Springer (2011)

Kuhlmann, M.: Quantum field theory. In: Zalta, E. N. (ed.) Stanford Encyclopedia of Philosophy (2020). https://plato.stanford.edu/entries/quantum-field-theory/

Meynell, L.: Picturing Feynman diagrams and the epistemology of understanding. Perspective on Science **26**, 459–481 (2018)

Peskin, M.E., Schroeder, D.V.: An Introduction to Quantum Field Theory. CRS Press, London (1995)

Plotnitsky, A.: The Principles of Quantum Theory, from Planck's Quanta to the Higgs Boson: The Nature of Quantum Reality and the Spirit of Copenhagen. Springer/Nature, New York, NY (2016)

Politzer, H.D.: Reliable perturbative results for strong interactions. Phys. Rev. Lett. **30**(26), 1346–1349 (1973). https://doi.org/10.1103/PhysRevLett.30.1346

Redhead, M.: A philosopher looks at quantum field theory. In: Brown, H.R., Harré, R. (eds.) Philosophical Foundations of Quantum Field Theory, pp. 9–24. Clarendon, Oxford (1990)

Ruetsche, L.: Interpreting Quantum Theories. Oxford University Press, Oxford (2011)

Schweber, S.S.: QED and the Men Who Made It: Dyson, Feynman, Schwinger, and Tomonaga. Princeton University Press, Princeton, NJ (1994)

Schwinger, J.: Particles, Sources, and Fields I. CRC Press, Boca Raton, FL (1970)

Shilladay, C., Busch, P.: Complementarity and uncertainty in Mach-Zehnder interferometry and beyond. Phys. Rep. **435**, 1–31 (2006)

Ulfbeck, O., Bohr, A., A.: Genuine fortuitousness: Where did that click come from? Found. Phys. **31**, 757–774 (2001)

Weinberg, S.: The search for unity: Notes for a history of quantum field theory. Daedalus **106**, 17–35 (1977)

Weinberg, S.: What is an Elementary Particle? (1996). http://www.slac.stanford.edu/pubs/beamline/27/1/27-1-weinberg.pdf

Weinberg, S.: The Quantum Theory of Fields, Volume 1: Foundations. Cambridge University Press, Cambridge (2005)

Weyl, H.: Theory of Groups and Quantum Mechanics, rpt. 1984. Dover, Mineola, NY (1931)

Wilczek, F.: Lightness of Being: Mass, Ether, and the Unification of Forces. Basic Books, New York, NY (2009)

Zee, A.: Quantum Field Theory in a Nutshell. Princeton University Press, Princeton, NJ (2010)

Chapter 9
From Circuits to Categories in Quantum Information Theory

> The quantal features of the phenomenon are revealed in the information about the atomic objects derived from the observations.
> —Niels Bohr, "Causality and Complementarity" (1958) (Bohr 1987, v. 3, p. 4).

Abstract The argument of this chapter builds on that of Chap. 4, by virtue of the relationships between Heisenberg's thinking, leading him to his discovery of QM, and reconstruction projects—projects aimed at (re)derivations of quantum theory, specifically for discrete variables, from a set of basic principles—in quantum informational theory, as exemplified in the work of Hardy and Mauro D'Ariano and coworkers, and the essential role of the situation is captured by the diagrams, introduced in Chap. 4:

QPA/QPAGG ($QUANTUMNESS \rightarrow PROBABILITY \rightarrow ALGEBRA$/
$QUANTUMNESS \rightarrow PROBABILITY \rightarrow ALGEBRA \rightarrow GENERALIZED\,GEOMETRY$)

in this work. This subject will be discussed in Sect. 9.2. Section 9.3 considers the role of category theory in these projects and quantum information theory in general. Finally, because more recently quantum information theorists moved beyond the reconstruction project, I shall, by way of conclusion, comment on these new agendas in Sect. 9.4.

Keywords Category · Category theory · Circuit · Functor · Morphism · Operational-probabilistic framework · Purification · Quantum information theory · Reconstruction projects

9.1 Introduction

The argument of this chapter builds on that of Chap. 4, by virtue of the relationships between Heisenberg's thinking, leading him to his discovery of QM, and reconstruction projects—projects aiming at the derivation of quantum theory, specifically

A. Plotnitsky, *Reality Without Realism*,
https://doi.org/10.1007/978-3-030-84578-0_9

quantum theory in finite dimensions (QTFD) from a set of basic (first) principles—in quantum informational theory. These relationships are defined by the shared role of the diagrams:

$$\text{QPA/QPAGG } (QUANTUMNESS \rightarrow PROBABILITY \rightarrow ALGEBRA/ \\ QUANTUMNESS \rightarrow PROBABILITY \rightarrow ALGEBRA \rightarrow GENERALIZED\,GEOMETRY)$$

Although Heisenberg's originality and inventiveness in his creation of QM were remarkable, and although it would be difficult to match the outcome, the creation of QM, his derivation of QM may not have been as rigorous a derivation from the first principles as some might ideally wish. The main reason is that, while borrowing the *form* of equations from classical mechanics was a logical deduction of part of the mathematical structure of QM, by the correspondence postulate, derived from Bohr's correspondence principle, Heisenberg essentially "guessed" the variables he needed. As discussed in Chap. 4, this guess was not unprepared by the preceding developments of quantum theory, from Bohr 1913 theory on. But it was ultimately a guess, nonetheless. The derivation of QM was made more rigorous by the later work of Born, Jordan, and Heisenberg himself; more, but not entirely (Born and Jordan 1925; Born et al. 1926). Even the derivation, using these and related developments, especially Dirac's work, offered by Heisenberg in his Chicago lectures (Heisenberg 1930), might still be seen as falling short of a rigorous derivation from the first principles. As discussed in Chap. 5, Schrödinger's derivation of both of his equations, time-independent and time-dependent also involved guesses at key junctures. Later on, Dirac, too, as much guessed his spinors as deduced them from the first principles in deriving his equation for the relativistic electron. Heisenberg's comment in his Chicago lectures indicates the nature of such concerns, whether Heisenberg himself shared them or not, given that QM itself was established, which was what counted most then and still might do for some now. As he said:

> It should be distinctly understood that this [the deduction of the fundamental equations of QM] cannot be a deduction in the mathematical sense of the word, since the equations to be obtained form themselves the *postulates* of the theory. Although made highly plausible by the following considerations [offered in his lectures], their ultimate justification lies in the agreement of their predictions with the experiment. (Heisenberg 1930, p. 108)

This is an important point in the context of reconstruction projects, aiming at rederiving quantum theory from fundamental principles or postulates. The term axiom is used sometimes as well, but for the reasons explained in Chap. 4, I prefer postulates, when dealing with physics. Even in the case of von Neumann's "axioms" (von Neumann 1932), which aims at establishing "the mathematical foundation of quantum mechanics," some of his grounding assumptions, which include Schrödinger's time-dependent equation, are physical postulates rather than mathematical axioms. The situation also opens the question concerning which principles or postulates are considered "natural" or "reasonable," or what constitutes a rigorous derivation. None of this has, thus, far been established with much consensus. One might argue, for example, that it is not sufficiently first-principle-like to see the equations of QM as postulates, and one might also prefer a less mixed derivation of quantum theory, QM, QTFD, or QFT. On the other hand, one might also question the necessity of a *fully* rigorous derivation. After all, with Heisenberg, especially as his scheme was developed by Born and Jordan, and then differently by Dirac, QM was established as a correct theory, as it was by Schrödinger. Then Dirac similarly invented QED. While a derivation of QM, or QTFD, might be made more rigorous than that offered by Heisenberg, one might also doubt that any such derivation, *from first principles*, could ever be as rigorous as "a deduction in the mathematical sense of the word." As Hardy suggested (Hardy 2001), it may be more a matter of what are "reasonable" initial axioms or postulates, although such a "reasonableness" is not an unconditional matter either.

These considerations do not, of course, diminish the significance of this line of research. Besides, as discussed in Sect. 9.4 below, more than merely rederiving QTFD is at stake in these programs, in particular, a deeper understanding of quantum phenomena and quantum theory, extendable to QFT and beyond via quantum information theory. So far, most attempts at such derivations along the lines of quantum information theory have concerned QTFD, if with the understanding that the principles involved may be extended (with a possible addition of other principles) to QM and QFT, and even to quantum gravity. Some of these efforts, however, have affinities with Heisenberg's approach, which, as stated in Chap. 3, exhibits a spirit of quantum informational thinking, along with the QPA/QPAGG structure:

$$QUANTUMNESS \rightarrow PROBABILITY \rightarrow ALGEBRA/QUANTUMNESS \rightarrow PROBABILITY \rightarrow ALGEBRA \rightarrow GENERALIZED\, GEOMETRY$$

I shall, in Sect. 9.2, consider two such approaches, the first is by D'Ariano and coworkers, Giulio Chiribella and Paolo Perinotti (hereafter DACP), and the second is by Hardy. Both are in accord with the QPA/QPAGG structure, although not necessarily with the RWR view, which was central to Heisenberg's thinking in his derivation. Hardy, in particular, appears to want an ontological view of the ultimate constitution of reality responsible for quantum phenomena.

Section 9.3 extends the analysis in Sect. 9.2 to consider the role of category theory in quantum information theory, a subject that acquired prominence in recent quantum informational approaches, in part in connection with computer science. The discussion offered in this section offers a new philosophical view on the subject. This view leads to a different, in terms of category theory, *functorial*, understanding of relationships between the algebraic structures defined by the arrangements of circuits (essentially measuring apparatuses), central to quantum information theory, and the mathematical formalism of quantum theory, rather than, as has been suggested by some quantum information theorists, by merely establishing an isomorphism or even only homomorphism between the first and the second structures. This new type of relationship is dictated by the RWR view, given that, in this view, the formalism of quantum theory, for either discrete or continuous variables, only offers probabilistic predictions concerning quantum experiments, without offering a representation or even conception of either quantum objects and behavior (or in the present interpretation the ultimate constitution of the reality responsible for quantum phenomena and quantum objects, both defined at the time of measurement) or measuring instruments, circuits, described by classical physics. Hence, there can be no such isomorphism or homomorphism. What may be possible instead is establishing what is known as a "functor" between two *multiplicities*, defined as "categories," of objects and the maps, "morphisms," between these objects. The first category would be defined by organized circuits and maps, "morphism," between them, and the second (which is already available) by Hilbert spaces over $\mathbb{C}$ and maps, "morphisms," between them. Such a functor would then relate each organized circuit to a Hilbert space. This relation would be different from a map or morphism (isomorphism or even homomorphism) between organized circuits and Hilbert spaces. Instead, it would allow one to predict, by using these Hilbert spaces (with the help of Born's or analogous rules), the probabilities of events observed in circuits, just as one does them now.

Section 9.4 briefly assesses the future of quantum information theory beyond QM.

9.2 Circuits, Operations, and Probabilities: Quantum Information Theories as Principle Theories

DACP's program, developed over the last decade and presented comprehensively in their book (D'Ariano et al 2017), belongs to the reconstructive trend in quantum information theory, and as most of the work there, it deals with discrete variables and the corresponding, finite dimensional, Hilbert spaces (e.g., Hardy 2001; Fuchs 2003; Chiribella and Spekkens 2016).[1] This is in part because, as DACP note (a view shared by others in this field), "the study of finite-dimensional systems allows one to decouple the conceptual difficulties in our understanding of quantum theory from the technical difficulties of infinite-dimensional systems" (D'Ariano et al 2017, p. 4). DACP's project is motivated by "a need for a deeper understanding of quantum theory [QTFD] in terms of fundamental principles," and by the aim of deriving QTFD from such principles, which, the authors contend, has never been quite achieved by their predecessors.

As elsewhere, the fluctuations of terms such as principles, axioms, postulates, commonly used in such reconstructive projects may be confusing and obscure their common aims. DACP use the term "axioms" as well, as does Hardy. In present terminology, established in Chap. 4, these axioms are postulates. On the other hand, while one can surmise DACP's understanding of the term "principle" from their use of it, they do not define the concept of "principle" either. As in Chap. 4, I

[1] These references only serve as influential examples. There have been numerous quantum informational attempts at reconstructing QTFD during the last decades. My main concerns in this chapter, however, are the role fundamental principles and postulates considered in quantum information theory, and the relationships among physics, mathematics, and philosophy there, from the RWR viewpoint, rather than reconstruction projects as such, and both DACP's and Hardy's work exemplify these concerns.

adopt the concept of principle defined by Einstein's view of principle theories. To restate his definition, principle theories: "The elements which form their [principle theories'] basis and starting point are not hypothetically constructed but empirically discovered ones, general characteristics of natural processes, principles that give rise to mathematically formulated criteria which the separate processes or the theoretical representations of them have to satisfy" (Einstein 1919, p. 228). This concept is, I would argue, in accord with DACP's use of principles, and their derivation of QTFD, or that of Hardy, may be seen as that of a principle theory in this sense.[2]

The main new feature of DACP's approach is adding to the view of QTFD as an extension of probability theory (a view found in earlier works and applicable to QM as well), "the crucial ingredient of *connectivity* among events" by using the operational framework of "circuits" (the term adopted in part from computer science and engineering, and also used by Wheeler, including in the quantum-informational context) and giving it a mathematics to the arrangements of circuits (D'Ariano et al. 2017, pp. 4–5). The framework of circuits has been used by others, including by Hardy, as considered later in this chapter. This addition allows DACP "to derive key results of quantum information theory and general features of quantum theory [QTFD]" without first assuming Hilbert spaces. Unlike von Neumann, to whom DACP refer for a contrast, Heisenberg, as we have seen, did not begin with a formalism either, although he used the equations of classical mechanics by the correspondence principle, made into a postulate. Thus, he *arrived* at his formalism from fundamental principles or postulates, even if not fully rigorously *derived* it from them, by more or less guessing new matrix variables. By contrast, von Neumann's aim was not to derive QM but to give a more rigorous formulation of the mathematical formalism of it as an already existing theory. In his more recent work, referenced in Chap. 2, D'Ariano, too, begins, axiomatically, with Hilbert spaces and their mathematical features, such as their tensor products, but his axioms are very different from those of von Neumann, especially, as concerns what he calls von Neumann's "unitarity dogma" (D'Ariano 2020).

Among the principles adopted by DACP, the purification principle plays a unique role as an essentially quantum principle, which distinguishes QTFD from classical probabilistic information theories, and in its mathematical form, as a postulate, is a form of symmetry. According to DACP: "The purification of mixed states is specifically quantum" (D'Ariano et al. 2017, p. 169). It is the single principle necessary for establishing this distinction, which may not be surprising given the history of quantum theory and attempts at its axiomatic derivations. Hardy's derivation also needed only one axiom, the continuity axiom, to do so (Hardy 2001). On the other hand, the fact that Hardy's continuity axiom is different from DACP's purification postulate suggests that there may not be a single system of postulates or principles from which QMDT could be derived, just as QM was originally established by Heisenberg and Schrödinger from very different principle and postulates.

[2] As I said, Hardy's use of "axioms" may be seen as somewhat different from the use of "principles" by DACP, but these differences do not affect my argument, insofar as in both cases one deals with the derivation of QTFD as a principle theory in Einstein's sense.

Whether one can define quantum phenomena as definitively different from classical phenomena remains a question, as indicated in Chap. 2 in connection with Planck's constant, *h*. While this question and literature concerning it are beyond my scope, I would like to comment on Spekkens's work, relevant here because it proceeds along the lines of quantum information theory. Spekkens introduced several toy models or theories, "epirestricted theories" (so-called because of epistemic restrictions on classical theory, assumed as a starting point), that reproduce many quantum phenomena and features of QM or QTFD, such as the presence of *h*, the uncertainty relations, noncommuting operator observables, entanglement, or the purification of mixed states (Spekkens 2007, 2016; Bartlett et al. 2012). Many but not all! These models expressly fail to reproduce some among the crucial features of quantum theory, specifically some of those dealing with correlations and entanglement, such as violations of Bell's inequalities and the existence of the Kochen–Specker theorem. This is conceptually important because these theorems reflect the essential features of quantum phenomena (in the case of Bell's inequalities independently of any theory, which of course must, if correct, satisfy them), including those related to the question of quantum nonlocality vis-à-vis realism (Spekkens 2016, p. 91). Thus, whether actual quantum phenomena can be captured by models of the type proposed by Spekkens remains an open question. According to Spekkens himself, this is unlikely:

> The investigation of epirestricted theories, therefore, need not—and indeed *should not*— be considered as the first step in a research program that seeks to find a ψ-epistemic ontological model [a realist model that assumes quantum states in the formalism to be epistemic, states of knowledge] of the full quantum theory. Even though such a model could always circumvent any no-go theorems by violating their assumptions, it would be just as unsatisfying as a ψ-ontic model [a realist model that assumes quantum states to be ontic, states of reality] insofar as it would need to be explicitly nonlocal and contextual. Rather, the investigation of epirestricted theories is best considered as a first step in a larger research program wherein the framework of ontological models—in particular the use of classical probability theory for describing an agent's incomplete knowledge—is ultimately rejected, but where one holds fast to the notion that a quantum state is epistemic. (Spekkens 2016, p. 95)

Ontological here has a different meaning from the one assumed in this study, which is synonymous with "realist" and thus is closer to Spekkens "ontic." As noted, the meaning of these terms fluctuates, and Spekkens's use is somewhat uncommon. This is, however, a secondary matter here. More pertinently, although the expression is common in quantum information theory, viewing quantum states (state vectors in the formalism) as "states of knowledge" could be misleading, at least, unless one refers to this knowledge as probabilistic or statistical. Even then quantum states are only part of the corresponding predictive machinery, in themselves they are not sufficient and one needs Born's or related rules to obtain these probabilities or statistics. They are not states of knowledge. States of knowledge are (mental) states of knowledge we form using quantum states and Born's rule. Besides, from the present, RWR-type, perspective, in which quantum states never refer to any actual knowledge (only obtained in measurements), our knowledge, actual or probabilistic, is incomplete only when compared with the kind of knowledge possible in classical physics or relativity, within their scope. In terms of this study, while QTFD (or QM or QFT), is *Einstein-incomplete*, it is as complete as possible for a theory of quantum phenomena to be, as things stand now, is *Bohr-complete.*

I am not sure whether, his skepticism concerning ontic, or in present terms, realist, models of quantum theory and the use of classical probability (based on "an agent's incomplete knowledge") notwithstanding, Spekkens would be willing to go as far as the RWR view, especially the strong RWR view, in which, as noted, this use of probability is inapplicable by definition. It appears, however, at least to the present reader, that his epirestricted theories and his argument suggest that the difference between the classical and the quantum may ultimately be irreducible, even though rigorously grounding reconstructive programs remains difficult. According to Spekkens, "it may be that there are particularly elegant axiomatic schemes that are currently in our reach and the road to program involves temporarily setting one's sight a bit lower," moving to partial reconstructive models of the kind he proposes (Spekkens 2016, p. 96). That may be, especially in the case of continuous variables, to which Spekkens's concept of an epirestricted theory would apply. DACP's or Hardy's approach may suggest otherwise, even while dealing only with QTFD.

There is also a question of whether such elegance should be important here. While Dirac, famously, as a proponent of such elegance or beauty, would have thought so, neither Heisenberg (in his derivation) nor Bohr was much worried about this elegance. It is true that an "elegant axiomatic scheme" is not the same as the mathematical beauty of physical laws, invoked by Dirac. Part of the ethos of reconstructive projects is defined by the imperative of a derivation of at least QTFD from more natural or reasonable first principles or postulates than those that grounded the original derivations of QM, which were moreover, semi-intuitive or even guess-like, by Heisenberg, Dirac, or Schrödinger, versus for example, the principles guiding Einstein's derivation of special relativity (Fuchs 2003). Whether, given the more complex (physically, mathematically, and philosophically) nature of QM, this is actually a viable comparison also remains a question, in my view. In fact, given that relativity could be derived otherwise, it might be a question, too, how natural Einstein's postulates were (e.g., D'Ariano 2017, p. 124). Such concepts as reasonable and natural are historically contingent. What now appears natural with quantum theory (or relativity) in hand was not natural before these theories were introduced. It is difficult to be certain which trajectories will lead us there, especially if this "there" is beyond QTFD, or QM and QFT, as it ultimately must be, for otherwise we are already there, insofar as we have the corresponding theories in place. There may not be one such trajectory. In the discovery of QM, there were two—Heisenberg's and Schrödinger's, the first of which appeared quite unnatural to most from the outset, and the second quickly lost its apparent greater naturalness (as a wave mechanics). As discussed in Chap. 5, there was, at the time, not the least for Schrödinger himself, nothing especially natural in the complex wave function.

I now return to DACP's derivation of QTFD and the purification principle, which they see as "specifically quantum." In nontechnical terms, the purification principle states that "every random preparation of a system can be achieved by a pure preparation of the system with an environment, in a way that is essentially unique"

(D'Ariano et al. 2017, p. 6). The principle originates in Schrödinger's response, in several papers, to the EPR paper and his concept of entanglement, as discussed in Chap. 7 (Sect. 7.6.). According to DACP:

> The purification principle stipulates that, whenever you are ignorant about the state of a system A, you can always claim that your ignorance comes from the fact that A is part of a large [composite] system AB, of which you have full knowledge. When you do this, the pure state that you have to assign to the composite system AB is determined by the state of A in an essentially unique way.
>
> The purification of mixed states is a peculiar feature—surely, not one that we experience in our everyday life. How can you claim that you know A *and* B if you don't have A alone? This counterintuitive feature has been noted in the early days of quantum theory, when Erwin Schrödinger famously wrote: "Another way of expressing the peculiar situation is: *the best possible knowledge of a whole does not necessarily include the best possible knowledge of all its parts.*" And, in the same paper: "I would not call that *one* but rather *the* characteristic trait of quantum mechanics, the one that enforced its entire departure from classical lines of thought" (Schrödinger 1935, p. 555). (D'Ariano et al. 2017, p. 168)

Technically, as implied by this statement, the situation needs to be clarified as regards measurement and thus circuits, only in relation to which one can speak of any knowledge or information here, actual or possible, but this subject could be put aside for the moment. DACP's purification principle offers a rigorous mathematical expression of this situation. They add, using the principle to link physics and information theory:

> The purification of mixed states is specifically quantum. But why should we assume it as a fundamental principle of Nature? At first, it looks like a weird feature—and it must look so, because quantum theory itself is weird and if you squeeze it inside a principle, it is likely that the principle looks weird too. However, on second thought one realizes that purification is a fundamental requirement: essentially, it is the link between physics and information theory. Information theory would not make sense without the notions of probability and mixed state, for the whole point about information is that there are things that we do not know in advance. But in the world of classical physics of Newton and Laplace, every event is determined and there is no space for information at the fundamental level. In principle, it does not make sense to toss a coin or to play a game of chance, for the outcome is already determined and, with sufficient technology and computational power, can always be predicted. In contrast, purification tells us that "ignorance is physical." Every mixed state can be generated in a single shot by a reliable procedure, which consists of putting two systems in a pure state and discarding one of them. As a result of this procedure, the remaining system will be a physical token of our ignorance. This discussion suggests that, only if purification holds, information can aspire to a fundamental role in physics. (D'Ariano et al. 2017, p. 169)

In the present view, the purification of mixed states is the principle arising in *our interactions with nature* by means of our experimental technology, rather than of nature itself, except insofar as we and our technologies are also nature. But then, in the present view, there are no other principles of nature, any more than any information, than those defined by us in our interactions with nature. DACP's formulation of the purification principle and their derivation of QTFD allows for an RWR-type interpretation, and I shall interpret it in this way, without claiming DACP's own interpretation of it is of the RWR-type. I should add that the purification principle also reflects a crucial feature of quantum physics, discussed in Chap. 7 in the context of the EPR experiment, the context of Schrödinger's assessment cited by DACP. This feature was emphasized throughout the work of D'Ariano and coworkers (but often missed by other quantum information theorists and others). In quantum theory, we always deal with compositions of systems, never individual systems, even in individual measurements, in which case, the composed system consists of the quantum objects under investigation and the measuring instruments, including their quantum strata. These strata enable the interaction between the instruments and the quantum objects, or in the present view, the ultimate constitution of the reality responsible for quantum phenomena.

Thinking in terms of "circuits" is close to Bohr's thinking concerning the role of measuring instruments in the constitution of quantum phenomena, as distinguished from quantum objects, which give rise to quantum phenomena by interacting with measuring instruments but which are never observable. Bohr, with the formalism of QM in place, was primarily concerned with interpretation rather than, as DACP (or Hardy), with giving the phenomena thus observed a mathematized structure, which could then be used (with certain additional postulates) to derive the formalism of QTFD. Circuits and their arrangements, too, embody those of measuring instruments capable of detecting quantum events, and thus enabling the probabilistic predictions of future events. Their arrangements and operations are enabled by rules that should ideally be derived from certain sufficiently *natural* assumptions (keeping in mind earlier qualifications concerning this naturalness). They are described classically, and represent (in this case in the usual realist sense) the structure of quantum information as a particular form of organization of classical information, an organization that cannot, however, be predicted by classical physics. This organization, supplemented by certain additional principles, can be used, as by DACP (or Hardy), in attempts to *derive* the mathematical formalism of QTFD. In fact, as discussed earlier, Heisenberg's derivation and then Bohr's interpretation were similarly based on the informational organization of the data manifested in classically observed quantum phenomena and a set of principles, under RWR-type assumptions, although not on the organization of measuring instruments, except implicitly, for example, in view of complementarity, as discussed in connection with Schwinger's argument concerning noncommutativity in Chap. 4.

While uniquely necessary in grounding the authors' derivation of QTFD, the purification postulate or principle (it functions in either capacity) is not sufficient to do so. They need five additional postulates (termed "axioms"): *causality* (essentially locality), *local discriminability*, *perfect distinguishability*, *ideal compression*, and *atomicity of composition* (D'Ariano et al. 2017, p. 6). These postulates define a large class of classical probabilistic informational theories, while the purification postulate distinguishes QTFD. The appearance of these additional postulates or principles is not surprising. Heisenberg's grounding principles, the quantum discreteness (QD) principle and the quantum probability or statistics (QP/QS) principle, were not sufficient for him to derive QM either. To do so, he needed the correspondence principle, which gave him half of the mathematical architecture of quantum theory. The other half was supplied by his matrix variables, not expressly derived from any specific principle, but enabling him to bring the formalism in accord with other principles and predict the data in question. There are, however, instructive parallels between DACP's and Heisenberg's approaches. As DACP say in their earlier article:

> The operational-probabilistic framework combines the operational language of circuits with the toolbox of probability theory: on the one hand experiments are described by circuits resulting from the connection of physical devices, on the other hand each device in the circuit can have classical outcomes and the theory provides the probability distribution of outcomes when the devices are connected to form closed circuits (that is, circuits that start with a preparation and end with a measurement). (Chiribella et al. 2011, p. 3)

This is similar to Heisenberg's thinking in his paper introducing QM, as these classical outcomes are discrete in both cases, even though Heisenberg was dealing with continuous variables. There are no other measurement outcomes in quantum physics, by the QD (quantum discreteness postulate). On the other hand, as noted above, the concept of "circuit" is closer to Bohr's view of the role of measuring instruments and his concept of phenomenon, defined by this role. Bohr's view was, arguably, the first operational view in this sense. As discussed in Chap. 4, however, the idea that in quantum theory we only deal with transition probabilities between the outcomes of the interactions between quantum objects and measuring instruments was introduced by Heisenberg, as part of his approach to QM, and was then adopted by Bohr. Heisenberg discovered that Bohr's frequency rules are satisfied by using, in general, noncommuting matrix variables with complex coefficients, from which one derives, by means of a Born-type rule, probabilities or statistics for transitions between stationary states, manifested in spectra observed in measuring devices. Thus, Heisenberg's derivation depended on measuring instruments as devices with classically describable observable parts, which are akin to "operational circuits," in his case dealing with continuous variables.

DACP aim to arrive at the mathematical structure of QTFD in a more first-principle-like way, in particular, independently of classical physics, which, because of the correspondence principle, was central to Heisenberg. Classical physics, to begin with, does not have discrete variables, such as spin, so there is no correspondence principle as concerns them. The rules governing the structure of operational devices, circuits, should, DACP argue, allow them to do so, because these rules are *empirical.* They are, however, *not completely empirical,* because, as discussed from the outset of this study, nothing ultimately ever is in physics, which is "experimental because of its mathematical project" (Heidegger 1967, p. 93). In this case, in particular, along with these rules, circuits *are given* a formal and, to some degree, mathematical structure by human agents, even though this structure may be partially defined by the organization, human in turn of the experimental arrangements in which circuits appear, for example, again, dues complementarity. The mathematized structure of circuits becomes a necessary condition for establishing the mathematical structure or algebra of QTFD, but it is not sufficient to do so. As noted in considering Schwinger's argument concerning quantum measurement, in Chap. 4, these two structures are not isomorphic or even homomorphic, for one thing, given the complexities of providing the mathematics to events of measurements, even if not to the organization of circuits. The mathematics of the formalism may be connected, both in terms of predictions and in terms of structure, to quantum measurement and thus to the organization of circuits, but it does not mathematically represent it. This means that additional structural entities, provided by additional postulates or axioms, are required to reach the (Hilbert-space or other) formalism over $\mathbb{C}$, while everything necessary for representing circles is defined over $\mathbb{R}$. DACP's derivation requires enormous technical work. It is next to impossible to do it justice here, with the preceding considerations, hopefully, sufficient to explain the main gradient of their thinking.[3] I shall instead close this section by considering, also as a bridge to the next section, the algebra of circuits in Hardy's work.

Hardy has a different set of main assumptions necessary to derive QTFD than those of DACP, but the main strategy is the same: establishing the architecture, "algebra," of circuits, technically of operations, that, with additional axioms, would allow one to derive the mathematical formalism of QTFD. According to Hardy:

> Circuits have:
>
> - A setting, s(H), given by specifying the setting on each operation.
> - An outcome set, o(H), given by specifying the outcome set at each operation (equals $o(A) \times o(B) \times o(C) \times o(D) \times o(E)$ in this case). We say the fragment "happened" if the outcome is in the outcome set.
> - A wiring, w (E), given by specifying which input/output pairs are wired together.
>
> (Hardy 2013, p. 7)

With this definition in hand, I shall comment on some of Hardy's fundamental assumptions discussed by him in a different paper. Hardy says:

> We will make two assumptions to set up the framework in this paper. ...
>
> **Assumption 9.1** The probability, Prob (A), for any circuit, A (this has no open inputs or outputs), is well conditioned. It is, therefore, determined by the operations and the wiring of the circuit alone and is independent of settings and outcomes elsewhere (Hardy 2010, p. 11).

This is a physical postulate, essentially that of locality, combined with probability or statistics, along the lines of the QP/QS principle. The task now becomes how to derive a QTFD that could correctly predict these probabilities. One needs another assumption:

> **Assumption 9.2** (*Operations are fully decomposable*) We assume that any operation $\mathrm{A}_{a1b2...c3}^{d4e5...f6}$ can be written, ... in a symbolic notation, $\mathrm{A}_{a1b2...c3}^{d4e5...f6} \equiv {}^{d4e5...f6}\mathrm{A}_{a1b2...c3}\mathrm{X}_{\mathrm{a1}}^{a1}\mathrm{X}_{\mathrm{b2}}^{b2}\cdots\mathrm{X}_{\mathrm{c3}d4}^{c3}\mathrm{X}_{e5}^{\mathrm{d4}}\mathrm{X}_{...f6}^{\mathrm{e5}}\mathrm{X}^{\mathrm{f6}}$. *In words we will say that any operationis equivalent to a linear combination of operations each of which consists of an effect for each input and a preparation for each output.* ... We allow the possibility that the entries in ${}_{d4e5...f6}\mathrm{A}_{a1b2...c3}$ are negative (and this will, indeed, be the case in quantum theory). Hence, in general, this cannot be thought of as physical mixing. ... (Hardy 2010, p. 19; emphasis added)

[3] One still needs a Born-*type* rule for getting from the formalism (defined over $\mathbb{C}$) to probabilities, which becomes an additional postulate, even if not expressly stated as a postulate, but still used as such, in DACP's scheme of derivation (D'Ariano et al. 2017, pp. 74–75). A version of Born's rule is expressly postulated, within D'Ariano's "minimal" set of axioms versus those of von Neumann, in D'Ariano (2020, pp. 1923–1924).

To make my main conceptual point, I only need the emphasized sentence, which means that one aims at constructing a suitable mathematized structure for each measuring or operational sequence. I cite the passage at a greater length to illustrate the algebraic structure of operations in Hardy's scheme, keeping in mind qualifications made above concerning the complexity of the connections between the operations of the formalism and the operations enacted in circuits, such as those defined by the order of successive measurement and its effect on the outcomes, in correspondence with noncommutativity of the corresponding *operators* in the formalism. Hardy then says:

> Assumption 9.2 introduces a *subtly* different attitude than the usual one concerning how we think about what an operation is. Usually we think of operations as effecting a transformation on systems as they pass through. Here we think of an operation as corresponding to a bunch of separate effects and preparations. We need not think of systems as things that preserve their identity as they pass through—we do not use the same labels for wires coming out as going in. This is certainly a more natural attitude when there can be different numbers of input and output systems and when they can be of different types. Both classical and quantum transformations satisfy this assumption. In spite of the different attitude just mentioned, we can implement arbitrary transformations, such as unitary transformations in quantum theory, by taking an appropriate sum over such effects and preparation operations. (Hardy 2010, pp. 19–20)

More natural or not, this rethinking of the concept of operation is important, especially if one adopts an RWR-type view. An "operation" is now defined in terms of observable "effects" of the interactions between quantum objects and measuring instruments, and not in terms of what happens, even in the course of these interactions (let alone apart from them), to the quantum objects or systems, *considered as independent systems*. In the present view, again, a quantum object can only be defined in and by this interaction. It is useful that we can treat classical systems in this way as well. In the classical case, however, we can, equivalently, use a more conventional concept of operation mentioned here, which is not the case in quantum theory. After a technical discussion of "duotensors," which I put aside, Hardy suggests a principle:

> *Physics to mathematics correspondence principle*. For any physical theory, there [exist] a small number of simple hybrid statements that enable us to translate from the physical description to the corresponding mathematical calculation such that the mathematical calculation (in appropriate notation) looks the same as the physical description (in appropriate notation).
>
> Such a principle might be useful in obtaining new physical theories (such as a theory of quantum gravity). Related ideas to this have been considered by category theorists. A category of physical processes can be defined corresponding to the physical description. A category corresponding to the mathematical calculation can also be given. The mapping from the first category to the second is given by a functor (this takes us from one category to another). (Hardy 2010, p. 39)

The language of correspondence should not mislead one into relating this principle to Bohr's correspondence principle, even in Heisenberg's *mathematical* form of his correspondence postulate. Bohr's correspondence principle deals with the correspondence between different physical theories (such as classical mechanics and quantum mechanics), insofar as their predictions would coincide in the regions where both theories could be used. By contrast, Hardy's "hybrid" construction implies that the category of physical processes could be "translated" into a proper formalism of QTFD, which could then, in terms of its mathematical calculations, be related to what is observed. For the reasons that I shall discuss in the next section, Hardy's principle may be better understood as or changed into the "physics to mathematics functoriality principle," as this translation, if possible, might and, in RWR-type interpretations, must, be seen as functorial in the sense of category theory. Hardy's suggestion is speculative and is not really worked out. It offers, however, an inviting starting point to address a possible role of category theory in quantum information theory or quantum theory in general because of Hardy's appeal to "a category of physical processes" and a possible functor from this category to that of mathematical calculation, versus recent suggestions along these lines by those using category theory in quantum information theory, to whom Hardy appears to refer here. As explained in the next section, these theorists (e.g., Robert Coecke and Samson Abramsky) only consider the second type of categories, those associated with the formalism.

Taking Hardy's suggestion as a point of departure, I shall, in the next section, address a possible role of category theory in the algebraization (QPA) and then re-geometrization (QPAGG) of physics. I do so without claiming that these comments are in accord with Hardy's view of the situation, because they are grounded in the RWR view. Hardy appears to prefer a realist view, "a search for what we hope to be the correct ontology of the world" (Hardy 2011), unless he assumes, as one would in adopting the RWR view, that the correct ontology of the world includes the strata of reality that is beyond thought and thus beyond any ontological representation. This, however, does not appear to be the case in Hardy, as there is no such qualification offered by Hardy on this occasion or elsewhere.

9.3 Thinking Categorically: From Morphisms to Functors

Category theory originated in algebraic topology in order to help study certain algebraic invariants, such as cohomology or homotopy groups, associated with topological spaces (MacLane 2013). It was then extensively used, especially thanks to Alexandre Grothendieck's work, in algebraic geometry, and eventually extended, via topos theory, introduced by Grothendieck as well, to mathematical logic (MacLane and Moerdijk 1994). Roughly, category theory considers multiplicities (categories) of mathematical objects conforming to a given concept, such as the category of topological spaces or geometrical spaces (say, Riemannian manifolds), and morphisms, also called arrows ($X \rightarrow Y$), which are the mappings between these objects that preserve this concept or structure. Studying morphisms allows one to learn about the individual objects involved more than we would by considering them just in themselves. Thus, in geometry, one does not have to start with a Euclidean space. Instead, the latter is just one specifiable object of a large categorical multiplicity, such as the category of Riemannian manifolds, an object marked by the particularly simple way we can measure the distance between points. Categories themselves may be viewed as multiplicities with arrows between them, and in this case, one speaks of "functors" rather than "morphisms."

Now, one can more easily think of properly defining mathematically the second category in Hardy's suggestion, say, as that of Hilbert spaces or some equivalent mathematical structures (such as C* algebras), by means of which the mathematical calculations in question in quantum theory, such QTFD, would be performed. On the other hand, the first category, that is, the structure of its objects and the morphisms between them, is a more complex matter, as is, it follows, that of a functor between these two categories. Hardy's formulation—"for any physical theory, there [exist] a small number of simple hybrid statements that enable us to translate from the physical description to the corresponding mathematical calculation such that the mathematical calculation (in appropriate notation) *looks the same* as the physical description (in appropriate notation)" (emphasis added)—might require qualification as concerns the meaning of the phrase "looks the same" and the relationships between "calculation" and "description." This formulation allows for different interpretations, especially given that "calculation" and "description" are different concepts. In particular, in what sense would they look the same? A similar question arises concerning his characterization of the functor in question as "*virtually* direct" (emphasis added), and even the statement "a category of physical processes can be defined *corresponding* to the physical description" (emphasis added).

I shall now sketch the reasons for requiring these qualifications, beginning with a possible meaning of "*physical processes*" in the statement that "a category of physical processes can be defined corresponding to the physical description." "Physical processes" may refer either to quantum processes or to circuits, which appears more likely because "the physical description" in Hardy's scheme is given only at the level of circuits. This view would also be suggested by the category theorists Hardy appears to refer to, those who work on using categories in quantum theory, specifically QTFD (considered by Hardy), such as Abramsky, Coecke, and others (e.g., Abramsky and Brandenburger 2011; Coecke and Kissinger 2017; Coecke 2017). Most of this work is primarily concerned with recasting the Hilbert-space language of QTFD formalism in a categorical framework by replacing Hilbert spaces, belonging to the category of the finite-dimensional Hilbert spaces over $\mathbb{C}$ (e.g., Hasegawa et al 2008), with objects of monoidal categories and the morphism between Hilbert spaces with morphisms between these objects, rather than with deriving the formalism from the first principles by using categories. There are some moves in this direction, in part following DACP's approach, for example, by starting with a suitable simple monoidal category and then considering how many additional pieces of structure one needs to arrive at QTFD (e.g., Coecke and Kissinger 2017). Essentially, however, at stake in this and related works is a recasting of formal "compositional" systems into a categorical framework (Coecke 2017), a project different from the one I am considering at the moment. This work further testifies to the dominant role of algebra in quantum theory, even though the work just mentioned uses a lot of diagrammatic operations. "Diagrams beat algebra," Coecke and Kissinger say (p. 50). In reality, however, while they make the algebra of QM easier to process for some, these diagrams still represent algebra and sometimes are in the way of geometrical thinking, say, of the type found in Heisenberg and Dirac, which the Hilbert-space formalism invites. Coecke and Kissinger's claim that "the traditional, Hilbert-space base language of quantum theory forms a major obstruction to discovering [the key quantum features]" is questionable, given that this is how these features were discovered in the first place (p. 9). It is also unclear whether their approach will be of much, if any, help in QFT, where the Hilbert space formalism is so effective, let alone beyond. More significantly, this approach manifestly amounts to a representation, helpful as it may be, of *logic* and so far, after two decades, it has not led to any really new physical concepts. This is also the difference between the use of category theory in algebraic topology (where it originated) or algebraic geometry, where this use led to an unending stream of new concepts, and mathematical logic, even though some new concepts did emerge there as well. I shall comment on this subject below. What is needed here is, I argue, a new concept, that of the category of circuits, as arrangements of measuring instruments.

As Coecke and Kissinger's title, "picturing *quantum processes*," indicates and as their argument suggests, they (and Coecke elsewhere) appear to hold a realist view in assuming that their diagrammatic picturalism represents actual quantum behavior. This is also suggested by their invocation of Whitehead (p. 251), whose views, as discussed earlier (Chap. 6, n. 50) are manifestly ontological. In fairness, "quantum processes" also refer to the formalism as such, which, however, still suggests a realist or ontological view. For otherwise, what do "processes" mean, unless something mental, which can be divorced from physical processes? Coecke and Kissinger do not appear to see the situation in this way or, in any event, do not consider this possibility. They do argue that quantum phenomena disallow local realism of the type Einstein wanted (pp. 656–661). On the other hand, their glosses of the views of Bohr or Heisenberg are misleading, as, for example, "any attempt to observe is bound to disturb" (p. 396). This "summation" obliterates the complexity of Bohr's and Heisenberg's thinking, given that, as discussed in this book, both are quite hesitant to speak of "disturbance." In any event, I would argue that, regardless of how much physical reality Coecke and Kissinger believe corresponds to their "diagrammatic picturalism," the latter is a diagrammatic mapping of quantum formalism and not of quantum phenomena, as effects (without classical causes), defined by the ultimate nature of physical reality responsible for them, effects that this formalism predicts. By the same token, their scheme only offers a categorical approach to the formalism and not to the functorial relationship between the categorical structures of the quantum formalism and those of measurement represented by circuits. In fact, although they maintain that observations are physically classical, their concept of circuit belongs to their diagrammatic representation of the formalism and does not refer to the arrangement of measuring devices, as does that of D'Ariano and coworkers. This is a crucial difference, especially in the present set of contexts, because the view of circuits as arrangements of measuring devices makes the situation open to an RWR-type interpretation, even if D'Ariano and coworkers do not adopt such an interpretation. They do not appear to take a strong stance on this point. On the other hand, Hardy, whose scheme is also defined by this type of relationship between the formalism and classical circuits, appears, as noted, to be guided by a realist or ontological view of the objects and behavior responsible for what is observed in circuits, by "a search for what we hope to be the correct ontology of the world" (Hardy 2011).

If the ultimate nature reality responsible for quantum phenomena is given a representation, the corresponding category may need to incorporate this representation in one way or another, by combining it with the effects of the ultimate constitution of the reality responsible for quantum phenomena, effects, thus, manifested in circuits and their arrangements. On the other hand, if one adopts an RWR-type view, one only deals with the physical description of circuits, which amounts to dealing only with the description of measuring instruments, given that the ultimate constitution of reality responsible for quantum phenomena, is beyond representation or even conception, and thus cannot be assumed to form a category either. Circuits and their organizations embody the arrangements of measuring instruments capable of detecting quantum events and enabling the probabilistic or statistical predictions of future events. These predictions, thus, relate to structures that may be mathematizable and as such *categorially*, which is to say, *functorially*, connected with the mathematical formalism of QTFD, which enables the necessary mathematical calculations. I speak of functorial connections, rather than, as Hardy does, "translations," because, as discussed in considering both Schwinger's argument in Chap. 4 and DACP's derivation of QTFD in this chapter, one need not and, in the RWR-type view, should not expect a representational correspondence ("translation") between the structures of circuits and the mathematical structures of the formalism of QTFD. For one thing, one, again, needs a formalism, a Hilbert-space, C*-algebra, or monoidal-category-one over $\mathbb{C}$, while as classical, the data manifested in circuits is over $\mathbb{R}$, and then one still needs a Born-type rule to get to probabilities.

If one adopts the RWR view, then, one could attempt to establish a category of circuits, each defined by some sort of algebraic structure of units and operations, and morphisms between them, and then, by using some reasonable axioms a category of mathematical objects defining the formalism, such as Hilbert spaces or some monoidal categories, and morphism between them (along, again, with a Born-like rule for probabilities). One would, then, connect these two categories by means of a well-defined functor. The advantage of this approach is that it preserves the role of the mathematical structure of circuits, which are observable. The derivation of the formalism could, then, consist in deriving the corresponding mathematical category from the category of the circuits. The difficulty is, again, that there does not appear to be a readily available categorical structure for circuits: their inherent mathematical structures, morphisms between them, and so forth. All of these need to be defined. This is, however, also true if one adopts a more realist view, which would in any event pertain to a representation of quantum objects and behavior by the formalism of QTFD rather than by the mathematical description of circuits. To build up a category of circuits, beginning with making them mathematical objects and giving them a mathematical structure would require additional entities and postulates.

In the approach just outlined, then, it is the question of *a functorial relationships between categories*, which, again, still needs to be established in the case of circuits (Hilbert spaces, used in the formalism, form a category), in the RWR view, in the absence of any *representational* mapping between the structure of circuits and the formalism of QTFD, or by implication, QM or QFT. Each category could consist of very different, but functorially related, types of mathematical objects, for example, one (the structures defining circuits) over $\mathbb{R}$ and the other (the formalism of QTFD) over $\mathbb{C}$. Mathematically, there are morphisms in each category, and there are functors between categories, and the mathematical-experimental situation of quantum physics would now be defined by these functors. Hence, as I said, from this perspective, Hardy's "physics to mathematics *correspondence* principle" would, if rigorously established, be replaced by the "physics to mathematics *functoriality* principle." It would be realized by the functorial relationships between the category based on circuits (the objects and morphisms of which need, again, to be defined) and the category of the mathematical objects defining the formalism, such as that of Hilbert spaces over $\mathbb{C}$ (finite-dimensional ones in QTFD) or some other category of objects and morphisms, still quite different from those of the first category.

This view is also more in accord with categorical thinking in mathematics. One of the most essential aspects of category theory, even its *raison d'être*, at least in its use in algebraic topology and algebraic geometry (mathematical logic is, again, a separate matter, although this is true there as well), consists in establishing the relationships between multiplicities, "categories," of objects of *different* types, for example, geometrical or topological objects and algebraic objects, or between algebraic objects of different types, rather than directly mapping the objects of the first category on those of the second. The concept of category was introduced in the field of algebraic topology and then extended to algebraic geometry to help study certain algebraic invariants, such as groups, associated with topological spaces. In contradistinction to geometry, defined, as a mathematical discipline, by the concept of measurement (geo-*metry*), topology, as a mathematical discipline, is defined by associating an algebraic structure or a set of structures, most especially groups, such as homotopy or cohomology groups (which came to play an important role in QFT and then string and M-brane theories), to a topological space. The structure of a topological space is defined by its continuities and discontinuities, and not by its geometry, even if it has a geometry, and not all topological spaces do. Insofar as one deforms a given figure continuously (i.e., insofar as one does not separate points previously connected and, conversely, does not connect points previously separated), the resulting figure or space is treated as mathematically equivalent. Thus, no matter how much one expands or continuously (without separating connected point or joining separated points) deforms a sphere the resulting spaces are topologically equivalent, while some of these objects are no longer spheres geometrically speaking. Such spaces are, however, topologically distinct from those of topologically deformed tori, because spheres and tori cannot be converted into each other without disjoining their connected points or joining the separated ones: the holes in tori make this impossible. In algebraic topology, mathematical objects of each type are arranged in categories and are related to one another by morphisms, while categories are related by functors. The category of topological spaces (or its subcategory, such as that of Riemannian manifolds) and their morphisms become related to a category of algebraic objects, such as groups and their morphisms. The relationships between these two categories, one topological or geometrical, and the other algebraic, allow one to extract an enormous amount of information concerning topological spaces and, although this is a subtler process, conversely, groups (through the structure of topological spaces to which these groups are associated). The approach also led to a cascade-like emergence and proliferation of new concepts, which still continues. A good categorical approach to, and a possible derivation of, quantum theory would similarly establish the functorial relationships between two categories, one based in circuits and the other defining the formalism, keeping in mind that at least the second one also allows one to establish a new form of geometry and, thus, extend the situation from QPG to QPAGG.

9.4 On Quantum Information Theory Beyond Reconstruction Projects

An important and difficult question, which need not depend on a categorical formulation of quantum theory, but which may be helped by it, is that of the relationship between the structure of the circuits, defined by the corresponding experimental arrangements, and the infinite-dimensional mathematical architecture of QM, or the same relationship in QFT. Consider the double-slit experiment, say, in the interference pattern setup. It is a circuit, which embodies, as operations, preparations, measurements, and predictions, all manifested in the emergence of the interference pattern. I would not presume to be able to mathematize it. But it is a circuit nevertheless, a complex one, albeit child's play in comparison to the circuitry found in high-energy quantum physics, such as that of the Large Hadron Collider (LHC).

Such questions will, however, need to be addressed if one is to extend the programs to derivations of QTFD to QM or to QED and QFT, or beyond. It is hardly sufficient to merely derive already established theories, or to use such informational derivation in order to offer a critique of previous derivations, admittedly, important tasks, which deepen our understanding of quantum theory, especially as based on the concept of information. The ultimate value of these programs, however, lies in what they can do for the future of fundamental physics. Thus, Hardy aims to rethink general relativity in operational informational terms, analogous to those of QTFD, and then to reach, in principle, quantum gravity, bypassing QFT or even QM, which would then be merely special cases of the ultimate theory (Hardy 2007, 2010). D'Ariano and coworkers, by contrast, appear first to move from QTFD to QFT. In their more recent work, they aim to develop a new approach to QFT, which, as based on the concept of quantum cellular automata, is different from the operational framework discussed thus far, but shares with it certain key informational features and the aim of developing QFT from fundamental principles.

D'Ariano and Perinotti's derivation of Dirac's equation is a step in this direction, for now in the absence of an external field, essential to the proper QED (D'Ariano and Perinotti 2014, 2017; D'Ariano 2017, 2020). Unlike Dirac's own or other previous derivations of the equation, their derivation only uses, along with other principles (homogeneity, isotropy, and unitarity), the principle of locality, rather than special relativity. The key principles of quantum theory and their quantum-theoretical mathematical expressions and consequences remain at work throughout—the probabilistic or statistical nature of the theory, locality, Hilbert spaces, operators, noncommutativity, the tensor structure, symmetry, group representation, and so forth (D'Ariano and Perinotti 2014, pp. 2–3). There is some important new mathematics as well, such as that of "quasi-isometric embedding," a profound recent development in geometry, primarily due to Misha Gromov and his geometric-group theory (D'Ariano and Perinotti 2014, p. 3; D'Ariano 2017, pp. 105–107). The QCA is the *quantum* cellular automaton operative at a larger scale. The derivation of Dirac's equation is achieved by finding Dirac's automata, which are "obtained by locally coupling Weyl automata" and give the Dirac equation at the relativistic limit from the informational principles proposed, *all of these principles*, rather than only linearity and unitarity, which are quantum in character (D'Ariano and Perinotti 2014, pp. 5, 6–9). This is a reflection and recasting of the fact that Dirac's spinors are composed of Weyl's spinors.

In this regard, the situation is unavoidably parallel to deriving QTFD from a larger (than only quantum) set of principles. Locality, homogeneity, and isotropy are additional principles here, which allow one to dispense with the principle of relativistic (Lorenz) invariance in deriving the equation, beginning with Dirac's original derivation. Locality merits special attention here. The locality requirement itself and the corresponding principle are defined by the requirement that the cardinality of the set, S_g, of sites g', interacting with the site g (both from a denumerable set G of systems, involved in the cellular automaton and described by Fermionic field operators) is "uniformly bound over G, namely, $|S_g| \leq k < \infty$ for every g" (D'Ariano and Perinotti 2014, p. 2). As noted earlier, this concept or principle is more general and deeper than the principle of (special) relativity, defined the Lorentz invariance, although relativity, special or general, conforms to the principle of locality. The Lorentz invariance emerges from this locality principle at the corresponding (relativistic) limit and the standard (Fermi) scale of quantum field theory, and hence of Dirac's equation. The significance of the locality requirement is, however, more general and is critical to the informational QCA framework proposed by the authors. This requirement is consistent with the view of locality adopted in this study (Einstein-locality): all physical interactions and indeed everything that essentially defines phenomena, information, and so forth is local. According to D'Ariano, while falsifiability is a crucial requirement to be satisfied by this framework, it is equally crucial that all falsifiability is *local*, which may be seen as a principle, the local falsifiability principle (D'Ariano 2018, pp. 8–9). The approach may, they hope, offer new possibilities for fundamental physics, even on Planck's scale.

It remains an open question whether this approach can reach beyond QFT and the standard model, and it is not certain either whether it can encompass even them. At the very least, however, it opens a trajectory beyond QTFD for quantum information theory and, thus, possibly an alternative to other current approaches aiming to reach beyond the standard model, especially in developing a theory of quantum gravity, thus far an elusive goal, with no consensus concerning what kind of theory it could be, for example, whether it will still be quantum.

9.5 Conclusion

Whether the QPA/QPAGG structure of quantum theory

$$QUANTUMNESS \rightarrow PROBABILITY \rightarrow ALGEBRA/$$
$$QUANTUMNESS \rightarrow PROBABILITY \rightarrow ALGEBRA \rightarrow GENERALIZED\,GEOMETRY$$

in place, as things stand now, regardless of interpretation, and the RWR view of it, which is interpretive, will remain viable, as Bohr thought it might (or even will), or will be defeated, fulfilling the hope of Einstein and those who followed him, is an open question as well. On the other hand, it appears likely, as no currently known attempts to move beyond QFT would indicate otherwise, that, as throughout the history of quantum theory and indeed all modern physics from Galileo on, we will continue to see the use of new mathematical structures, either already available in mathematics itself or yet to be invented by mathematicians or, as has happened before, by physicists. This mathematics may continue to be dominated by QPA/QPAGG structures, as the mathematics of quantum theory has been thus far, or it may, as Einstein hoped, return fundamental physics to more geometrically representing the ultimate nature of physical reality. This return would still involve algebra, because, while there might be algebra without geometry, there is no geometry without algebra. As discussed in this study, from the Pythagoreans on, our thinking in mathematics and physics alike has always been defined by both geometry and algebra, and the multi-faceted, and sometimes love–hate, relationships between them. This is, however, a good thing. It helps to move mathematics and physics forward, and will continue to do so, assuming that physics will remain a *mathematical*-experimental science of nature, as it has been from Descartes and Galileo to our own time, with mathematics, again, defining this conjunction. Philosophical thinking in physics itself and in the philosophy of physics has been equally defined by the mathematical-experimental character of modern physics—a remarkable way in which thought relates to matter, via technology, even when the ultimate constitution of matter is beyond thought.

References

Abramsky, S., Brandenburger, A.: The sheaf-theoretic structure of non-locality and contextuality (2011). arXiv:1102.0264 [quant-ph]

Bartlett, S.D., Rudolph, T., Spekkens, R.W.: Reconstruction of Gaussian quantum mechanics from Liouville mechanics with an epistemic restriction. Phys. Rev. A **86**, 012103 (2012)

Bohr, N.: The Philosophical Writings of Niels Bohr, vol. 3. Ox Bow Press, Woodbridge, CT, USA (1987)

Born, M., Jordan, P.: Zur quantenmechanik. Z. Phys. **34**, 858–888 (1925)

Born, M., Heisenberg, W., Jordan, P.: On quantum mechanics. In: Van der Waerden, B.L. (eds.) Sources of Quantum Mechanics, 1968, pp. 321–385. Dover, New York, NY, USA (1926)

Chiribella, G., D'Ariano, G.M., Perinotti, P.: Informational derivation of quantum theory. Phys. Rev. A **84**, 012311-1–12339 (2011)

Chiribella, G., Spekkens, R.: Quantum Theory: Informational Foundations and Foils. Springer/Nature, New York, NY, USA (2016)

Coecke, R.: From quantum foundations via natural language meaning to a theory of everything. In: Cooper, S.B., Soskova, M.I. (eds.) The Incomputable: Journeys Beyond the Turing Barrier, Theory and Applications of Computability, pp. 63–80. Springer/Nature, Berlin, Germany (2017)

Coecke, R., Kissinger, A.: Picturing Quantum Processes. Cambridge University Press, Cambridge, UK (2017)

D'Ariano, G.M.: Physics without physics. Int. J. Theor. Phys. **56**(1), 97–138 (2017)

D'Ariano, G.M.: Causality re-established. Philos. Trans. R. Soc. A **376**, 20170313 (2018). https://doi.org/10.1098/rsta.2017.0313

D'Ariano, G.M.: No purification ontology, no quantum paradoxes. Found. Phys. **50**, 1921–1933 (2020). https://doi.org/10.1007/s10701-020-00398-6

D'Ariano, G.M., Chiribella, G., Perinotti, P.: Quantum Theory from First Principles: An Informational Approach. Cambridge University Press, Cambridge, UK (2017)

D'Ariano, G.M., Perinotti, P.: Derivation of the Dirac equation from principles of information processing. Phys. Rev. A **90**, 062106 (2014)

D'Ariano, G.M., Perinotti, P.: Quantum cellular automata and free quantum field theory. Front. Phys. **12**(1), 1–11 (2017)

Einstein, A.: What is the theory of relativity? In: Einstein, A. (ed.) Ideas and Opinions, 1954, pp. 227–231. Bonanza Books, New York, NY, USA (1919)

Fuchs, C.A.: Quantum mechanics as quantum information, mostly. J. Mod. Opt. **50**, 987–1023 (2003)

Hardy, L.: Quantum mechanics from five reasonable axioms (2001). arXiv:0101012v4 [quant-ph]

Hardy, L.: Towards quantum gravity: a framework for probabilistic theories with non-fixed causal structure. J. Phys. A **40**, 3081–3099 (2007)

Hardy, L.: A formalism-local framework for general probabilistic theories, including quantum theory (2010). arXiv:1005.5164 [*quant-ph*]

Hardy, L.: Foliable operational structures for general probabilistic theory. In: Halvorson, H. (ed.) Deep Beauty: Understanding the Quantum World Through Mathematical Innovation, pp. 409–442. Cambridge University Press, Cambridge, UK (2011)

Hardy, L.: Reconstructing quantum theory (2013). arXiv:1303.1538 [quant-ph]

Hasegawa, M., Hofmann, M., Plotkin, G.: Finite dimensional vector spaces are complete for traced symmetric monoidal categories. LNCS [Lecture Notes in Computer Science] **4800**, 367–385 (2008)

Heidegger, M.: What is a thing? (tr. Barton, WB, Jr., Deutsch, V), Gateway, South Bend, IN, USA (1967)

Heisenberg, W.: The Physical Principles of the Quantum Theory (tr. Eckhart, K., Hoyt, F.C.), rpt. 1949. Dover, New York, NY, USA (1930)

MacLane, S.: Categories for the Working Mathematician. Springer, Berlin, Germany (2013)

MacLane, S., Moerdijk, I.: Sheaves in Geometry and Logic: A First Introduction to Topos Theory. Springer, Berlin, Germany (1994)

Schrödinger, E.: Discussion of probability relations between separated systems. Proc. Camb. Philos. Soc. **31**, 555–563 (1935)

Spekkens, R.W.: Evidence for the epistemic view of quantum states: a toy theory. Phys. Rev. A **75**(3), 032110 (2007)

Spekkens, R.W.: Quasi-quantization: classical statistical theories with an epistemic restriction. In: Chiribella, G., Spekkens, R.W. (eds.) Quantum Theory: Informational Foundations and Foils, pp. 83–136. Springer/Nature, New York, NY, USA (2016)

Von Neumann, J.: Mathematical Foundations of Quantum Mechanics (tr. R.T. Beyer), rpt. 1983. Princeton University Press, Princeton, NJ, USA (1932)

Index

A. Plotnitsky, *Reality Without Realism*,
https://doi.org/10.1007/978-3-030-84578-0